ABOUT ISLAND PRESS

Since 1984, the nonprofit Island Press has been stimulating, shaping, and communicating the ideas that are essential for solving environmental problems worldwide. With more than 800 titles in print and some 40 new releases each year, we are the nation's leading publisher on environmental issues. We identify innovative thinkers and emerging trends in the environmental field. We work with world-renowned experts and authors to develop cross-disciplinary solutions to environmental challenges.

Island Press designs and implements coordinated book publication campaigns in order to communicate our critical messages in print, in person, and online using the latest technologies, programs, and the media. Our goal: to reach targeted audiences—scientists, policymakers, environmental advocates, the media, and concerned citizens—who can and will take action to protect the plants and animals that enrich our world, the ecosystems we need to survive, the water we drink, and the air we breathe.

Island Press gratefully acknowledges the support of its work by the Agua Fund, Inc., Annenberg Foundation, The Christensen Fund, The Nathan Cummings Foundation, The Geraldine R. Dodge Foundation, Doris Duke Charitable Foundation, The Educational Foundation of America, Betsy and Jesse Fink Foundation, The William and Flora Hewlett Foundation, The Kendeda Fund, The Forrest and Frances Lattner Foundation, The Andrew W. Mellon Foundation, The Curtis and Edith Munson Foundation, Oak Foundation, The Overbrook Foundation, the David and Lucile Packard Foundation, The Summit Fund of Washington, Trust for Architectural Easements, Wallace Global Fund, The Winslow Foundation, and other generous donors.

The opinions expressed in this book are those of the author(s) and do not necessarily reflect the views of our donors.

ABOUT THE SOCIETY FOR ECOLOGICAL RESTORATION INTERNATIONAL

The Society for Ecological Restoration (SER) International is an international nonprofit organization comprising members who are actively engaged in ecologically sensitive repair and management of ecosystems through an unusually broad array of experience, knowledge sets, and cultural perspectives.

The mission of SER is to promote ecological restoration as a means of sustaining the diversity of life on Earth and reestablishing an ecologically healthy relationship between nature and culture.

The opinions expressed in this book are those of the author(s) and are not necessarily the same as those of SER International. Contact SER International at 285 W. 18th Street, #1, Tucson, AZ 85701. Tel. (520) 622-5485, Fax (270) 626-5485, e-mail, info@ser.org, www.ser.org.

RIVER FUTURES

Society for Ecological Restoration International

The Science and Practice of Ecological Restoration

James Aronson, editor
Donald A. Falk, associate editor
Margaret A. Palmer
Richard J. Hobbs

Wildlife Restoration: Techniques for Habitat Analysis and Animal Monitoring, by Michael L. Morrison

Ecological Restoration of Southwestern Ponderosa Pine Forests, edited by Peter Friederici, Ecological Restoration Institute at Northern Arizona University

Ex Situ Plant Conservation: Supporting the Survival of Wild Populations, edited by Edward O. Guerrant Jr., Kayri Havens, and Mike Maunder

Great Basin Riparian Ecosystems: Ecology, Management, and Restoration, edited by Jeanne C. Chambers and Jerry R. Miller

Assembly Rules and Restoration Ecology: Bridging the Gap Between Theory and Practice, edited by Vicky M. Temperton, Richard J. Hobbs, Tim Nuttle, and Stefan Halle

The Tallgrass Restoration Handbook: For Prairies, Savannas, and Woodlands, edited by Stephen Packard and Cornelia F. Mutel

The Historical Ecology Handbook: A Restorationist's Guide to Reference Ecosystems, edited by Dave Egan and Evelyn A. Howell

Foundations of Restoration Ecology, edited by Donald A. Falk, Margaret A. Palmer, and Joy B. Zedler

Restoring the Pacific Northwest: The Art and Science of Ecological Restoration in Cascadia, edited by Dean Apostol and Marcia Sinclair

A Guide for Desert and Dryland Restoration: New Hope for Arid Lands, by David A. Bainbridge

Restoring Natural Capital: Science, Business, and Practice, edited by James Aronson, Suzanne J. Milton, and James N. Blignaut

Old Fields: Dynamics and Restoration of Abandoned Farmland, edited by Viki A. Cramer and Richard J. Hobbs

Ecological Restoration: Principles, Values, and Structure of an Emerging Profession, by Andre F. Clewell and James Aronson

River Futures: An Integrative Scientific Approach to River Repair, edited by Gary J. Brierley and Kirstie A. Fryirs

River Futures

An Integrative Scientific Approach to River Repair

Edited by

Gary J. Brierley and Kirstie A. Fryirs

SOCIETY FOR ECOLOGICAL RESTORATION INTERNATIONAL

Washington · Covelo · London

Brierley, Gary J.
River futures : an integrative scientific approach to river repair / Gary J. Brierley, Kirstie A. Fryirs.
p. cm.
ISBN-13: 978-1-59726-112-8 (cloth : alk. paper)
ISBN-10: 1-59726-112-2 (cloth : alk. paper)
ISBN-13: 978-1-59726-113-5 (pbk. : alk. paper)
ISBN-10: 1-59726-113-0 (pbk. : alk. paper)
1. Rivers–Regulation. I. Fryirs, Kirstie A. II. Title.
TC530.B73 2008
627'.12–dc22 2007042296

Printed on recycled, acid-free paper 

Manufactured in the United States of America

10 9 8 7 6 5 4 3 2 1

Keywords: river restoration, river repair, integrative river science, river rehabilitation, river science management, river management policy

CONTENTS

PREFACE

Human activities have significantly altered aquatic ecosystems across most of our planet. Recognizing the damage done, many conservation and rehabilitation programs have kick-started an era of river repair. In this book we examine strategies that can be adopted to promote healthier river futures. Although issues discussed extend across social, cultural, and economic dimensions, the primary emphasis is on scientific considerations with which to guide the process of river repair through management practice. Key messages from the book include:

- The importance of having a future focus for river rehabilitation, rather than a focus on the past (through restoration).
- A paradigm shift in the science and management of river systems, marked by the emergence of crossdisciplinary applications with an increasingly practical, ecosystem-based focus.
- Visions that shape what we seek to achieve in the process of river repair, linking what is biophysically achievable in relation to what is socially acceptable for any given catchment.
- Issues of scale and approach that must be addressed in moving beyond discipline-bound knowledge structures and toward an integrative scientific understanding of river systems.
- Respect for the inherent diversity, complexity, and variability of river systems, framing crossdisciplinary approaches to river science and management upon a catchment-scale landscape template.
- Analysis of contemporary river condition in relation to river dynamics and evolution, which aids our ability to forecast likely future trajectories of river adjustment (including system responses to rehabilitation treatments and climate change).
- Interpretation of controls on ecosystem structure and function for each river system, developing management actions that target key processes that fashion ecosystem functionality.
- Application of carefully targeted measures to monitor the condition of a river system and appraise the success of management efforts in the process of river repair.

- Dedication to remembering that our relationship to any given river system is a key determinant of prospects for healthier river futures, never underplaying the importance of place in designing rehabilitation initiatives.

Case studies highlight the use of integrative river science in river rehabilitation practice in the United States, Europe, Japan, South Africa, and Australia. As landscape and climatic settings, population densities, and land uses vary markedly in these countries (among many other factors), they face differing management issues and priorities. However, in each instance there is a clear trend toward an ecosystem-based approach to river repair. This marks a fundamental transition in the way we view and manage river systems. Increasingly, management practices recognize and in some instances even embrace inherent uncertainties. Adoption of adaptive management principles is considered a key to the success of the era of river repair.

In facilitating the uptake of scientific insight in this book, we hope to promote direct collaboration among researchers, stakeholders, managers, decision-makers, and the community at large, striving to enhance prospects for healthier river futures. Ultimately, ecological rehabilitation is about people and process. In many instances, effective management does not require the design and implementation of sophisticated, high-cost measures. Indeed, overzealous technical applications could distract us from the work of protecting ecosystems. In many instances, such practices emphasize concerns for corporate objectives but fail to bring about substantive shifts in the attitudes, behavior, and practices that underpin more authentic approaches to river repair. Finding a way to link environmental and social goals offers the most likely pathway to rehabilitation success. Unless applications are "owned" by the communities involved, prospects for long-term, sustainable river health are likely to be compromised.

This book is written for a wide range of river practitioners, targeting science students (especially final-year undergraduates and postgraduate students in hydrology, geomorphology, aquatic/terrestrial ecology, aquatic geochemistry, resource/environmental science, and/or management), river managers and consultants, resource managers, and environmentalists concerned with the health of river systems. Framed in a constructive manner, with minimal use of jargon, the book can be used by scientists and managers alike.

The origins of this book can be traced back to a workshop in South Africa in 1994, where a range of ecosystem scientists and social scientists discussed the validity of various approaches to river classification and the uptake of these tools by river managers. Findings from the workshop highlighted resoundingly the lack of comparability in results derived from these various research tools. This primarily reflected the differing forms and scales of data used within differing discipline-bound perspectives, fundamentally limiting their application in guiding coherent management plans (Uys 1994). In response, a workshop on "Geomorphology and River Health" was orchestrated at Macquarie University (Brierley and Nagel 1995). The subsequent phase of research was dedicated to the development of a geomorphic template with which to guide coherent, crossdisciplinary river management. This research tool is called the River Styles Framework (Brierley and Fryirs 2005). There has been extensive uptake of this tool in Australia and elsewhere, aided by the teaching of a professional short course that has been attended by a wide range of river practitioners (see www.riverstyles.com).

Various postgraduate students at Macquarie University completed research projects that demonstrated the nature and extent of river response to human disturbance in southeastern Australia. One of these students, Andrew Brooks, went on to lead major research on reinstatement of wood as a component of river rehabilitation plans. As part of this work, Brooks initiated the emergence of UHRRI (Upper Hunter River Rehabilitation Initiative), with significant support from the former Hunter Catchment Trust Board, the former NSW Department of Land and Water Conservation, and Macquarie University. The employment of a research manager, Craig Miller, triggered a major shift in the scale of operation that culminated in a successful ARC Linkage Grant application LP0346918 (2003–2007). This multi-institutional, crossdisciplinary project linked five PhD theses and postdoctoral research with a major rehabilitation project. On-the-ground rehabilitation occurred with enormous support from various agencies including the NSW Department of Lands and NSW Fisheries. Industry support by Bengalla, Mt Arthur Coal, and Macquarie Generation provided not only financial support, but also access to eight kilometers of riverfront land near Muswellbrook. The chief investigators for this research when the project started were: Gary Brierley, Andrew Boulton, Andrew Brooks, Kirstie Fryirs, Michelle Leishman, and Darren Ryder.

Craig Miller, as research manager, was tasked with bringing together the skills and experiences of the researchers while coordinating rehabilitation efforts that met various environmental goals *and* the requirements of five PhD projects. These were demanding and challenging times. To kick-start the process, a workshop was held to develop a conceptual model outlining how the study reach "works." The workshop highlighted very starkly the differences in approach, scale of enquiry, and familiarity with the literature across disciplines ranging from geomorphology, aquatic ecology, plant ecology, hydrology, and social science, as well as a host of other impediments that hindered our ability to develop such an ecosystem model. In response, an appeal was launched to a broader audience, and we hosted an international workshop on "Integrative River Science" as part of the International Geophysical Union (IGU) Commission entitled "Geomorphic Challenges for the Twenty-First Century." This workshop was very successful, prompting the development of the proposal for this book and subsequent contributions by various authors (Angela Arthington, Richard Hobbs, Tomomi Marutani, Larissa Naylor, and Hervé Piégay). We would like to thank staff at Auckland General Hospital for helping to get Gary to this conference after his bike accident.

Richard Hobbs provided the initial connection with Island Press. We thank James Aronson, editor of the Restoration Ecology series, for providing initial guidance on the book prospectus and significantly tightening various parts of the final product (though deficiencies are entirely our responsibility). Barbara Dean and Barbara Youngblood prompted changes to several chapters and clarified our presentation of the text. We are extremely grateful to all authors for providing such substantive input while generally keeping within our timelines. Special thanks go to the lead authors in coordinating these efforts. All chapters were sent for review, typically to one author from the book and an external reviewer. Review comments by Paul Angermeier, Angela Arthington, Emily Bernhardt, Ian Boothroyd, Andrew Boulton, Brad Coombes, Carola Cullum, Steve Darby, Barbara Downes, Peter Downs, Brian Finlayson, Gordon Grant, Ken Gregory, Mick Hillman, Richard Hobbs, Matt Kondolf, Malcolm Newson, Takashi Oguchi, Hervé Piégay, Geoffrey Poole, Darren Ryder, Martin Thoms, Stephen Tooth, Robyn Watts, Joe Wheaton, and Ellen Wohl provided enormous support to us in our role as editors, and we truly appreciate these helpful and constructive contributions.

Several additional themes warrant mention. Our work in developing this book has been enhanced by our collaboration with Mick Hillman. His PhD work on environmental justice in river rehabilitation has been an inspiration for many elements of this book. Every conversation with him prompted new and intriguing lines of inquiry—it's a minor miracle that we've been able to bring the book to a close! Various other research colleagues provided insightful comments whether in conversation or their reading of various chapters—our thanks go to Claire Gregory, Susan Owen, Helen Reid, Nadine Trahan, and Ines Winz. The title for this book was prompted by a comment in a conversation with Mark Taylor many years ago. We also thank Emmy Macdonald for her sterling efforts in editing, proofreading, and reference checking the book and Dean Oliver Graphics for drafting many figures in the book.

To close, it's worth pointing out the outcome from the ARC Linkage Project. We held a workshop in December 2006 to bring together the findings from the various PhD projects. Perhaps the initial expressions of frustration and alarm a few years ago prompted this group of researchers to reframe their PhD projects in a broader crossdisciplinary context. Perhaps our initial goals were overly ambitious. Regardless of the underlying causes, we were eventually able to produce a conceptual model of the study reach. The key lesson here is the need to spend time together to consider a problem before becoming hung-up on the practicalities and time constraints of PhD or contract research commitments. As noted by the PhD students themselves, twelve months of working together prior to commencing their PhD projects would have enabled them to adopt a crossdisciplinary starting point; instead, that unified starting point always seems to be slightly beyond our grasp. These various research projects would not have been completed without assistance from the three universities involved in the project (Macquarie University, University of New England, and Griffith University), and the commitment of the research manager and research officer employed through the ARC Project—a vote of thanks to Mark Sanders and Dan Keating, respectively.

The front cover image was photographed by Christopher Fryirs. It depicts a river red gum on the Murray River just downstream of Echuca, Victoria, Australia. We chose this image to depict the vibrant colors of the Australian landscape . . . just one reason why we do the jobs we do!

We dedicate this book to the next generation of river scientists and practitioners, hoping that this contribution assists in the development of more coherent crossdisciplinary applications that make effective use of integrative river science.

References

Brierley, G. J., and K. A. Fryirs. 2005. *Geomorphology and river management: Applications of the River Styles framework*. Oxford, UK: Blackwell.

Brierley, G. J., and F. Nagel, eds. 1995. *Geomorphology and river health in New South Wales*. Proceedings of a one-day conference held at Macquarie University, October 7, 1994, Graduate School of the Environment Working Paper 9501.

Uys, M., ed. 1994. *Classification of rivers and environmental health indicators*. Proceedings of a joint South African/Australian Workshop. February 7–14 1994, Cape Town, South Africa. Water Research Commission Report No. TT 63/94.

PART I

The Emerging Process of River Repair

> You can't roll back history. You can create a sort of zoo or museum for people to glimpse the way things were—but you can't return to another time.
>
> Wheeler 1998, 17

Rivers and streams represent some of Earth's most altered ecosystems. Across much of our planet, rivers bear little relation to the way they looked—and functioned—in the not-too-distant past. Flow regulation and river diversion have fragmented almost all major river networks, with devastating ecological consequences. Few rivers retain their natural riparian vegetation or loading of wood. Tens of thousands of river kilometers have been channelized. Most rivers now operate under fundamentally different conditions from those that existed prior to human disturbance.

A growing awareness of human-induced change and damage to river systems has prompted a shift in management programs toward strategies that strive to improve river health. A paradigm shift is underway, marking the transition away from "command and control" perspectives that engineered river systems for human purposes and toward an ecosystem perspective that strives to balance human needs with environmental values. This new way of thinking has a sustainability focus, striving to meet biodiversity needs while maintaining ecosystem services that meet human needs. Rather than trying to restore or reconstruct the past, river management in this era of river repair has a future focus, aiming to *rehabilitate* river systems.

The move toward an era of river repair is discussed in chapter 1. There is now significant investment in river conservation and rehabilitation projects across the world. Many of these efforts are characterized by a move away from projects that manage for a single species, involve engineering methods to alter channel morphology, or focus on a specific site/reach, and toward a holistic view that integrates crossdisciplinary themes within an ecosystem perspective. Today, effective river rehabilitation programs emphasize catchment-scale linkages, recognizing the multidimensional processes that are inherent components of functioning, dynamic, and self-sustaining ecosystems. Inevitably, these environmental perspectives must be balanced with socioeconomic and cultural values, the nature of which varies in differing parts of the world (see part III). Part II outlines the key components of integrative river science that are used to guide the process of river repair.

The visions we create express what we seek to achieve in the process of river repair. In chapter 2, ecosystem perspectives that give a voice for the environment, rather than merely expressing human and/or managerial aspirations, are framed in terms of the structural and functional attributes of river systems. However, these scientific perspectives cannot be meaningfully integrated into management plans unless they are appropriately related to social values. Hence, the rehabilitation process explicitly entails the merging of scientific and cultural values and aspirations, setting targets that are biophysically possible and socially acceptable. Our success in such exploits requires appropriate commitment to monitoring and auditing procedures, appraising and reframing the effectiveness of our efforts to improve river health.

Chapter 3 considers two major challenges that limit prospects for the process of river repair: "turbulence" and "train wrecks." Turbulence refers to the challenges inherent in meaningfully connecting scientific perspectives within managerial pursuits, while train wrecks refers to prospective disasters that may ensue from inappropriately framed scientific perspectives, or inappropriate use of these insights. Mismatches of scale (both space and time), the mindsets associated with discipline-bound practices, and steps that can be taken to move beyond such thinking receive special attention. We hope that the integrative approach to the analysis of river systems promoted in this book will assist the next generation of river practitioners in moving beyond discipline-bound thinking and management applications.

Reference

Wheeler, R. S. 1998. *The buffalo commons*. New York, NY: Tom Doherty Associates.

Chapter 1

Moves Toward an Era of River Repair

Gary J. Brierley and Kirstie A. Fryirs

> Despite our dependence on healthy ecosystems, society has made the decision to continue life as usual until a loss of valued goods and services is realized; then, society will expect and rely on science to clean up the mess and make it look natural.
>
> Hilderbrand et al. 2005, 1

Rivers are part of society's lifeblood. We live along these natural arteries of the landscape, and they provide fundamentally important services: ready access to potable water, an easy means of transportation, fertile and replenished lands readily irrigated for agricultural use, and a reliable source of renewable energy, among many other applications. In many parts of the world, cultural associations are tied intimately to these biophysical and economic values. Throughout history, many peoples have developed a strong emotive and psychological connection to river systems. Rivers are also extremely important in ecological terms. Researchers estimate that, although aquatic ecosystems occupy only 0.8 percent of the surface area of the planet, 12 percent of all animal species live in fresh water (Abramovitz 1996).

Despite these values and associations, ill-conceived, conflicting, and unsustainable practices litter the history of human exploitation of land and water resources. The damage inflicted upon natural ecosystems is no longer in dispute (Boon et al. 2000; Millennium Ecosystem Assessment 2005), and there is severe pressure on the ecosystem services that we and future generations ultimately depend on (Daily 1997). Efforts to sustain biodiverse and functional river ecosystems represent one of the greatest environmental challenges for the twenty-first century (Bernhardt et al. 2006; Dudgeon et al. 2006).

The damage inflicted upon river systems, and prospects for river recovery, vary markedly across the globe. In environmental terms, there is no turning back the clock to former conditions and relationships—though we may be able to slow or reverse degradation trends. In most instances, reinventing the past simply isn't practical, possible, or desirable. In engaging with prospects for river futures, we need to move beyond the rose-tinted impressions of the past that are framed in terms of idyllic rural lifestyles and notionally harmonious human-environmental relationships. Such unrealistic images convey a misleading sense of former practices, the impacts of which were diluted by lower population densities and the capacity of societies to abandon unsustainable practices at any given locality by simply moving elsewhere (and proceeding to do the same thing, until "collapse" ensued once more; e.g., Diamond 2005; Wright 2005).

Across much of the Western world, rivers that

were physically transformed and ecologically ruined to facilitate industrial and agricultural developments are now receiving increasing societal demands for rehabilitation (e.g., Postel and Richter 2003; Wohl 2004; our preference for the term "rehabilitation," rather than "restoration," is discussed later in this chapter). Significant investment in large-scale rehabilitation initiatives has triggered notable recovery of many aquatic ecosystems, such as Chesapeake Bay and the Everglades in the United States and the Danube and Rhine rivers in Europe. Protection of high-value ecological remnants is now a key focus for many national and international agencies. Perhaps the greatest source of hope for healthier river futures, however, lies in the growth and success of small-scale rehabilitation initiatives that promote river recovery and reconnect local communities to their river systems. The process of river repair is underway.

Prospects for ecosystem recovery reflect societal values and the priority we place upon such issues. Among other factors, the amount that society is prepared to pay for such applications (i.e., what is socially acceptable), and societal attitudes toward maintenance and upkeep of rehabilitation measures, are major influences on these prospects. The quest for healthy and sustainable river futures reflects our capacity to develop harmonious relationships between human values and environmental needs (Wang 2006).

The emerging process of river repair extends across managerial, societal, and scientific dimensions. A paradigm shift is underway, as we realize the unsustainable consequences of past actions, and we adjust our perspectives to meet the reframed perspectives of this new era. In this book, we examine four key attributes of this emerging approach to river management:

1. The importance of a future focus for setting visions for river rehabilitation, and challenges faced in developing crossdisciplinary understanding with which to approach the process of river repair (part I).

2. The development of integrative, crossdisciplinary river science with which to facilitate the process of river repair (part II).

3. The primacy of regional and catchment-scale considerations as determinants of differing priorities and strategies that shape river futures in various parts of the world (part III).

4. The development of adaptive management frameworks that respect the inherent diversity, variability, and complexity of river systems (part IV).

In this chapter, we outline various components of the shift in thinking that underpins the emergence of the era of river repair. We then highlight the importance of coherent scientific information with which to guide this process. Finally, we define several key concepts used in this book, and provide an overview of the structure of the book as a whole.

The Emerging Process of River Repair

Access to, and use of, water resources has profoundly influenced the emergence of industrial society. In meeting societal needs, demands for guaranteed water supply for agricultural, domestic, navigational, and industrial applications were addressed with limited regard for ecosystem values (Hillman and Brierley 2005). The dominant mindsets were security of supply and minimization of risk. Notional progress through development in this era of "command and control" management brought about pervasive modification of environmental systems (Holling and Meffe 1996). Some of the key factors that affected rivers included the construction of dams and irrigation schemes, channelization and flood control programs, and a myriad of activities that sought to make drylands wetter and wetlands drier.

Although industrial society was not oblivious to the environmental consequences of its actions, it took some time to shift priorities away from the sole concern for economic considerations and toward an effort to address the health and well-being of society *and* natural ecosystems. However, initial attempts at repair emphasized issues that directly affected humans, such as water quality and disease (i.e., sanitation facilities). Eventually, this was viewed as an incomplete solution—one that ultimately fails to fix the problem.

The failure of management systems to deliver environmental goods, or to remedy environmental harm, renders the command and control approach vulnerable—if not obsolete—and open to replacement by new ways of thinking. Such transitions, if and when they occur, are the result of push and pull factors. "Push factors" refer to the apparent failure of traditional science to explain or predict controls on ecosystem functionality, and consequently a failure of management intervention. In part, this reflects the failure of reductionist, discipline-bound knowledge to adequately inform policy and management, thereby failing to reverse the pervasive degradation of aquatic ecosystems. Pull factors include demands from voters and influential community members for greater emphasis on environmental protection and a more substantive say in natural resource management.

In many parts of the world, a fundamental shift in river management practice is underway, marking a move away from a deterministic focus that endeavors to control nature and toward an ecosystem perspective that strives to work with nature, viewing humans as part of the ecosystem (table 1.1, figure 1.1). The emerging "ecosystem approach" to river management strives to establish healthy, productive, and resilient ecosystems that are able to recover from, rather than resist, disturbance. This new approach views ecosystem values and human needs side by side, placing biodiversity management and the sustainability agenda at the heart of the era of river repair.

Early developmental approaches to river management endeavored to meets society's needs through the application of engineering skills to create stable, predictable, and reliable channels. These imperatives were met with considerable flair. However, a suite of unintentional, largely unconsidered consequences ensued, inflicting enormous damage to aquatic ecosystems. Attempts to redress the environmental shortcomings of former practices through discipline-bound engineering applications have often further exacerbated the problems. A wider, crossdisciplinary knowledge base is required to inform the process of river repair.

A shift in scientific practice and uptake has accompanied the reframed needs of the process of river repair. Two interrelated trends are particularly noteworthy: the move beyond discipline-bound thinking, and an increased emphasis upon the use of scientific insight to address practical, real-world issues. As noted by Ziman (2000), most practical problems do not emerge ready-made in the middle of existing research specialities—they are essentially crossdisciplinary. Ecosystems do not operate within the boundaries we place on understanding through discipline-based teaching and learning. Failure to integrate the complex interplay of linkages across discipline boundaries constrains our capacity to deliver effective guidance to environmental managers. Fragmented science and/or management can only yield partial solutions. Informing political/social debate and stakeholder negotiations requires coherent scientific guidance. Meeting this new scientific agenda requires numerous adjustments in perspective. The critical responsibility of researchers is to merge conflicting perspectives, rather than expecting managers to do so.

Crossdisciplinary applications do not set out to replace or exclude traditional modes of research. Rather, these two approaches are complementary. Generating integrative knowledge requires the combined skills of specialists and integrators. In generating more holistic knowledge, effective use must be made of available discipline-bound insights. However, collaborative research is required to address the big questions that lie outside the comfort zone and conservatism of discipline-bound thinking. Through effective cooperation, parties that see different aspects of a problem are

TABLE 1.1

Attributes of command and control and ecosystem-based approaches to river management (modified from Hillman and Brierley 2005)

Principle or Theme	Command and Control Approach	Ecosystem-Based Approach
Goals/aims	• Outcome-driven, goal-oriented • Single purpose, discipline-bound (engineering focus), reductionist • Perceives problems as technical issues requiring technical solutions, emphasizing the desire for certainty in outcomes	• Emphasizes processes and outcomes, means and ends • Multi-objective, holistic, cross-disciplinary • Perceives problems as symptomatic of wider socioeconomic, cultural, and biophysical considerations
Perceptions	• Strives to create simple and predictable water management systems, viewing rivers as conduits with which to maximize the conveyance of water, sediments, and environmental "waste products" through uniform, stable, hydraulically smooth (homogenous) channels • Strives to stabilize, train, or improve rivers • Views human activities as separate from ecosystems	• Strives to (re)instigate natural variability in river structure, recognizing that many channels are naturally messy, irregular, and rough, while other areas may have no channel (e.g., wetlands or discontinuous watercourses) • Explicitly recognizes inherent complexity and uncertainties, emphasising concerns for living, vibrant, dynamic, and evolving ecosystems • Views people as part of ecosystems.
Scientific approach	• Applies deterministic (positivist), cause-and-effect science using engineering principles (fluid mechanics and hydraulics). Builds upon experimental procedures performed under controlled sets of conditions • Generates and applies general theories and principles	• Applies probabilistic reasoning, recognizing that ecosystems are emergent, nonlinear phenomena that are not amenable to reductionist explanations • Frames system-specific knowledge in relation to generalized principles
Institutional framework	• Top-down, politically driven approach designed and enforced by government agencies	• Bottom-up approach, applying participatory frameworks that integrate managerial, stakeholder, researcher, and community perspectives
Management approach	• Applies prescriptive (cookbook) approaches to river repair • Site-specific or reach-scale applications, typically framed in the quest for stability over decadal timeframes • Construction focus, with high level of intervention. Often embellished under labels such as "environmentally sympathetic," "soft," "sensitive," or "ecologically sound" engineering practices	• Promotes flexible (system-specific) approaches to ecosystem management (e.g., space to move programs) • Catchment-framed rehabilitation programs recognizing the range of natural variability over centuries or millennia • Considers a continuum of interventions, including conservation programs, the "do nothing" option, and strategic interventions that strive to enhance recovery. Minimizes the use of "hard" engineering to protection of key infrastructure and assets
Approach to prioritization	• Reactive. Focuses attention upon sites of greatest societal alarm, typically located in the most degraded reaches. Such strategies may accentuate damage or transfer problems elsewhere • Typically considers only a part of the problem, commonly addressing symptoms rather than underlying causes of degradation	• Proactive, conservation-first approach that strategically targets reaches with high recovery potential. Uses "whole of system" thinking to prioritize actions, recognizing system connectivity and the potential for lagged, off-site responses • Addresses causes rather than symptoms of degradation

TABLE 1.1

(Continued)

Principle or Theme	Command and Control Approach	Ecosystem-Based Approach
Auditing and monitoring	• Limited accountability • Monitoring is externalized, with maintenance divorced from design	• Long-term commitment • Monitoring is internalized and maintenance is a core management (community) activity
Approach to social engagement	• Extension science educates people about the environment	• Action research promotes mutual learning by all practitioners

Command and control approach | **Ecosystem-based approach**

FIGURE 1.1. Engineering- and ecosystem-based approaches to river management.
(a) Sabo dams and sediment management—Hokkaido, Japan (source: Kirstie Fryirs)
(b) "Stabilizing" rivers through urbanization and stormwater management—Cessnock, NSW (source: Kirstie Fryirs)
(c) Reinstating braided and anabranching channels along previously channelized rivers in "space to move" programs in Europe—© Jürgen Petutschnig (see chapter 11)
(d) Maintenance and enhancement of valley bottom swamps and wetlands—Barbers swamp, upper Shoalhaven catchment, NSW (source: Kirstie Fryirs)

able to explore their differences constructively and search for (and implement) plans and solutions that go beyond their individual visions of the possible (Morrison et al. 2004). Such collegial initiatives should not compromise the capacity for individual endeavor. Indeed, avoiding the limitations of conformist "groupthink" is imperative. As noted by Wilson (1998, 269): "We are drowning in information, while starving for wisdom. The world henceforth will be run by synthesizers, people able to put together the right information at the right time, think critically, and make important decisions wisely."

In the process of integration, researchers must move beyond personal biases, prejudices, and the inherent suspicions that disciplinary groups seem to hold of each other (Palmer and Bernhardt 2006). Recognition of diverse approaches *within* disciplines and avoidance of stereotyping or misrepresentation are key points of departure for cross-disciplinary practice (Pawson and Dovers 2003). Adoption of a whole-of-system approach at the outset, rather than retrospectively striving to connect threads between disciplines, provides the most effective way to promote integrative thinking.

To enhance prospects for environmental repair, researchers must work with managers, stakeholders, and decision-makers to address the most important questions, rather than focusing their attention on those questions to which they can readily provide answers (Rogers 2006). As noted by Ravetz (1999, 652): "The management of complex natural and social systems as if they were simple scientific exercises has brought us to our present mixture of triumph and peril. We are now witnessing the emergence of a new approach to problem-solving strategies in which the role of science, still essential, is now appreciated in its full context of the uncertainties of natural systems and the relevance of human values."

Funtowitz and Ravetz (1991) use the term "post-normal" science to describe these adjustments in scientific practice. These developments reflect enhanced appreciation of the methodological, societal, and ethical issues raised by scientific activities and their application (Chalmers 1999). Post-normal practice inverts the traditional opposition of "hard" facts and "soft" values. Decisions are "hard" in every sense, but the scientific inputs are irremediably "soft," as facts are uncertain, values in dispute, stakes high, and decisions urgent (Funtowitz and Ravetz 1991). Such applications recognize that science is neither value-free nor ethically neutral.

The Emergence of Integrative River Science

Ultimate success in river management depends implicitly upon our efforts to conceptualize river systems in a clear, systematic, and organized manner. In a sense, efforts to synthesize our knowledge of river systems revisit the proposal by Penck (1897) for discourse in "potamology" as the science of flowing waters. A fundamental shift in perspective lies at the heart of this process—that of seeing and conceptualizing river systems as dynamic wholes rather than static collections of parts. Rather than consider elements from ecology, geomorphology, hydrology, or aquatic geochemistry in isolation, *integrative river science* builds upon holistic crossdisciplinary analyses of aquatic ecosystems (all terms in italics are defined in table 1.2). While formal recognition of integrative river science is seldom stated explicitly, convergence of perspectives from differing disciplinary backgrounds is increasingly informally expressed. The emergence of notions such as riverscape, ecohydrology, ecohydraulics, and geodiversity is testimony to the adoption of more holistic approaches to river science.

In recent years there has been a remarkable convergence of thinking in the development and application of catchment-based scientific frameworks with a managerial focus, which have been used to conceptualize the structure and function of river systems (e.g., North America—Frissell et

TABLE 1.2

Definition of terms used in this book
(modified from SER 2004, Brierley and Fryirs 2005, Clewell and Aronson 2006, 2007)

Term	Definition
Integrative river science	Holistic, crossdisciplinary analysis of aquatic ecosystems that integrates physical and ecological integrity as a platform to analyze controls upon ecosystem integrity. This approach encourages practitioners and researchers to "think like an ecosystem."
Physical integrity	The interaction of abiotic factors (primarily water, sediment, and vegetation) that shapes the physical structure and function of a river. Differing river types have characteristic sets of form-process interactions that determine *available habitat* for flora and fauna.
Ecological integrity	The operation of biotic factors (i.e., biological and chemical components) that sustain the natural range of flora and fauna for that system. The *viability of available habitat* is determined by the collective effect of these biotic interactions.
Ecosystem integrity	The combined role of abiotic factors (physical integrity) and biotic factors (ecological integrity) that maintains viable, self-sustaining ecosystems. The state or condition of an ecosystem expresses attributes of biodiversity (e.g., composition and structure) within normal/expected ranges, relative to its ecological state of evolution/development.
Ecological resilience	The capacity of an ecosystem to sustain all critical measures of functionality, such that the self-sustaining capacity of the system is regained or maintained. Loss of ecological resilience, in whatever form or capacity, makes a system increasingly vulnerable, increasing the likelihood of catastrophic change and ecosystem collapse.
River condition	A measure of the capacity of a river to perform ecosystem functions that are expected for that type of river within the environmental setting that it occupies. Dynamic attributes of ecosystems are expressed within normal/expected ranges of activity, relative to the ecological stage of evolution/development.
River health	A societal perspective that frames biophysical concerns for river condition in relation to human associations/values for river goods and associated ecosystem services.
River rehabilitation	An intentional activity that enhances/assists the recovery of an ecosystem that has been degraded, damaged, or destroyed through manipulation of its structure and function. Management activities aim to promote the recovery of ecosystem processes so as to regain normal/expected function and self-sufficiency without necessarily aiming to recover all indigenous biota.
River creation	A process of river repair that seeks to improve river condition through adjustments toward the "best attainable" state for a site given contemporary biophysical interactions and constraints on the system. Human disturbance has instigated irreversible changes to these rivers.
River restoration	A process of river repair that strives to promote recovery toward a pre-disturbance state (near-intact condition). Reversible geomorphic change has occurred following human disturbance.

al. 1986; Naiman et al. 1992; Bohn and Kershner 2002; Europe—Petts and Amoros 1996; South Africa—Rogers and O'Keefe 2003; Australasia—Brierley and Fryirs 2005; Snelder and Biggs 2002). In noting the coherency of crossdisciplinary thinking in these frameworks, it is interesting to note their varying disciplinary origins, which extend across geomorphology, hydrology, and aquatic/terrestrial ecology. Inevitably, it's one thing to have these insights, but quite another to consider what we do with them!

Effective approaches to river management address concerns for the key drivers and relationships that determine *ecosystem integrity* for any given system. Such endeavors meaningfully frame the *ecological integrity* of the system in relation to its abiotic setting (i.e., its *physical integrity*; see figure 1.2). Abiotic interactions exert a direct influence

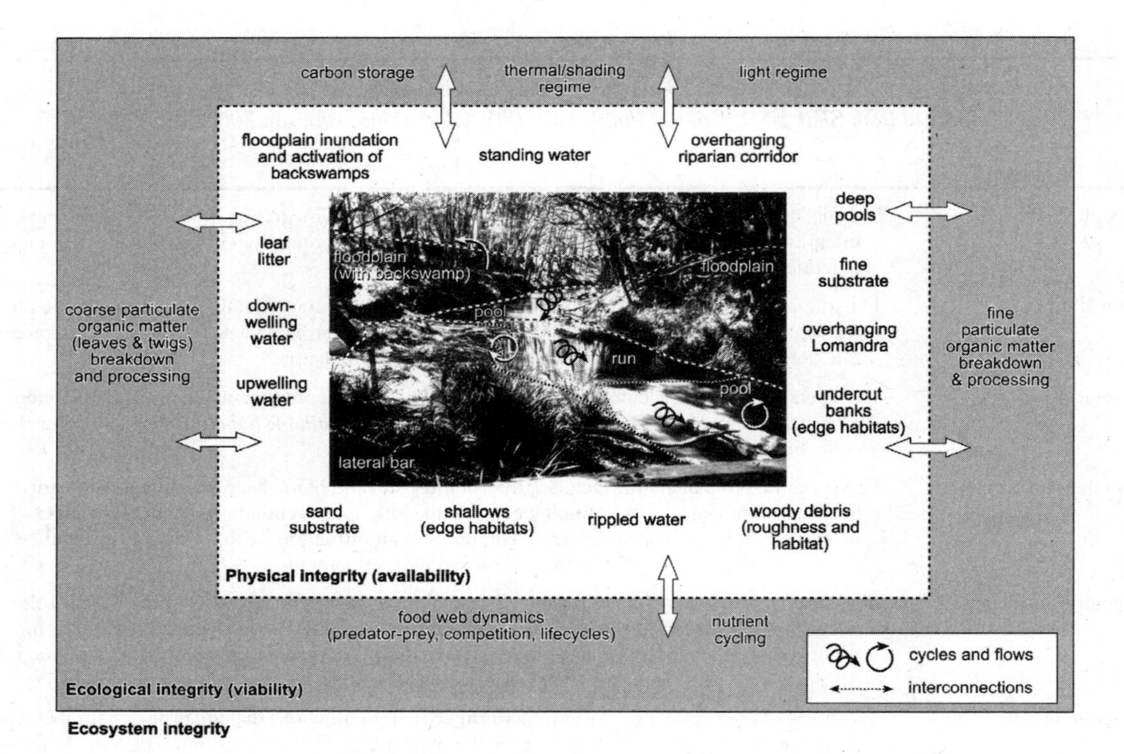

FIGURE 1.2. A conceptualization of the attributes that comprise the physical and ecological integrity of a river system and how these interact to determine ecosystem integrity. Physical integrity includes the structure of the physical template and the processes involved in the formation and reworking of that physical template. The structure of this template establishes the availability of habitat for aquatic flora and fauna. The physical template incorporates the abiotic interactions between sediment, water, and vegetation in a river system. Ecological integrity includes the processes and functions that make the physical habitat viable for aquatic flora and fauna. Biotic interactions associated with organic matter processing, nutrient spiralling, and food web dynamics are just some of the functions required for maintenance of ecosystem integrity.

upon the three-dimensional space, or habitat, in which river organisms live and complete their life processes. Biotic factors atop or within this template determine whether the physical habitat is viable for colonization, enabling flora and fauna to complete their life cycles. Measures of ecosystem functionality shape the viability of the aquatic habitat at any given locality. In framing management initiatives to maintain ecosystem integrity, measures must target those elements of *ecological resilience* that are vulnerable or under stress, striving to enhance the self-sustaining capacity of the system.

Despite the uniformity of the underlying physics that shapes river behavior, river systems demonstrate a remarkable array of biophysical interactions and evolutionary trajectories. Complex arrays of abiotic and biotic functions, processes, and structures are largely the result of system-inherent, dynamic genesis and development (Jungwirth et al. 2002). Researchers have developed a sophisticated understanding of the primary controls upon this diversity, variability, and complexity. In general terms, we have sound knowledge of the key abiotic and biotic interactions that fashion ecosystem functionality for any given ecoregion (Cushing et al. 1995). However, we should not oversimplify natural variability and complexity by using unduly rigid and categorical approaches to river research and management (Hilderbrand et al. 2005; Simon et al. 2007). Each catchment has its own configuration and history of disturbance events. Indeed, in many situations, exceptions to generalized relationships may be the very things we should target in management programs, as these unique or rare attributes may represent critical components of biodiversity.

An authentic, precautionary approach to river management strives to "work with nature," seeking to protect and enhance the values of a given place (Brierley et al. 2006; van der Velde et al. 2006). Sustainable approaches to river rehabilitation apply process-driven and strategic frameworks, with inputs from a wide range of specialists (Clarke et al. 2003; Wohl et al. 2005). Coherent insight into how a particular system works is required in order to frame local situations in their broader spatial and temporal context. Ideally, coherent and "owned" information bases merge scientific insights with local knowledge (de Groot and Lenders 2006). In this book, we adopt the term "transdisciplinary approach" to refer to the integration of social science perspectives with ecosystem science, and the collective use of these insights to guide management practice.

Framing Our Goals in the Process of River Repair

The labeling of what we seek to achieve in the process of river repair has engendered significant debate. In simple terms, this discussion revolves around the distinction between *restoration* and *rehabilitation* perspectives. In this book, we purposefully promote a move away from a restoration focus for river management activities, as we feel that this term implies an undue connection to the past, when boundary conditions under which rivers operated were fundamentally different from the conditions that prevail today. In a general sense, use of the term *river restoration* implies returning an aquatic ecosystem to a state as similar as possible to that which prevailed prior to human disturbance. In this sense, ecological restoration refers to the process of repairing damage caused by humans in order to reestablish the diversity and dynamics of indigenous ecosystems. From a purist perspective, such thinking is naïve and unrealistic for two fundamental reasons. First, as nature is a product of never-to-be-repeated conditions, reversing the clock is nonsensical. Second, many human changes to aquatic ecosystems are irreversible or only partially reversible (Jungwirth et al. 2002). To us, restoration is something that you do to a piece of furniture, a painting, or a building with a known history.

In contrast, river rehabilitation programs seek to achieve healthy and sustainable aquatic ecosystems under prevailing conditions, striving to minimize the need for ongoing maintenance while balancing social, economic, and environmental needs. Human pressures upon the environment determine the "best attainable condition" that can be achieved. Our intent here is not to view rehabilitation in terms of the replacement of past conditions and processes, though recovery mechanisms may regenerate former conditions and processes. Rather, we seek to maximize the capacity of the system to function in ecological terms while maintaining the provision of ecosystem services. In many instances, efforts to improve river condition under prevailing conditions may entail the *creation* of a "new" state, as human disturbance has instigated irreversible changes to the structure and function of aquatic ecosystems (Brierley and Fryirs 2005).

In this book, we use the term *river condition* to describe the integrity of an ecosystem and the term *river health* to relate these attributes to societal perspectives of rivers, framed in terms of societal needs/desires for ecosystem goods and services (Vugteveen et al. 2006). This distinction is important for determining the needs for a functioning ecosystem relative to human-values. The concept of river health provides an integrating theme with which to frame concerns for the process of river repair. Although it is a value-laden concept with widely debated relevance to scientific perceptions of ecological systems (Boulton 1999; Karr 1999), the metaphorical parallel between river health and human health provides an effective means to communicate concepts, processes, and values among scientists, managers, and the community. Our bodies function with broken limbs and artificial or replaced organs. We may choose to undergo cosmetic surgery to selectively transform our appearance. However, our least-healthy attribute usually determines the way we perform (and feel). Alternatively, if a vital organ breaks down and critical functioning is lost, we may die. So it is with rivers. Whatever factor performs least effectively limits the performance of the system as a whole.

River health is a useful concept in another sense. As rivers link different parts of the landscape, river health provides a barometer of catchment health and hence environmental health at the landscape scale. River health also provides a meaningful barometer with which to gauge human relationships to the environment. In general terms, healthy rivers are products of healthy societies and communities. In light of these definitions, this book adopts the term "rehabilitation" to reflect intentional acts that strive to improve river health.

Structure of the Book

This book is about preparing for the future in terms of management timeframes of the next fifty to one hundred years. What can we do, in a proactive sense, to maximize prospects for improving river health while maintaining or enhancing the provision of ecosystem services? In thematic terms, the book has four parts. Part I sets the scene for "river futures," highlighting the emergence of an era of river repair and implications for scientific practice in providing coherent insights with which to guide this process. Part II outlines an approach to integrative river science. Part III presents case studies from differing parts of the world that examine the use of integrative river science to guide rehabilitation efforts relative to other river management priorities. Finally, part IV considers the crossover between science and management, highlighting issues that must be addressed to promote healthier river futures.

Management efforts require a clear sense of what we are trying to achieve. As outlined in chapter 2, coherent scientific guidance is required to frame our visions for river protection and repair. In framing visions for river futures, knowledge of "what is biophysically achievable" is tied to an understanding of "what is socially acceptable." Effective uptake of scientific knowledge requires that scientists, practitioners, and the community "speak the same language," breaking down the barriers that limit interaction between these spheres and their embedded disciplines. Chapter 3 considers issues faced in developing and applying integrative river science in management practice.

The first two chapters of part II consider the importance of spatial context (chapter 4) and temporal dynamics (chapter 5) in developing an approach to integrative river science that facilitates the process of river repair. The spatial configuration of landscapes provides a physical platform with which to appraise the changing character, pattern, and linkages of river forms and processes in differing topographic and climatic settings. Increasingly, biophysical relationships are analyzed in relation to their landscape (geomorphic) setting, recognizing the importance of position within a catchment. The heterogeneity of river structure fashions the range of aquatic habitat. Ecosystems continually adjust to prevailing fluxes, disturbance events, lagged responses, and altered boundary conditions (such as climate and land use changes). Chapter 5 discusses the importance of understanding system dynamics and evolutionary tendencies, examining prospective river futures in relation to the range of variability of the system, the trajectory of change, and the prospect for threshold-induced shifts in state. Longer-term understanding of system evolution is required to unravel the variable role of multiple disturbance events, placing human impacts in light of natural variability.

Throughout this book, efforts to conceptualize river systems emphasize concerns for individual catchments, rather than an overly generalized understanding. Exactly what is required in rehabilitation depends on what is wrong in the system under investigation. As discussed in chapter 6, proactive river management strategies identify key controls on ecosystem integrity and functionality for any given system. Place-based conceptual models (discussed in chapter 5) enable us to test our understanding of ecosystem functionality. These models can forecast how ecosystems are likely to respond to disturbance events, including management interventions, and to appraise system sensitivity and prospects for ecological recovery. Integrative approaches to measuring river condition are critical for monitoring improvement or deterioration and assessment of the effectiveness of management actions, as outlined in chapter 7.

The relation to place identity is also fundamental to community engagement in river management programs, building social capital, and fostering a sense of commitment to the process of river repair. If properly conceived, the process of river repair brings people to a new, enlightened awareness of their relationship with place. Chapter 8 explores the relationships between biophysical and social connectivity.

Part III presents case study applications from around the world: Australia (chapter 9), the United States (chapter 10), Europe (chapter 11), Japan (chapter 12), and South Africa (chapter 13). Each chapter highlights the use of integrative river science in management practice, framing concerns for river rehabilitation in relation to other management priorities.

Finally, part IV highlights implications for managing the process of river repair. Chapter 14 frames prospects for river repair in terms of our relationship to the environment. The move beyond an era of command and control management and toward the adoption of ecosystem-based applications marks a shift in worldview from one shaped by a deterministic outlook to one that respects the inherent diversity, variability, and complexity of the natural world. Implicitly, this reframed outlook recognizes that efforts to promote healthier river futures must work with the inherent uncertainty of natural, socioeconomic, and managerial interactions. Rather than striving to control such issues, the outlook proposed in this chapter strives to restore uncertainty. Chapter 15 considers lessons from the book, and implications for the management of healthier river futures.

Acknowledgments

We are grateful to Mick Hillman for many discussions relating to the themes of this chapter. Comments by James Aronson significantly tightened the definition of terms used in this chapter, but we are responsible for any remaining deficiencies. Barbara Dean significantly improved the structure and presentation of the chapter. Themes presented here were focal points for discussion at an International Geophysical Union (IGU) workshop on integrative river science held at Muswellbrook, New South Wales, in February 2005. We thank all participants for their contributions to this workshop and hope that we have done justice to opinions expressed at that time. Australian Research Council Linkage Grant LP0346918 supported the workshop.

References

Abramovitz, J. N. 1996. *Imperiled waters, impoverished future: The decline of freshwater ecosystems*. Washington, DC: Worldwatch Paper 128, Worldwatch Institute.

Bernhardt, E., S. E. Bunn, D. D. Hart, T. M. Malmqvist, R. J. Naiman, C. Pringle, M. Reuss, and B. van Wilgen. 2006. Perspective: The challenge of ecologically sustainable water management. *Water Policy* 8:475–79.

Bohn, B. A., and J. L. Kershner. 2002. Establishing aquatic

restoration priorities using a watershed approach. *Journal of Environmental Management* 64:355–63.

Boon, P. J., B. R. Davies, and G. E. Petts, eds, 2000. *Global perspectives on river conservation: Science, policy, and practice*. Chichester, UK: Wiley.

Boulton, A. J. 1999. An overview of river health assessment: Philosophies, practice, problems and prognosis. *Freshwater Biology* 41:469–79.

Brierley, G. J., and K. A. Fryirs. 2005. *Geomorphology and river management: Applications of the river styles framework*. Oxford, UK: Blackwell.

Brierley, G. J., M. Hillman, and K. A. Fryirs. 2006. Knowing your place: An Australasian perspective on catchment-framed approaches to river repair. *Australian Geographer* 37:133–47.

Chalmers, A. 1999. *What is this thing called science?* St. Lucia, Australia: University of Queensland Press.

Clarke, S. J., L. Bruce-Burgess, and G. Wharton. 2003. Linking form and function: Towards an eco-hydromorphic approach to sustainable river restoration. *Aquatic Conservation: Marine and Freshwater Ecosystems* 13:439–50.

Clewell, A. F., and J. Aronson. 2006. Motivations for the restoration of ecosystems. *Conservation Biology* 20: 420–28.

Clewell, A. F., and J. Aronson. 2007. *Ecological restoration*. Washington, DC: Island Press.

Cushing, C. E., K. W. Cummins, and G. W. Minshall, eds. 1995. *River and stream ecosystems*. Amsterdam, Netherlands: Elsevier.

Daily, G. C., ed. 1997. *Nature's services: Societal dependence on natural ecosystems*. Washington, DC: Island Press.

De Groot, W. T., and H. J. R. Lenders, 2006. Emergent principles for river management. *Hydrobiologia* 565:309–16.

Diamond, J. M. 2005. *Collapse: How societies choose to fail or succeed*. New York, NY: Viking.

Dudgeon, D., A. H. Arthington, M. O. Gessner, Z. Kawabata, D. Knowler, C. Lévêque, R. J. Naiman, A. H. Prieur-Richard, D. Soto, and M. L .J. Stiassny. 2006. Freshwater biodiversity: Importance, threats, status, and conservation challenges. *Biological Reviews* 81: 163–82.

Frissell, C. A., W. J. Liss, C. E. Warren, and M. D. Hurley. 1986. A hierarchical framework for stream classification: Viewing streams in a watershed context. *Environmental Management* 10:199–214.

Funtowitz, S. O., and J. R. Ravetz. 1991. A new scientific methodology for global environmental issues. In *Ecological economics: The science and management of sustainability*, ed R. Costanza, 136–52. New York, NY: Columbia University Press.

Hilderbrand, R. H., A. C. Watts, and A. M. Randle. 2005. The myths of restoration ecology. *Ecology and Society* 10 (1): Article 19. http://www.ecologyandsociety.org/vol10/iss1/art19/ (accessed December 16, 2007).

Hillman, M., and G. J. Brierley. 2005. A critical review of catchment-scale stream rehabilitation programmes. *Progress in Physical Geography* 29:50–70.

Holling, C. S., and G. K. Meffe. 1996. Command and control and the pathology of natural resource management. *Conservation Biology* 10:328–37.

Jungwirth, M., S. Muhar, and S. Schmutz. 2002. Reestablishing and assessing ecological integrity in riverine landscapes. *Freshwater Biology* 47:867–87.

Karr, J. R. 1999. Defining and measuring river health. *Freshwater Biology* 41:221–34.

Millennium Ecosystem Assessment. 2005. *Ecosystems and human well-being: Synthesis*. Washington, DC: Island Press.

Morrison, T. H., G. T. McDonald, and M. B. Lane. 2004. Integrating natural resource management for better environmental outcomes. *Australian Geographer* 35:243–58.

Naiman, R. J., D. G. Lonzarich, T. J. Beechie, and S. C. Ralph. 1992. General principles of classification and the assessment of conservation potential in rivers. In *River conservation and management*, ed. P. J. Boon, P. Calow, and G. E. Petts, 93–123. Chichester, UK: Wiley.

Palmer, M. A., and E. S. Bernhardt. 2006. Hydroecology and river restoration: Ripe for research and synthesis. *Water Resources Research* 42:W03S07.

Pawson, E., and S. Dovers. 2003. Environmental history and the challenges of interdisciplinarity: An antipodean perspective. *Environment and History* 9:53–75.

Penck, A. 1897. Potamology as a branch of physical geography. *The Geographical Journal* 10:619–23.

Petts, G. E., and C. Amoros, eds. 1996. *Fluvial hydrosystems*. London: Chapman and Hall.

Postel, S., and B. Richter. 2003. *Rivers for life: Managing water for people and nature*. Washington, DC: Island Press.

Ravetz, J. 1999. What is post-normal science? *Futures* 31:647–54.

Rogers, K. H. 2006. The real river management challenge: Integrating scientists, stakeholders and service agencies. *River Research and Applications* 22:269–80.

Rogers, K. H., and J. O'Keefe. 2003. River heterogeneity: Ecosystem structure, function and management. In *The Kruger experience: Ecology and management of savanna heterogeneity*, ed. J. T. du Toit, K. Rogers, and H. C. Biggs, 189–218. Washington, DC: Island Press.

SER (Society for Ecological Restoration Science and Policy Working Group). 2004. *The SER primer on ecological restoration*. www.ser.org (accessed March 30, 2006).

Simon, A., M. Doyle, G. M. Kondolf, F. D. Shields Jr.,

B. Rhoads, G. Grant, F. Fitzpatrick, K. Juracek, M. McPhillips, and J. MacBroom. 2007. Critical evaluation of how the Rosgen Classification and associated "natural channel design" methods fail to integrate and quantify fluvial processes and channel response. *Journal of the American Water Resources Association* 43:1–15.

Snelder, T. H., and B. J. F. Biggs. 2002. Multi-scale river environment classification for water resources management. *Journal of the American Water Resources Association* 38:1225–40.

van der Velde, G., R. S. E. W. Leuven, A. M. J. Ragas, and A. J. M. Smits. 2006. Living rivers: Trends and challenges in science and management. *Hydrobiologia* 565:359–67.

Vugteveen, P., R. S. E. W. Leuven, M. A. J. Huijbregts, and H. J. R. Lenders. 2006. Redefinition and elaboration of river ecosystem health: perspective for river management. *Hydrobiologia* 565:289–308.

Wang, S. 2006. *Resource-oriented water management: Towards harmonious coexistence between man and nature*. Singapore: World Scientific Press.

Wilson, E. O. 1998. *Consilience: The unity of knowledge*. New York, NY: Knopf.

Wohl, E. 2004. *Disconnected rivers: Linking rivers to landscapes*. New Haven, CT: Yale University Press.

Wohl, E., P. L. Angermeier, B. Bledsoe, G. M. Kondolf, L. MacDonnell, D. M. Merritt, M. A. Palmer, N. L. Poff, and D. Tarboton. 2005. River restoration. *Water Resources Research* 41:AW10301.

Wright, R. 2005. *A short history of progress*. New York, NY: Carroll and Graf Publishers.

Ziman, J. 2000. *Real science: What it is, and what it means*. Cambridge, UK: Cambridge University Press.

Chapter 2

Vision Generation: What Do We Seek to Achieve in River Rehabilitation?

DARREN RYDER, GARY J. BRIERLEY, RICHARD HOBBS, GARRETH KYLE, AND MICHELLE LEISHMAN

> If you don't know where you are going all roads lead there.
>
> Roman Proverb

Given the intimate link between rivers and their catchments, river rehabilitation is a complex and difficult task involving practitioners from multiple scientific disciplines, managers, and community stakeholders. Developing a vision or a guiding image in the early stages of a rehabilitation project provides a mechanism with which to successfully direct all stakeholders toward a common goal. As outlined in chapter 1, rehabilitation efforts target realistic improvements in river condition, framing the most desirable outcome in relation to available resources.

Generations of short-term thinking have extended or even exacerbated environmental degradation, rather than promoting a sustained commitment to a process of river repair. A different, more strategic and inclusive approach to river rehabilitation is required. The development of realistic but visionary management plans is an integral part of this process. The visioning process entails the identification of the best achievable condition and services, framed in terms of social considerations and physical constraints, and the articulation of steps that must be taken in working toward this state. A balanced yet strategic approach to the targeting of rehabilitation initiatives is necessary. Integrative rehabilitation incorporates the current state of knowledge about how rivers work and societal demands on rivers. The process of vision generation provides a powerful tool with which all stakeholders can negotiate the steps that must be taken to meet human needs while minimizing negative impacts upon aquatic ecosystems and maximizing environmental functions, benefits, and services.

This chapter focuses on the role of developing a guiding image to unify scientists, managers, and the community to a common and achievable rehabilitation goal. If the a priori generation of a guiding image for rehabilitation success is an important and initial step in river rehabilitation projects, then the ability to unambiguously measure the attainment of this vision is equally imperative. In reviewing the current procedures for setting rehabilitation goals, this chapter starts by appraising the development and use of guiding images. Derivation of this guiding image is based on an assessment of what it is biophysically possible to achieve for any given reach or catchment. This is framed in terms of both physical and ecological integrity. The appropriate application and development of indicators to assess rehabilitation success are critiqued in light of these considerations. Ultimately, however, these scientific aspirations must be tied to societal expectations, presenting a unified platform for community-based projects. The chapter concludes with an assessment of the role of a guiding image in bridging the gap between the science of rehabilitation ecology and the practice of ecological rehabilitation.

Using a Guiding Image to Set Rehabilitation Goals

With any activity, there must be a goal or vision—something to work toward. A broad vision for ecosystem rehabilitation provides an overall goal or desired future state, along with an outline of the steps needed to get there. The broad vision thus answers the question "Where do we want to go?" The strategy answers the question "How do we want to get there?" Tactics provide the answer to the question "What actions do we need to take?" (Hobbs and Saunders 2001). Although setting goals for management and rehabilitation is perhaps one of the most important steps in designing and implementing a project or program, it is often either overlooked entirely or not done very well (Hobbs 2003). There is a tendency to jump straight to the "doing" part of a project without clearly articulating the reasons why things are being done and what the intended outcome should be. Ensuring that goals are both explicit enough to be meaningful and realistic enough to be achievable is a key to successful projects. Whether projects are developed as a bottom-up societal activity or a top-down government initiated program, rehabilitation goals should be decided among all stakeholders. In this light, the process of vision generation is a system-specific exercise that reflects social, cultural, and biophysical attributes of a given place (Hobbs 2003). An inclusive vision is therefore the product of iteration and consensus-building between researchers, managers, and the community.

Historically, guiding images, visionary targets, or *leitbild* have been used to assemble a vision of a natural reference condition without being concerned for the economic or political aspects that influence the realization of the scheme (Kern 1992). This view of a guiding image has great practical advantages for rehabilitation efforts, primarily because it provides an objective benchmark. However, there is a need to move beyond the "operational guiding image" and toward a concept of a guiding image that is truly visionary; one that is optimal under present ecological conditions, but framed in light of social and economic opportunities and constraints (McDonald et al. 2004).

A guiding image provides the opportunity to explicate what society wants from rivers while it enables scientists to influence those expectations. Rather than use this process to recreate an unachievable or unknown historical condition, Palmer et al. (2005) argue for a more pragmatic approach in which goals should move the rehabilitation toward the least degraded and most ecologically dynamic state possible. This is defined as a state where biota vary in abundance and composition over space and time, and channel shape and configuration change in response to flow variability that is characteristic of the region (recognizing the imperative to retain linkages to floodplains whenever practicable). The process of developing a guiding image should include the use of historical information such as aerial photographs, maps, and biological surveys (e.g., Brierley and Fryirs 2005), the use of relatively undisturbed or already restored sites as space-for-time analogues that represent a reference point for channel condition and biological composition (e.g., Rheinhardt et al. 1999), and a modeling approach to assess whether specific rehabilitation actions are appropriate (e.g., Skidmore et al. 2001). These approaches to the development of a guiding image are not mutually exclusive and must be complementary, and are best surmised by the "lines of evidence" approach advocated by Downes et al. (2002) that is inclusive of input from scientists, managers, and the community.

Scientific Considerations in Vision Generation

By definition, the success of ecological rehabilitation is measured against what we set out to achieve. Prospects for optimal environmental outcomes are maximized if emphasis is placed upon efforts to achieve tangible, short-term goals (timeframes of a few years) that are set within longer-term management timeframes of fifty to one hundred years (Jackson et al. 1995). To be effective, a guiding image must encapsulate the abiotic and biotic components of a river system in a manner that integrates both structural and functional attributes. In a sense, it combines concerns for both physical and ecological integrity, striving to enhance recovery mechanisms in working toward the best achievable condition under prevailing conditions (see chapters 1 and 5).

In recent years, it has become increasingly evident that the most effective way to approach the process of river rehabilitation lies in efforts to improve ecological resilience, enhancing the ability of a system to bounce back after disturbance. Such endeavors maximize prospects to generate self-sustaining systems (i.e., promoting ecosystem functions that help the river to look after itself). In developing such initiatives, geomorphic considerations that fashion the physical integrity of a reach must be linked to analyses of biotic and abiotic controls on the ecosystem functionality that determine the viability of habitat (see chapters 4, 5, and 6).

Assessing Rehabilitation Success

There is broad consensus among restoration ecologists that measuring success using ecologically based criteria is an essential ingredient for all ecological rehabilitation projects (Hobbs and Harris 2001; Hobbs 2003; Lake 2005; Ryder and Miller 2005). Specific performance goals and learning objectives must be related to ecological outcomes and must be measurable in terms such as increases in health indicators (e.g., increasing similarity of species or structure with reference community) or decreases in degradation indicators (e.g., active erosion, salinity extent or impact, non-native plant cover; Hobbs 2003). We can only measure success if we have some idea of what it looks like (i.e., we know what we are aiming for *and* we have ways of assessing condition and detecting change in condition; see chapter 7). The choice of parameters and processes to be monitored must go hand in hand with the setting of goals, ensuring that assessment criteria provide a meaningful measure of ecosystem functionality.

Recent reviews of approaches to river rehabilitation have been at the project level (Downs and Kondolf 2002), regional level (Moerke and Lamberti 2004), national level (Bernhardt et al. 2005), and international level (Giller 2005). Several recurring themes have emerged from these publications:

1. Only 10 to 50 percent of projects included post-project monitoring and evaluation.
2. Of these, nearly all the monitoring was short-term, inadequately funded, and, as a consequence, predominantly voluntary.
3. Monitoring programs were frequently based on an inadequate design and included inappropriate indicator species by which to judge success of the stated rehabilitation goals.
4. The reporting of river rehabilitation projects appeared to be biased toward successful projects, limiting the potential to learn from mistakes.
5. The setting of realistic goals and development of standardized monitoring methods were advocated in all reviews.

Unfortunately, measures of performance are rarely appraised in a systematic and appropriate manner, limiting our capacity to assess success (Grayson et al. 1999; Brooks et al. 2002; Bernhardt et al. 2005; Palmer et al. 2005). In those instances when performance is measured, attributes that are recorded may provide limited insight into ecosystem functionality. Through not knowing whether these projects are ecologically successful or not—either because attempts at measuring success were never made or because success was measured using inappropriate criteria or indicators—resources will be wasted and ecosystems will remain degraded. Despite the consensus among rehabilitation ecologists on the need for goal setting, debate continues on the appropriate level of organization at which goals should be specified and their success measured (Cairns 2000; Ehrenfeld 2000). At present, many rehabilitation projects set goals in relation to function or sustainability yet measure their success using structural ecosystem attributes. Similarly, greater attention should be given to assessment of ecosystem services such as water purification, recreation, and fish stocks (Grayson et al. 1999).

Typically, a range of biophysical indicators is used to assess the current health of a river (Boulton 1999). These indicators provide managers with a measure of which elements of the system should be tackled for the best possible rehabilitation outcome, ecologically, socially, and financially. Due to constraints on time and funding, rapid assessments are often performed over a relatively short period of time using simple *structural attributes* of the physical habitats, and biological and chemical environments (Bunn and Davies 2000; see chapter 7). The relative ease of measurement and interpretation of structural ecosystem attributes is perhaps the major advantage for their use in setting rehabilitation goals and evaluating success. Keddy (1999) suggests desirable properties of such structural variables including ease of sampling and processing, relative low cost, lack of ambiguity such as taxonomic certainty, high sensitivity to rehabilitation measures, and direct relevance to the hypothesis being tested. The rehabilitation of ecosystem structure is based on an understanding of the biology (e.g., genetic structure, population dynamics) of an organism or population and their habitat requirements, such as geomorphic setting or hydrologic requirements. For example, structural attributes are frequently used as indicators of the success of riparian rehabilitation, such as the population attributes of vegetation (Galatowitsch and Richardson 2005) or birds (Jansen 2005). A major problem associated with goals focusing on a particular attribute of ecosystem structure is the implicit knowledge needed of ecosystem or landscape-level interactions and processes influencing the organisms or populations (Ehrenfeld 2000). This approach is even more complicated in streams, as fluxes of water, nutrients, energy, and organisms between geomorphic features result in a mosaic of interdependent habitats, each one suitable for different species and communities (Pedroli et al. 2002). Any attempt to rehabilitate rivers in favor of biodiversity needs to focus on these intrinsic attributes of flowing water.

The production of organic matter, establishment of food webs, and movement of material and energy are important *functional aspects* of stream ecosystems. The changing relationships over time among biomass, productivity, respiration, and nutrient turnover have formed the basis for the theory of ecosystem function (*sensu* Odum 1969). Rehabilitation ecologists can use spatial and temporal changes to these patterns to evaluate the successful achievement of a shared vision. The use of ecosystem processes as an approach to setting goals and measuring success in rehabilitation projects recognizes that the viability of populations of all species, including rare and endangered species, depends on the maintenance of both large- and small-scale ecological processes, the presence of a characteristic mosaic of community types over a broad area, and the movement of individuals and material within a landscape (Ehrenfeld 2000). As such, critical measures of ecosystem function may also vary from system to system (see chapter 6). Identification of key drivers and relationships, and the weakest links that may compromise ecosystem sustainability, may require the development of catchment-specific conceptual models (see chapter 5).

Measures of ecosystem processes require an appreciation of the inherent *temporal variability* for each variable and system when developing a guiding image. Derivation of these insights requires an assessment of system evolution, and determination of fundamental controls exerted by combinations of geological, climatic, and human-induced factors. This requires extensive historical and field-based investigations. Interpretations of geo-hydraulic interactions with biotic processes, and how they change over time, provide a basis to assess system resilience, prospects for recovery, and the likely response trajectories. Analysis of the ecosystem functions of a river system requires researchers to consider the placement of the project in the landscape: its boundaries, its connections or lack thereof to adjoining ecosystems, and its inputs and losses of materials and energy from its physical surroundings (Ehrenfeld and Toth 1997; see chapter 8). Quantifying the fluxes of energy and materials into, out of, and within ecosystems is critical for analysis of river function. For example, the energy of flowing water provides the means by which the in-stream and riparian habitats are structured and altered (Friedman et al. 1996). The mosaic of habitats provides sites where the fluxes of energy may subsidize the growth of organisms (e.g., low flows promoting benthic autotrophic production; Ryder 2004), or prevent their growth (e.g., high velocity flows scouring benthic autotrophs, Ryder et al. 2006).

The relationship between the structure and function of an ecosystem is a vital issue in developing a guiding image for a rehabilitation project and being able to measure its success. Although conditions that allow the replacement of plants, animals, and even specific habitats can be created and maintained, the rehabilitated system may still not perform the critical ecosystem functions required for ecosystem sustainability. In many instances, the return of key processes may lag behind the reinstatement of ecosystem structure (e.g., Zedler 1996; Findlay et al. 2002). Patterns and processes in ecosystem attributes are, however, not the same, and rehabilitation of one does not necessarily equate to the rehabilitation of the other (Grayson et al. 1999; Cairns 2000). Therefore, measuring processes rather than patterns may provide better indicators of ecological integrity (*sensu* Karr 1996) for assessing river health (Bunn et al. 1999; Jansson et al. 2005). By combining these two types of variables, our understanding of inputs, storage, transformations, and exports of material and energy in ecosystems is enhanced considerably (Dahm et al. 1998). Recognition of the dynamic and interconnected nature of ecological entities underpins efforts to formulate a substantive guiding image for rehabilitation projects.

However, despite well-developed theories, practical implementation of scientific principles in rehabilitation projects is still far from a reality. This is where the terms ecological rehabilitation (the actual practice and implementation of projects) and rehabilitation ecology (scientific principles upon which these actions should be based) provide the pivotal link between scientists and practitioners, with the guiding image providing the common currency.

Socioeconomic Considerations: An Inclusive Approach to Vision Generation

What we seek to achieve in the process of river rehabilitation is a matter of choice. Typically, this choice reflects social and cultural values, and what we feel we can afford, rather than scientific insights into what is biophysically achievable. Ecological rehabilitation can refer as much to a movement and mode of living as it does to a scientific pursuit. From this premise, any definition of good rehabilitation should include social, cultural, aesthetic, economic, political, and moral values (Higgs 2003). Our sense of "what is biophysically achievable" must be linked with "what is socially acceptable, desirable, and affordable."

A truly integrative rehabilitation approach

needs to bridge the gaps between what biophysical scientists know about how rivers work and what society may want or expect from rivers. The latter reflects a mix of social, economic, cultural, and ethical factors. Hence, the rehabilitation process needs to integrate both the biophysical and sociological understandings of rivers. Prospectively, the process of vision generation provides a powerful sociological tool with which to enable river stakeholders to come to grips with the consequences of routinely discounting sustainability in determination of management options. Although the functional aspects of rivers are crucial to scientific understanding of rivers, they are often not highly valued by stakeholders. For stakeholders to care about functional aspects, scientists need to link river functions to valued ecosystem services in compelling ways. In the future, development of a guiding image could provide a crucial link between the sociological and ecological dimensions of river rehabilitation. Development of a guiding image provides the opportunity to explicate what society wants from rivers while enabling scientists to influence those expectations.

Scientists must recognize that their role in river management is secondary (Rogers 2006). Societal demands on rivers, which stem from value systems, are the root causes of management problems. Societal refusal or reluctance to apply available science and fully consider the costs and benefits of preventative measures in river management reflects a major disconnect between what river scientists do and advocate, and what society thinks is important (see chapter 15). Although societal leaders do not openly advocate for unsustainable outcomes, neither do they typically advocate for sustainability. Perpetual growth and prosperity in the human economy is a much more common (and politically appealing) societal mantra. In many instances, even when societal leaders are advised about the consequences of ecologically unsustainable tactics, they still choose an unsustainable trajectory because of other (e.g., social, economic, political) perceived advantages. It is only when those whose actions that engender unsustainable outcomes take responsibility for their decisions (or are held accountable for them) that there are genuine prospects for reform.

The decision to manipulate ecosystem processes and components involves not only a scientific judgment that a restored or rehabilitated condition is achievable, but also a value judgment that this condition is more desirable than the status quo. Perhaps the key contribution that scientists can make entails clear communication of the implications of differing management scenarios, promoting initiatives that maximize ecological prospects under prevailing environmental and social conditions. Ultimately, this is a process of negotiation, hopefully performed through informed debate of potential impacts of differing management strategies. These can be used to frame rehabilitation trajectories in terms of a range of potential ecological outcomes that accommodate variability in controlling factors.

Although there is a clear need for rehabilitation projects to incorporate appropriate measures of success into their work, numerous impediments inhibit this process. As most projects are undertaken by community-based and volunteer organizations, participants may not be familiar with the concept of ecological outcomes and success, or the broad range of scientific and technical issues surrounding the effective measurement of them (Freeman and Ray 2001). Nor are they likely to have major funding to support the large-scale quantitative research that is needed to measure the ecological outcomes of their projects. Consequently, they may see the inclusion of science and ecological theory as an unnecessary complication adding to the already substantial temporal, financial, staffing, and motivational challenges (Safstrom and O'Byrne 2001). Harris et al. (1996) proposed a conceptual model highlighting the level of community influence on ecological rehabilitation at differing spatial scales (see figure 2.1). It is not surprising that an increased weight from communities is given to local-scale projects, as the increasing

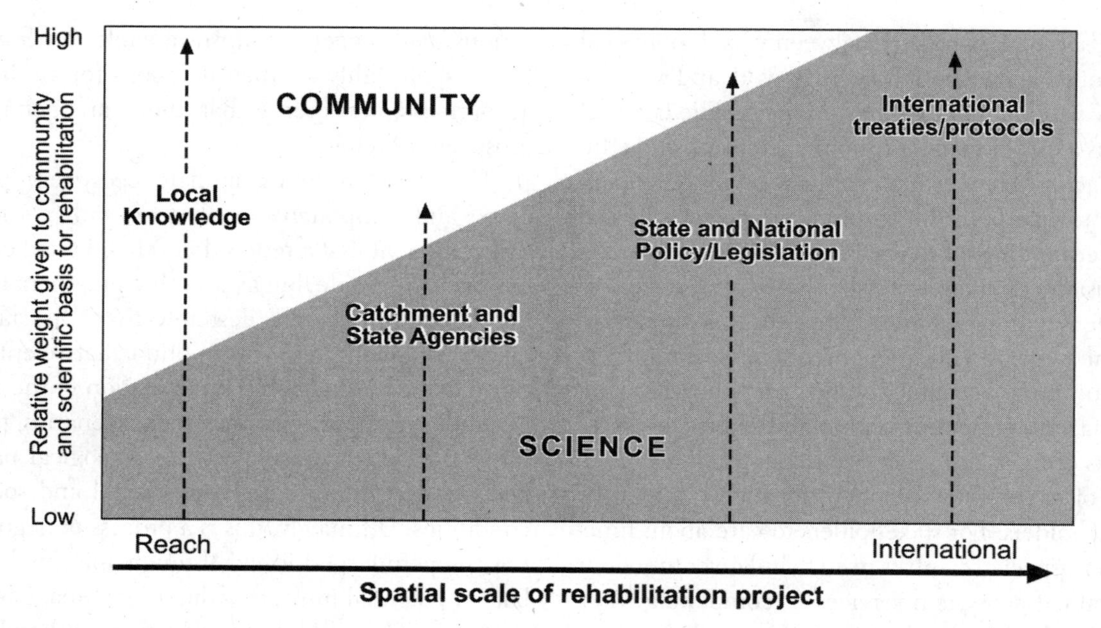

FIGURE 2.1. A conceptual model highlighting the level of community and scientific influence on ecological rehabilitation projects at local to international spatial scales (adapted from Harris et al. 1996).

spatial scale (and complexity) of projects results in a decrease in the direct influence of the communities involved, and a concomitant increase in the influence of scientific opinion (Harris and van Diggelen 2006).

Developing and implementing a vision for river rehabilitation is a catchment-scale task that integrates perspectives across multiple disciplines, scales and dimensions (Boon 1998; Rutherfurd et al. 2000; Brierley and Fryirs 2005). Authentic transdisciplinary perspectives extend beyond the undue focus on "single immediate pressures" (Finlayson 2002), such as particular species or water uses or river reaches, and attempt to link these concerns in programs that focus on the big questions that must be addressed in any given system. A usable vision is sufficiently *flexible and adaptive* such that it can be modified in light of changing circumstances, perceptions, and understanding, or as targets are achieved. It is also *generative and catalytic*, galvanizing and maintaining action around broadly identified themes (Hillman and Brierley 2005).

Incorporating a Guiding Image into Successful River Rehabilitation Practice

The development of a guiding image and appropriate measures of success for rehabilitation projects should be a priority not only to the rehabilitation industry, but also to the community as a whole. Measuring success must include the possibility of measuring failure, enabling mid-course corrections, or even complete changes in direction. Although not all rehabilitation projects require quantitative measures of their success, the question of ecological success and how to measure it is pivotal for both rehabilitation ecology and ecological rehabilitation, extending across the spectrum from community-based, action-orientated groups to government-initiated programs (Possingham 2001). The process of river rehabilitation is a learning experience that requires ongoing and effective scientific evaluation within the context of societal expectations and values (Pelley 2000). Adoption of adaptive management principles enhances flexibility in this process, recognizing and communicating to stakeholders that scientific understanding of ecosystems is uncertain and constantly changing (see chapters 14 and 15).

The reality of the current situation is that there is often little effective communication between rehabilitation ecologists and on-ground practitioners regarding issues of goal-setting, monitoring, and measuring success. Thus, community-based groups often fail to see how monitoring and goal-setting go hand in hand, as they are often faced with a confusing array of different ecological variables that could be monitored. Moreover, the monitoring of ecological variables comes at a great expense of time, money, and effort. If monitoring is to be done by community-based groups, there is little point in ecologists proposing complicated schemes that involve expert identification or analyses (Hobbs 2003). Given this, it is hardly surprising that appropriate goals are rarely set, monitoring schemes are not maintained, and ecological success is seldom quantified in community-based rehabilitation projects.

What does the future hold for the closer alignment of ecological rehabilitation and rehabilitation ecology, and the setting of rehabilitation goals and measuring of their success in stream ecosystems? As proposed by Hobbs (2003), the process of setting rehabilitation goals through scientific, management, and community forums is paramount for ensuring that there are clear target conditions, each with a timeframe for achievement. As noted by Harris and van Diggelen (2006), the arena in which rehabilitation ecology and ecological rehabilitation meet is societal.

Rehabilitation Ecology

Prospects for biodiversity will be maximized through management initiatives that comprise a genuine balance of conservation and rehabilitation strategies, framed within a landscape-scale plan of action. This provides the most rational basis with which to prioritize efforts to achieve the best geo-ecological outcomes for effort and cost. Second, although guiding principles for rehabilitation practice can be characterized (e.g., effective use of science, ethical behavior, commitment to environmental justice, and so on), the process of vision generation is ultimately a system-specific exercise, reflecting the social, cultural, and biophysical attributes of a given place.

Efforts to achieve and sustain success in rehabilitation projects are enhanced when multi-stakeholder communication lies at the heart of the vision generation process. This process also facilitates the advancement of rehabilitation ecology as a science, as it requires the effective selection of indicators to measure rehabilitation success by identifying stressors on the ecosystem that are valued by society and measurable by scientists. For example, Chapman and Underwood (2000) suggest that if measurements of ecosystem structure are to be used to set goals and provide information on the success of projects, then there is value in undertaking research into where and under what circumstances structure and function are linked. Any relationships that may facilitate links between structure and function, or the development of surrogate measures for ecosystem services that are more easily communicated to communities, will require substantial development and communication among scientists from different disciplines. For example, the combination of structural attributes such as geomorphic structure (Brierley and Fryirs 2005), biotic habitats (Rohde et al. 2005), and fish and invertebrate populations (Brooks et al. 2002; Lepori et al. 2005), with functional attributes such as decomposition and retention of organic matter (Lepori et al. 2005) and food web structure (Hession et al. 2000) providing an integrated information base for assessing rehabilitation success. However, this also requires effective communication among scientists from divergent disciplines, as well as scientists with divergent views from the same discipline (see chapter 3).

Ecological Rehabilitation

In many instances, ecological rehabilitation is trying to do in a matter of years what takes decades or centuries under natural conditions. All too often, it is assumed that systems will return to a "natural" state in fairly short order if they are nudged in the right direction through adjustments to structural attributes or by regulating species composition. In striving to be as efficient and effective as possible, it is critical that we do not take shortcuts that can potentially undermine the integrity of aquatic ecosystems or result in an inability to measure success against goals.

Clearly, in order to incorporate ecological principles into community-based rehabilitation projects and ensure positive environmental outcomes, there need to be stronger links and better communication between rehabilitation ecologists and on-ground practitioners (e.g., Rhoads et al. 1999; Lake 2001; Safstrom and O'Byrne 2001; Hobbs 2003). By becoming involved in local projects, rehabilitation ecologists can facilitate communication between scientists and nonscientists about ecological theory and the ways it can be used both to create new knowledge and to solve real-world problems. This provides opportunities for practitioners to give scientists a firsthand view of the on-ground problems and practical realities faced by community-based rehabilitation groups (Rhoads et al. 1999).

To facilitate communication between scientists and practitioners, new methods are needed to deliver scientific principles to practitioners in a simple, accessible, and cost-effective way (Norton 1998). A study of nonprofit conservation groups in the United States by Freeman and Ray (2001) found that such groups were accepting of increased assistance from rehabilitation ecologists, but any assistance or recommendations must be clear, succinct, and easily applied to frontline initiatives. Hence, the solution to truly integrative rehabilitation projects lies in scientists condensing ecological concepts and principles into more accessible formats, such as the development of a guiding image or vision for how the rehabilitated ecosystem might be structured and how it might function. With this guiding image in hand, both scientists and practitioners can develop, implement and assess the success of rehabilitation projects based on sound scientific principles, appropriate indicators and community involvement. Only then can river management, founded on the conceptual basis that a rehabilitated site should be self-sustaining (i.e., is structurally and functionally viable), be successful.

Conclusion

There is growing consensus among river scientists that an improved scientific basis for rehabilitation projects must include the setting of goals and establishing standards against which success can be assessed. The development of an a priori guiding image by all stakeholders to set rehabilitation goals, and the ability to measure success through pre- and post-monitoring, are key elements of rehabilitation ecology. Bridging the gap between rehabilitation science and the practice of rehabilitation will require more explicit links between the scientific knowledge of how rivers work and societal demands on rivers. For example, although functional aspects of rivers are crucial to our scientific understanding of systems, they need to be linked to valued ecosystem services and expressed in a compelling way if these attributes of rivers are to be valued by stakeholders. Thus, the development of a guiding image to set rehabilitation goals provides an opportunity to influence societal expectations through improved communication and interaction among scientists from different disciplines. Once developed, a guiding image can be used to communicate the implications of differing management scenarios, promoting initiatives that maximize ecological prospects under prevailing environmental and social conditions. Our success in these endeavors is constrained by our capacity to meaningfully link scientific perspectives beyond traditional discipline boundaries, as discussed in the following chapter.

Acknowledgments

Many of the ideas expressed in this chapter arose from a workshop on integrative river science in Muswellbrook in March 2005. We thank many colleagues who have discussed this chapter with us, particularly Andrew Boulton, Kirstie A. Fryirs, Paul Frazier, Robyn Watts, Ben Wolfenden, and Sarah Mika. Constructive review comments by Paul Angermeier and Emily Bernhardt enhanced this chapter. The authors acknowledge the Australian Research Council, Macquarie Generation, Mt. Arthur Coal, Bengalla Mining Company, the NSW Department of Natural Resources, and the Hunter-Central Rivers Catchment Management Authority for funding and support.

References

Bernhardt, E. S., M. A. Palmer, J. D. Allan, G. Alexander, K. Barnas, S. Brooks, J. Carr, S. Clayton, C. Dahm, J. Follstad-Shah, D. Galat, S. Gloss, P. Goodwin, D. Hart, B. Hassett, R. Jenkinson, S. Katz, G. M. Kondolf, P. S. Lake, R. Lave, J. L. Meyer, T. K. O'Donnell, L. Pagano, B. Powell, and E. Sudduth. 2005. Synthesizing U.S. river restoration efforts. *Science* 308:636–37.

Boon, P. J. 1998. River restoration in five dimensions. *Aquatic Conservation: Marine and Freshwater Ecosystems* 8:257–64.

Boulton, A. J. 1999. An overview of river health assessment: Philosophies, practice, problems and prognosis. *Freshwater Biology* 41:469–79.

Brierley, G. J., and K. A. Fryirs. 2005. *Geomorphology and river management: Applications of the River Styles Framework*. Oxford, UK: Blackwell.

Brooks, S. S., M. A. Palmer, B. J. Cardinale, C. M. Swan, and S. Ribblett. 2002. Assessing stream ecosystem rehabilitation: Limitations of community structure data. *Restoration Ecology* 10:156–68.

Bunn, S. E., and P. M. Davies. 2000. Biological processes in running waters and their implications for the assessment of ecological integrity. *Hydrobiologia* 422/423: 61–70.

Bunn, S. E., P. M. Davies, and T. D. Mosisch. 1999. Ecosystem measures of river health and their response to riparian and catchment degradation. *Freshwater Biology* 41:333–45.

Cairns, J. 2000. Setting ecological restoration goals for technical feasibility and scientific validity. *Ecological Engineering* 15:171–80.

Chapman, M. G., and A. J. Underwood. 2000. The need for practical scientific protocols to measure successful restoration. *Wetlands (Australia)* 19:28–49.

Dahm, C. N., N. B. Grimm, P. Marmonier, H. M Valett, and P. Vervier. 1998. Nutrient dynamics at the interface between surface waters and groundwaters. *Freshwater Biology* 40:427–51.

Downes, B. J., L. A. Barmuta, P. G. Fairweather, D. P. Faith, M. J. Keough, P. S. Lake, B. D. Mapstone, and G. P. Quinn. 2002. *Monitoring ecological impacts: Concepts and practice in flowing waters*. Cambridge, UK: Cambridge University Press.

Downs, P. W., and G. M. Kondolf. 2002. Post-project appraisals in adaptive management of river channel restoration. *Environmental Management* 29:477–96.

Ehrenfeld, J. G. 2000. Defining the limits of restoration: The need for realistic goals. *Restoration Ecology* 8:2–9.

Ehrenfeld, J. G., and L. A. Toth. 1997. Restoration ecology and the ecosystem perspective. *Restoration Ecology* 5:307–17.

Findlay, S. E., G. E. Kiviat, W. C. Nieder, and E. A. Blair. 2002. Functional assessment of a reference wetland set as a tool for science, management and restoration. *Aquatic Sciences* 64:107–17.

Finlayson, M. 2002. Ecosystem assessment: links between science and community—The common ground? *Ecological Management and Restoration* 4:3–4.

Freeman, R. E., and R. O. Ray. 2001. Landscape ecology practice by small scale river conservation groups. *Landscape and Urban Planning* 56:171–84.

Friedman, J. M., W. R. Osterkamp, and W. M. Lewis Jr. 1996. Channel narrowing and vegetation development following a Great-Plains flood. *Ecology* 77:2167–81.

Galatowitsch, S. M., and D. M. Richardson. 2005. Riparian scrub recovery after clearing of invasive alien tress in headwater streams of the Western Cape, South Africa. *Biological Conservation* 122:509–21.

Giller, P. S. 2005. River restoration: Seeking ecological standards. *Journal of Applied Ecology* 42:201–7.

Grayson, J. E., M. G. Chapman, and A. J. Underwood.

1999. The assessment of restoration of habitat in urban wetlands. *Landscape and Urban Planning* 43:227–36.

Harris, J. A., P. Birch, and J. Palmer. 1996. *Land restoration and reclamation: Principles and practice*. Harlow, UK: Longman Higher Education.

Harris, J. A., and R. van Diggelen. 2006. Ecological restoration as a project for global society. In *Restoration ecology: The new frontier*, ed. J. van Andel and J. Aronson, 3–15. Oxford, UK: Blackwell.

Hession, W. C., T. E. Johnson, D. F. Charles, D. D. Hart, R. J. Horwitz, D. A. Kreeger, J. E. Pizzuto, D. J. Velinsky, J. D. Newbold, C. Cianfrani, T. Clason, A. M. Compton, N. Coulter, L. Fuselier, B. D. Marshall, and J. Reed. 2000. Ecological benefits of riparian reforestation in urban watersheds: Study design and preliminary results. *Environmental Monitoring and Assessment* 63:211–22.

Higgs, E. 2003. *Nature by design: People, natural process and ecological restoration*. Cambridge, MA: MIT Press.

Hillman, M., and G. J. Brierley. 2005. A critical review of catchment-scale stream rehabilitation programmes. *Progress in Physical Geography* 29:50–70.

Hobbs, R. J. 2003. Ecological management and restoration: Assessment, setting goals and measuring success. *Ecological Management and Restoration* 4 (supplement):S2–S3.

Hobbs, R. J., and J. A. Harris. 2001. Restoration ecology: Repairing the Earth's ecosystems in the new millennium. *Restoration Ecology* 9:239–46.

Hobbs, R. J., and D. A. Saunders. 2001. Nature conservation in agricultural landscapes: Real progress or moving deckchairs? In *Nature conservation 5: Nature conservation in production landscapes*, ed. J. Craig, N. Mitchell, and D. A. Saunders, 1–12. Chipping Norton, Australia: Surrey Beatty & Sons.

Jackson, L. L., N. Lopoukhine, and D. Hillyard. 1995. Ecological restoration: A definition and comments. *Restoration Ecology* 3:71–75.

Jansen, A. 2005. Avian use of restoration plantings along a creek linking rainforest patches on the Atherton Tablelands, North Queensland. *Restoration Ecology* 13:275–83.

Jansson, R., H. Backx, A. J. Boulton, M. Dixon, D. Dudgeon, F. M. R. Hughes, K. Nakamura, E. H. Stanley, and K. Tockner. 2005. Stating mechanisms and refining criteria for ecologically successful river restoration: A comment on Palmer et al. (2005). *Journal of Applied Ecology* 42:218–22.

Karr, J. R. 1996. Ecological integrity and ecological health are not the same. In *Engineering with ecological constraints*, ed. P. C. Schulze, 97–109. Washington, DC: National Academy Press.

Keddy, P. 1999. Wetland restoration: The potential for assembly rules in the service of restoration. *Wetlands* 19:716–32.

Kern, K. 1992. Restoration of lowland rivers: The German experience. In *Lowland floodplain rivers: Geomorphological perspectives*, ed. P. A. Carling and G. E. Petts, 279–97. Chichester, UK: Wiley.

Lake, P. S. 2001. On the maturing of restoration: Linking ecological research and restoration. *Ecological Management and Restoration* 2:110–15.

Lake, P. S. 2005. Perturbation, restoration and seeking ecological sustainability in Australian flowing waters. *Hydrobiologia* 552:109–20.

Lepori, F., D. Palm, and B. Malmqvist. 2005. Effects of stream restoration on ecosystem functioning: Detritus retentiveness and decomposition. *Journal of Applied Ecology* 42:228–38.

McDonald, A., S. N. Lane, N. N. Haycock, and E. A. Chalk. 2004. Rivers of dreams: On the gulf between theoretical and practical aspects of an upland river restoration. *Transactions of the British Institute of Geographers* 29:257–79.

Moerke, A. H., and G. A. Lamberti. 2004. Restoring stream ecosystems: Lessons from a Midwestern state. *Restoration Ecology* 12:327–34.

Norton, B. G. 1998. Improving ecological communication: The role of ecologists in environmental policy formation. *Ecological Applications* 8:350–64.

Odum, E. P. 1969. The strategy of ecosystem development. *Science* 164:262–70.

Palmer, M. A., E. S. Bernhardt, J. D. Allan, P. S. Lake, G. Alexander, S. Brooks, J. Carr, S. Clayton, C. Dahm, J. Follstad Shah, D. J. Galat, S. Gloss, P. Goodwin, D. H. Hart, B. Hassett, R. Jenkinson, G. M. Kondolf, R. Lave, J. L. Meyer, T. K. O'Donnell, L. Pagano, P. Srivastava, and E. Sudduth. 2005. Standards for ecologically successful river restoration. *Journal of Applied Ecology* 42:208–17.

Pedroli, B., G. de Blust, K. van Looy, and S. van Rooij. 2002. Setting targets in strategies for river restoration. *Landscape Ecology* 17:5–18.

Pelley, J. 2000. Restoring our rivers. *Environmental Science and Technology* 34:86A–90A.

Possingham, H. P. 2001. The business of biodiversity: Applying decision theory principles to nature conservation. *Tela Paper, Issue* 9. Australian Conservation Foundation and Earthwatch Institute.

Rheinhardt, R. D., M. Craig, M. M. Brinson, and K. E. Faser. 1999. Application of reference data for assessing and restoring headwater ecosystems. *Restoration Ecology* 7:241–51.

Rhoads, B. L., D. Wilson, M. Urban, and E. Herricks. 1999. Interaction between scientists and non-scientists in community-based watershed management: emergence of the concept of stream naturalization. *Environmental Management* 24:297–308.

Rohde, S., M. Schütz, F. Kienast, and P. Englmaier. 2005. River widening: An approach to restoring riparian habitats and plant species. *River Research and Applications* 21:1075–94.

Rogers, K. H. 2006. The real river management challenge: Integrating scientists, stakeholders and service agencies. *River Research and Applications* 22:269–80.

Rutherfurd, I. D., K. Jerie, and N. Marsh. 2000. *A rehabilitation manual for Australian streams, volumes 1 and 2.* Canberra, Australia: Cooperative Research Centre for Catchment Hydrology, and the Land and Water Resources Research and Development Corporation.

Ryder, D. S. 2004. Response of epixylic biofilm metabolism to water level variability in a regulated floodplain river. *Journal of the North American Benthological Society* 23:214–23.

Ryder, D. S., and W. Miller. 2005. Setting goals and measuring success: Linking patterns and processes in stream restoration. *Hydrobiologia* 552:147–58.

Ryder, D. S, R. J. Watts, E. Nye, and A. Burns. 2006. Can flow velocity regulate biofilm structure in a regulated lowland river? *Marine and Freshwater Research* 57:29–35.

Safstrom, R., and M. O'Byrne. 2001. Community volunteers on public land need support. *Ecological Management and Restoration* 2:85–86.

Skidmore, P. B., F. D. Shields, M. W. Doyle, and D. E. Miller. 2001. A categorization of approaches to natural channel design. In *Wetlands Engineering and River Restoration Conference*, ed. D. F. Hayes. Reston, VA: American Society of Civil Engineers.

Zedler, J. B. 1996. Coastal mitigation in southern California: The need for a regional restoration strategy. *Ecological Applications* 6:84–93.

Chapter 3

Turbulence and Train Wrecks: Using Knowledge Strategies to Enhance the Application of Integrative River Science to Effective River Management

ANDREW BOULTON, HERVÉ PIÉGAY, AND MARK D. SANDERS

> Conducting an epistemological analysis of the scientific disciplines involved in environmental problems can help establish realistic expectations about the role of science, the adequacy of information, and the effectiveness of management policies in dealing with particular problems.
>
> Benda et al. 2002, 1134

The research literature is rife with tales of woe in the development and application of crossdisciplinary science in environmental management—failed scientific projects, disenchanted teams, unsatisfied managers, an upset public, and so on (e.g., Hobbs 1997; Kondolf 1998; Chenoweth et al. 2002; Wohl et al. 2005). Borrowing the graphic metaphor from Benda et al. (2002), these failures, in terms of time, personal grief, and money, can be referred to as "train wrecks." In this chapter, we consider steps that can be taken to avoid these train wrecks. Part of the solution relates to sharing "common ground" among all stakeholders at the outset (Allen and Kilvington 2005). Scientific questions that address environmental problems must be framed in relation to their human dimension, considering social, political, and economic issues (Wu and Hobbs 2002; Graf 2005; Hillman and Brierley 2002, 2005). This involves smoothing what Cullen (1990) describes as the "turbulent boundary" between environmental scientists and resource managers. Such turbulence arises because of a lack of effective communication that arises from a misunderstanding of the cultures within which each works. This can generate mutual distrust, stifle information flow, and hamper problem resolution (Allen and Kilvington 2005).

Benda et al. (2002) argued that an explicit analysis of the knowledge structures of various disciplines should be conducted at the outset of crossdisciplinary collaborations addressing environmental problems. In this chapter, we advocate extending this approach to include *all* stakeholders, especially the public, because public support is essential for successful research toward sustainable solutions to environmental problems (Golet et al. 2006). We review ten tenets we think should be held as shared goals and twelve commitments that promote behavior or actions to help integrative river science occur. We conclude by urging practitioners to try this approach, documenting its failings as well as its successes so that we can improve the rate of progress of truly integrative river science.

Sources of Turbulence

Several sources of "turbulence," as defined by Cullen (1990), characterize transdisciplinary interactions. The first is the mistaken perception of polarity of attitudes, reflecting the broad spectrum of views and attitudes between scientists and managers (Halse and Massenbauer 2005). The second relates to the rapidity of recent advances in our environmental awareness, with changes in emphasis in river management from engineering needs to knowledge needs associated with ethics, conservation, sustainable development, and social justice (Piégay et al. 2002; Hillman and Brierley 2002, 2005). Turbulence occurs because issues change at different paces and places. These changes are often so rapid and cross such a broad sector (from technology to social science) that senior scientists and managers may not keep pace, which leads to discord in communication and progress (Richter et al. 2006).

A third source of turbulence reflects prior experiences. Managers, disillusioned by a slow rate of progress or the questionable relevance of scientific work commissioned to tackle a specific project (Halse and Massenbauer 2005), may be reluctant to work with subsequent scientists (Allen and Kilvington 2005). University researchers becoming embroiled in time-consuming court cases or protracted issues of access to land or data may retreat from future opportunities to work on applied issues. To resolve these causes of turbulence—or, even better, to avoid them at the outset—it is best to begin with shared points of agreement. Prospects for meaningful collaboration are enhanced by establishing ground rules and common goals, building relationships with all relevant participants, and adopting appropriate spatial and temporal scales (box 3.1).

Reducing Turbulence with Shared Beliefs: Tenets and Commitments

There is a surprising amount of crossdisciplinary agreement among geomorphologists, hydrologists, and ecologists in understanding how rivers function, at least at a general level (Benda et al. 2002, chapter 1). Much of this agreement emerged in the 1980s from the increased emphasis in systems thinking in the environmental sciences (e.g., Roux 1982; Diamond and Case 1986) and the enhanced interactions that resulted from exercises such as expert panel assessments for flow allocations (e.g., Arthington 1998; Richter et al. 2006). In addition to a common, broad understanding of river function, we suggest that scientists, practitioners, river managers, and other stakeholders should share a common set of tenets about the rationale for integrative river science and management. These tenets (table 3.1) could provide a useful entry point for obtaining consensus and initiating the dialogue that underpins successful sharing of knowledge.

The first two tenets acknowledge the importance of science in river management and that it is valued by managers (table 3.1). The third, fourth, and fifth tenets establish the benefits of crossdisciplinary research (especially in solving environmental problems) at an appropriate scale (box 3.1) and with due recognition of the emergent properties that typify complex ecosystems (example in box 3.2). The next two tenets consider the uncertainty of prediction associated with scale of study and how knowledge structures from differing disciplines affect approaches to data interpretation (Benda et al. 2002). These differences in knowledge structure can lead to "turbulence" (*sensu* Cullen 1990), the eighth tenet. The final two tenets emphasize the significance of communication and the relevance of social scientists (table 3.1).

If all or most of these tenets are shared, the crucial early stages of trust, acknowledgment, respect, and direction should result (Allen and Kilvington 2005, box 3.3). But *actions* are needed as well. A workshop on integrative river science (see acknowledgments, below) generated a dozen "commitments" (table 3.2), expressed in terms of actions and behaviors that participants had found to promote success in crossdisciplinary research and problem-solving. Some relate directly back to the tenets: effective communication, breadth of experience, responsibility, and passion. Others underpin

Box 3.1. Focusing on System Understanding Rather Than an Immediate Solvable Question: The Long Term Ecological Research Basin of the Rhône (ZABR Rhône), France (http://www.graie.org/zabr/index.htm)

Even when working in an interdisciplinary framework, the goals of scientists and river managers often diverge (Cullen 1990). This is most likely when an applied question must be solved in a limited period of time. Typically, scientists contribute as reluctant "super" environmental consultants whereas managers do not really profit from the potential benefits of the science because scientists must adapt to the managers' time scale. In such cases, managers would have more reliable data if they flagged issues earlier and scientists had more time to collect adequate information to advise the managers. To address this problem, the ZABR group focuses collectively on specific geographic areas (single reaches up to catchments of 90,000 square kilometers) so that the scientific community can provide basic knowledge on environmental system trajectories to solve applied questions.

Two parallel programs operate in the ZABR: one focusing on fundamental interdisciplinary programs providing knowledge about system parameters and technological approaches; the second centered on applications where scientific objectives are discussed by a steering committee comprising scientists, river basin managers, and end-users. The whole project is funded by different agencies depending on the focus of the programs. Nonetheless, fundamental research funded from its own sources provides meaningful elements that can be developed for management purposes with other strategies and sources of funds. Within this type of framework, scientists can adopt long-term perspectives and sustain high-quality research while managers can profit from the cumulative knowledge in their region, enhancing management actions and funding effective applications of the science at appropriate spatial and temporal scales.

TABLE 3.1

Ten Tenets of Successful Integrative River Science and Management. These tenets could provide a starting point from which to derive common ground for discussions of bridging the gaps between scientists and managers.

1	Science can provide generalizations that may be usefully applied to river management. Therefore, scientists should play a role in river management.
2	Scientific input is valued and sought by most nonscientists, such as managers, policy-makers, and the public.
3	There are benefits of crossdisciplinary approaches from fields such as hydrology, geomorphology, and ecology to integrative river science.
4	There is a need for integrative studies of whole river systems at the catchment scale, if not broader.
5	Because rivers are large, complex systems, they have crucial emergent properties that cannot be understood when investigated in a reductionist, discipline-specific mode; "the whole exceeds the sum of the parts."
6	Uncertainties about outcomes (scientific predictions and management expectations) increase with spatial and temporal scale.
7	Scientists from different disciplines have different research histories and traditions. These dictate what variables are measured and how data are analyzed and interpreted.
8	Philosophical differences exist between managers and scientists although these may be along a continuum.
9	Scientists should put more effort into effective technology transfer to the general public, decision-makers, and field practitioners. The success of crossdisciplinarity hinges on scientists' abilities to communicate, share knowledge, and define collective aims.
10	Stakeholder participation in integrative river management is becoming more widespread and better informed. Social scientists are essential to evaluate and enhance stakeholder participation and the adoption of proposed recommendations.

Box 3.2. The Upper Hunter River Rehabilitation Initiative (UHRRI), Australia (further information www.hcr.cma.nsw.gov.au/uhrri)

This project on the Hunter River in New South Wales, Australia, sought to integrate an extensive river rehabilitation program (UHRRI) with crossdisciplinary scientific research spanning vegetation ecologists, geomorphologists, and stream biologists. However, it soon became clear that the scientists and industry partners had different expectations of what science might achieve. Industry sponsors were looking for clear answers to practical, site-specific questions such as "what are the best methods for growing trees at this site?" In contrast, the scientists on the project expected to make small, incremental gains in knowledge that would be more broadly applicable, such as "how are geomorphic site conditions related to plant functional traits?" In our experience, such mismatches in expectations are common, but do not have to be problematic if addressed early so that all stakeholders are clear about what science can deliver (table 3.1).

This project was also a good example of mismatches of scale that needed to be considered early. Ecologists were attempting to assess the influence of woody debris on microhabitat availability to macroinvertebrates or localized exchanges of water between the stream and the hyporheic zone (10^{-2} to 10^{0} meters) whereas the geomorphologists were focusing on information collected at the catchment-scale (10^{2} to 10^{4} meters). Although these latter data provided a useful context for the study at a river-reach or river style scale (sensu Brierley et al. 2002), the ecologists needed to collect their own geomorphic data on changes in sediment size and erosion/deposition patterns in response to the introduction of wood at this finer scale.

successful activities: precise goals, practicality and efficiency, creativity and flexibility. Yet others are behavioral: reflective and willing to learn, ethical and social (table 3.2). Principles on this list could help readers from a wide range of backgrounds and knowledge structures address crossdisciplinary concerns in both scientific and management practice, thus helping to smooth turbulence and avoid train wrecks.

Seeking Solvable Problems: Comparative Analysis of Knowledge Structures

Within the various disciplines that comprise environmental science, there are marked differences in research culture, theories of knowledge, spatial and temporal scales of research, technical language, and accuracy of predictions. These differences or mismatches are sometimes only identified late in the collaborative process, frequently when data analysis fails because of a fatal mismatch of spatial scale or some problem in the precision of measurement of a driving variable (Boulton et al. 2005). These failures seldom see print or are discussed openly, representing missed opportunities to learn from our mistakes (Brierley et al. 2005; Tress et al. 2005). To address these crises before they occur, Benda et al. (2002) propose a method for evaluating knowledge structures as a way of revealing potential biases or mismatches (table 3.3). These insights can then be used to explore approaches to either resolving the mismatch or redefining the research question into a problem that can be solved given the lowest common denominator of expertise or knowledge—that is, a "solvable problem."

TABLE 3.2

Twelve Commitments of Action and Behavior That Will Facilitate Integrative River Science. These were drawn from the experiences of participants of the workshop held in 2005, and are phrased to be inspirational and active.

1	Be communicative. Use clear and integrative language to establish common grounds and shared knowledge structures.
2	Be precise. An agreed-upon definition of clear objectives and a dynamic common vision help determine the resolution and precision needed to study necessary processes as well as unite players in a common goal.
3	Be a student. Be willing to learn and share collective wisdom and previous experiences/context.
4	Be practical. Aim for practical management outcomes.
5	Be reflective and humble. Learn from mistakes. Measure the effectiveness of research. Allocate time for reflection. Recognize limitations and gaps in our personal or discipline-specific understanding.
6	Be creative and flexible. Multiple lines of evidence may be needed to illustrate the success of integrative river science and new tools may arise that can be used.
7	Be efficient. Time and money are limited resources. A team approach can often identify efficiencies after some discussion.
8	Be myrioramic (multifaceted). Maintain a positive appreciation of other fields and disciplines, respecting their contributions.
9	Be responsible and passionate. Passion and commitment from the group often needs a responsible champion to drive and maintain impetus.
10	Be contumacious. Don't be constrained by institutional structure and be prepared to think (and act) "outside the box."
11	Be social. Share ideas and talk to everyone about your work and thoughts, and listen to all suggestions and approaches.
12	Be scientifically ethical. Rigor, peer review, transparency, and intellectual honesty are critical components of scientific integrity and credibility.

The typical approach to environmental investigations is to begin with a broad research question and then embark on a set of research studies that addresses specific aspects of the question (Rogers and Biggs 1999). For example, one research question in the field of integrative river science and rehabilitation might be "How does the introduction of woody debris into the Hunter River, New South Wales, alter the hydrology and sediment dynamics of the system, and what are the repercussions of these alterations on aquatic biota and ecosystem processes?" One likely scenario for a team of collaborating scientists would be for the hydrologists to first assess changes in flow patterns within the system, possibly even exploring localized and reach-scale changes in groundwater–surface water exchange. The geomorphologists then would consider how these changes in flow would affect erosion, sediment transport, and deposition in the channel and through the system. Ecologists might then explore how the presence of wood and alterations in channel shape and flow influenced the aquatic biota and associated ecosystem processes.

Although this approach may seem logical, it has several serious flaws. First, there is unlikely to be a sequence where the hydrologists first make their measurements, followed by the geomorphologists, and finally the ecologists, because projects have time constraints and many measurements must be taken simultaneously so that data matching can occur. Further, some sampling methods may impact the success of others. For example, the removal of large volumes of sediment from the stream bed for geomorphic analysis may disrupt benthic invertebrate samples collected by the ecologist and vice versa. Second, the broad conceptual model of river function, especially from the ecological aspect, has considerable uncertainty (Wu and Hobbs 2002; Nilsson et al. 2003) and the expectations of predictive power in response to any experimental manipulation across these different research fields must be realistic. Finally, in all of this, some fundamental sampling decisions about spatial and temporal scale must be made (Downes et al. 2002), even across different groups within the same discipline (e.g., biofilms versus fish).

Box 3.3. Project River Recovery in New Zealand

While the Project River Recovery (PRR) in the central South Island, New Zealand, experienced turbulence in its early years, it avoided becoming a train wreck and is now considered an excellent example of ecological restoration (Caruso 2006). This research-based river conservation program commenced with a strong focus on endangered birds, and tackled several difficult problems that generated much public controversy. It has now matured into a whole-ecosystem project that protects or manages more than 30,000 hectares of braided rivers and associated wetlands, and enjoys wide stakeholder and interdisciplinary involvement.

One of us (Mark D. Sanders) worked as manager and scientist on this project and considers several key factors that were instrumental in its success. The first was employing staff with strong science training, committed to living locally and achieving on-the-ground results rather than seeking publication, yet maintaining close links with full-time research scientists. A second was secure, long-term external funding and relative autonomy, but within a strong management framework requiring clear planning, budgeting, and quantitative performance and financial reporting. Finally, a strategic planning approach involved stakeholder input in the development of a clear rationale and mandate for the project, setting clear guiding principles and realistic operational plans (Woolmore and Sanders 2005). PRR's guiding principles are:

1. Integrated habitat, ecological community, and species management
2. Science-informed work
3. Monitoring to enable cost-benefit analyses
4. Close liaison with the community
5. Excellence

PRR has operated in close accord with the tenets and commitments outlined in tables 3.1 and 3.2. PRR's extensive consultation with stakeholders since its inception in 1991, in diverse formal and informal settings, has resulted in a deep understanding of the knowledge structures of multiple disciplines and stakeholder groups, including university and research institute scientists, field staff, internal management staff, indigenous people, local farmers, community members and businesses, and members of local, regional, and national government and nongovernmental organizations.

TABLE 3.3

Broad Components of a Knowledge Structure. The first five are proposed by Benda et al. (2002) and we suggest that many social scientists would advocate the sixth.

1	Disciplinary history and the forms of scientific knowledge that are currently available within that discipline
2	Spatial and temporal scales of that knowledge and its applications
3	Knowledge precision (i.e., qualitative versus quantitative forms), and how this affects our understanding across scales
4	Accuracy of predictions, again scale dependent
5	Availability of data to construct, calibrate, and test predictive models to refine knowledge
6	Contrasting theories of knowledge and resulting research paradigms

Benda et al. (2002) acknowledge these problems as common in river science. They suggest that when faced with a research question, all collaborators begin by explicitly stating their particular discipline's knowledge structure (see table 3.3). By getting each participant to evaluate these features for their own specific discipline and explain how they relate to the research objective, the team is able to collectively advance their understanding and to identify potential mismatches early in the process.

Four Logical Steps to Evaluate Knowledge Structures

The evaluation of knowledge structures follows four logical steps (Benda et al. 2002). The first step is the selection of a specific question (for instance, the Hunter River example above) or a general class of question (e.g., "what might be the effects of woody debris introduction into any river?"). The next step, far more challenging, is for each discipline to define its knowledge structure as defined in table 3.3. The research traditions thus exposed will reveal what techniques are available, what variables are measured, what levels of precision exist, and how data will be analyzed. The third step is to identify potential difficulties that derive from comparison of the various knowledge structures of the different disciplines, including the absence of scientific knowledge, mismatch of scale or precision, or lack of specific data. The final step entails determining strategies to circumvent these difficulties and to construct solvable problems.

This four-step process of analyzing knowledge structures for each discipline is a real challenge for its proponents because it requires a good working knowledge of the discipline's origins and history as well as the ability to describe this in relation to the research topic. For various reasons, scientists may have only limited familiarity with past literature, a problem that is coupled with inadequate integration of empiricism and theory (at least in ecology, Belovsky et al. 2004). Another challenge is specifying spatial and temporal scales of the various forms of knowledge of each discipline. Benda et al. (2002) give an excellent example of incompatibilities arising across hydrology, geomorphology, and ecology, and suggest a trend across the three disciplines. Hydrology, with the fewest parameters, makes relatively accurate predictions at small scales; geomorphology, with more parameters, makes relatively inaccurate quantitative predictions and so is pressed to be qualitative; and ecology, with the greatest number of parameters spanning several spatial and temporal scales, makes predictions mostly in qualitative or conceptual terms. This trend in increasing quantitative uncertainty with an increasing number of parameters and associated scales poses significant challenges to successful crossdisciplinary collaboration (chapter 14).

This raises the question as to the most appropriate scales of analysis for each of the disciplines involved (box 3.3). Is it more appropriate for one group to work at one scale while the other group works at another scale, or should the team try to concentrate all of its effort at a single scale? Most often, it comes down to a question of cost-benefits, time, management priorities, and the expertise and interest of the individual researchers. While these are all good practical reasons for choice of scale, they may be less defensible when it comes to actually solving the problem at an appropriate scale, especially since the answers probably lie at multiple scales. For all of these reasons, elements of knowledge structures must be drawn together in a practical context to construct a solvable problem. Critically, this should be done early in the process.

Strategies for Constructing Solvable Problems: Difficulties and Potential Solutions

Having compared the various knowledge structures across the collaborating disciplines, the team must next explore ways to address these difficulties and mismatches such that a solvable problem can be generated. All too often, mismatches are perceived late in a project and then either become caveats to the final analysis or, worse, are glossed over completely. Obviously, it is not enough to simply agree that the scales of assessment differ between ecologists and geomorphologists and move on—strategies to redefine the problem and sampling approach are needed to address the issue (boxes 3.1 and 3.3).

There are at least four potential difficulties that arise during the comparative analysis of knowledge structures. The first and rather obvious one is where scientific knowledge or specific field data are lacking. In the Hunter River example (box 3.2), although it was known from other studies that nutrients could be generated in the hyporheic zone and transported into the surface waters in upwelling zones (e.g., Hancock and Boulton 2005), it was not known what nutrients limited surface primary productivity and whether there would be any algal response to changes in surface-subsurface water exchanges in response to wood addition. Primary data collection is the most direct way to solve the problem if time and resources allow.

An alternate approach could involve suitably cautious extrapolation of data from other studies of similar systems along with other "levels-of-evidence" approaches (table 3.2, Downes et al. 2002). In cases where this is impossible, Benda et al. (2002) suggest eliminating or bypassing the discipline. They give an example from "watershed analysis" used to address effects of timber harvesting on fishery resources in the Pacific Northwest. Land use impacts were evaluated as effects on physical habitats under the simplifying assumption that physical habitat conditions are tightly linked to biological responses; the biological discipline was bypassed completely. When primary data cannot be obtained, levels of evidence can be gleaned from databases and case studies that are now being collated as readily accessible resources (e.g., Jenkinson et al. 2006).

Another difficulty is when the knowledge is available at the wrong spatial or temporal scale. This is exemplified by our earlier example of geomorphology and ecology in the research question in the Hunter River (box 3.2). Unfortunately, the hierarchical relationships of ecological responses to spatial scale are seldom clear. For example, Boulton et al. (1998) adopted three "scales" (sediment, reach, and catchment) for discussing the functional significance of the hyporheic zone to stream processes but were unable to indicate how sediment-level responses, for example, were related to reach-level processes. Another example of rescaling problems is when efforts are used to scale down complexity by constructing model systems such as flumes where environmental conditions can be controlled (Merritt and Wohl 2002). Of course, extrapolating these results back to the field can be problematic, but experimental manipulations of scale-dependent features in model systems could help resolve scaling issues if the behavior of these features could be understood.

A popular way to approach this problem is to collectively develop a conceptual model describing how the system works (box 3.1). From this, we can assess the potential effects of a given manipulation or infer the expected evolution of the river (chapter 5). Not only are these models useful for predicting outcomes, but it is also possible to rewrite each contributor's questions as hypotheses to be tested. Broad generalities can then be simplified so that the model can help combine all the stakeholders' hypotheses to provide an integrated answer to a specific management problem. This model can continue to be revised throughout the project, providing a valuable tool for focusing management options as well as understanding system function (Sime 2005).

The third common difficulty that arises is the unequal ability to quantify parameters across disciplines. The most logical solution is to modify the analysis to incorporate these differences yet still address the solvable problem. Ecologists, not surprisingly, are familiar with these kinds of approaches where they need to make decisions about the "biological significance" of a difference compared to its "statistical significance"; sometimes the conventional probability level of 0.05 is unwarranted when quantitative data are hard to obtain (Downes et al. 2002). Often, questions can be modified in a way that renders their answers useful, even though the responses being sought are only semiquantitative. In the Hunter River example, it would be impossible to quantify exactly the degree to which sediments would be redistributed in a reach from a given volume of woody debris introduced into the river, and prospective changes in aquatic invertebrate community composition. Therefore, a solvable research question could be phrased in terms of small (less than 10 centimeters), moderate (10 to 100 centimeters), or large (greater than 100 centimeters) changes in channel sediment redistribution, qualitatively assessing the changes in invertebrate community composition along this gradient of the magnitude of change. As a manager is likely to be more interested in relative degrees of change in channel morphology, and perhaps in the health of the stream as indicated by aquatic community composition, exact quantitative data are not going to be needed. Thus, the mismatch in precision is not an issue.

Related to the unequal ability across disciplines to quantify parameters is the low accuracy of predictions that might arise from being obligated to use the lowest common denominator of qualitative information, the fourth problem. Benda et al. (2002) advocate "coarse grain" analysis to tackle this, combining empirical knowledge and theoretical reasoning through mathematical synthesis and computer simulations to improve the accuracy of the predictions. We seek to extend this to crossdisciplinary collaboration when the accuracy of scientific predictions must be harnessed to value judgments by managers and the general public (Edgar et al. 2001). For example, when evaluating impacts, all stakeholders (not just scientists) must nominate what they consider as "unacceptable" change (Rogers and Biggs 1999; Downes et al. 2002). This step must be done before variables are selected or research designs developed. Using the Hunter River example once more, it may be possible to predict that introducing woody debris will enable viable populations of native fish to persist even if the accuracy of where, when, and what species is lacking because it may be that the value judgment by the public is simply that they want a river that has some native fish that can persist in it. A prediction that this will occur given a certain timeframe and river rehabilitation strategy could be all that is needed, solving the problem and going some way toward helping ensure ecologically successful river rehabilitation (Palmer et al. 2005).

Conclusions

The perspectives advocated here promote the clear articulation of "problems" at the outset of a project, allowing the research team and other stakeholders to be aware of the difficulties, mismatches, and strategies so that they can redefine the problem to make it "solvable." Explicit commitment to transparency at the outset reveals the limitations of science to nonscientists, obligates managers to help restate problems in order to make them solvable yet useful, and fosters collaboration from the beginning in a way that should smooth some of the turbulence described by Cullen (1990). Identification of knowledge structures, formal documentation of assumptions, data types, scales, and precision of measured variables, and compromises that have been made in recognition of mismatches among disciplines are invaluable to people who join collaborative teams later in the process. In long-term projects particularly, staff come and go. Therefore, it is crucial that detailed records are maintained so that new participants can quickly become familiar with reasoning behind decisions and the evolution of the solvable questions. Changes to the knowledge base, research, and management questions, and composition of the team must be aligned with redefinition of milestones and targets. Effective management of data is crucial, often including Geographical Information

Systems (GIS) implementation, shared data-bases on the Web, evolving graphic tools to develop and illustrate conceptual models, and frequent public consultation.

Although the approach proposed by Benda et al. (2002), extended to include social scientists and issues of ethics and policy, addresses many of the concerns for turbulence at the interface between science and management identified by Cullen (1990), there remains a need for more applications of this knowledge structure approach and assessments of its success in integrative river science. The trend toward transdisciplinary research has many compelling arguments in its favor, so we need to identify and resolve problems in this type of collaboration and share our experiences (including the failures) to improve the efficiency of this approach. Comparing knowledge structures as a framework for identifying and addressing disciplinary mismatches to yield solvable research questions is just a beginning. Alternative tools and approaches can be borrowed and adapted from other disciplinary fields (e.g., network governance, Imperial 2005). Individual personalities and circumstantial serendipity play crucial but often unacknowledged roles in collaboration and creativity may be stifled by an overrigid framework. Yet, we urge scientists, managers, decision-makers, and the general public to adopt our commitments and to consider the benefits of the knowledge structures approach. Otherwise, lack of communication early in the process about common ground and potential mismatches can lead to turbulence and train wrecks.

Acknowledgments

Many of the ideas and commitments expressed here arose from a workshop in Muswellbrook in March 2005, and we thank Gary J. Brierley and Kirstie A. Fryirs for organizing it and editing this book. We also thank the many colleagues who have shared their ideas, particularly Darren Ryder, Richard Hobbs, Bill Johnson, Gary Brierley, and Julian Prior. Andrew Boulton and Mark Sanders acknowledge the Australian Research Council, Macquarie Generation, Mt. Arthur Coal, Bengalla Mining Company, the NSW Department of Natural Resources, and the Hunter-Central Rivers Catchment Management Authority for funding and support. Hervé Piégay is grateful for the cross-disciplinary discussions and experiences of the group of the French Drôme catchment working site of the Rhône Basin LTER (http://www.graie.org/zabr/index.htm). We thank Darren Ryder, Gary J. Brierley, Mick Hillman, and Gordon Grant for valuable comments on drafts of this manuscript, and F. M. Sunion for support during its preparation.

References

Allen, W. J., and M. J. Kilvington. 2005. Getting technical information into watershed decision making. In *The farmers' decision: Balancing economic successful agriculture production with environmental quality*, ed. J. L. Hatfield, 45–61. New York, NY: Soil and Water Conservation Society.

Arthington, A. H. 1998. *Comparative evaluation of environmental flow assessment techniques: Review of holistic methodologies*. LWRRDC Occasional Paper 26/98. Canberra, Australia: Land and Water Resources Research and Development Corporation.

Belovsky, G. E., D. B. Botkin, T. A. Crowl, K. W. Cummins, J. F. Franklin, M. L. Hunter, A. Joern, D. B. Lindenmayer, J. A. MacMahon, C. R. Margules, and J. M. Scott. 2004. Ten suggestions to strengthen the science of ecology. *BioScience* 54:345–51.

Benda, L. E., N. L. Poff, C. Tague, M. A. Palmer, J. Pizzuto, S. Cooper, E. Stanley, and G. Moglen. 2002. How to avoid train wrecks when using science in environmental problem solving. *BioScience* 52:1127–36.

Boulton, A. J., S. Findlay, P. Marmonier, E. H. Stanley, and H. M. Valett. 1998. The functional significance of the hyporheic zone in streams and rivers. *Annual Review of Ecology and Systematics* 29:59–81.

Boulton, A. J., D. Panizzon, and J. Prior. 2005. Explicit knowledge structures as a tool for overcoming obstacles to interdisciplinary research. *Conservation Biology* 19:2026–29.

Brierley, G., K. Fryirs, D. Outhet, and C. Massey. 2002. Application of the River Styles framework as a basis for river management in New South Wales, Australia. *Applied Geography* 22:91–122.

Brierley, G., C. Miller, A. Brooks, K. Fryirs, A. Boulton, D. Ryder, M. Leishman, D. Keating, and J. Lander. 2005. Making integrative, cross-disciplinary research happen: Initial lessons from the Upper Hunter River Rehabilitation Initiative. In *Linking rivers to landscapes*, ed. I. D. Rutherfurd, I. Wiszniewski, M. Askey-Doran, and R. Glazik, 125–33. Hobart, Australia: Department of Primary Industries, Water and Environment.

Caruso, B. S. 2006. Project River Recovery: Restoration of braided gravel-bed river habitat in New Zealand's high country. *Environmental Management* 37:840–61.

Chenoweth, J. L., S. A. Ewing, and J .F. Bird. 2002. Procedures for ensuring community involvement in multijurisdictional river basins: A comparison of the Murray-Darling and Mekong River basins. *Environmental Management* 29:497–509.

Cullen, P. 1990. The turbulent boundary between water science and water management. *Freshwater Biology* 24:201–9.

Diamond, J., and T. J. Case. 1986. *Community ecology*. New York: Harper & Row.

Downes, B. J., L. A. Barmuta, P. G. Fairweather, D. P. Faith, M. J. Keough, P. S. Lake, B. D. Mapstone, and G. P. Quinn. 2002. *Monitoring ecological impacts: Concepts and practice in flowing waters*. Cambridge, UK: Cambridge University Press.

Edgar, B., N. Schofield, and A. Campbell. 2001. Informing river management policies and programs with science. *Water Science and Technology* 43:185–95.

Graf, W. L. 2005. Geomorphology and American dams: The scientific, social, and economic context. *Geomorphology* 71:3–26.

Golet, G. H., M. D. Roberts, R. A. Luster, G. Werner, E. W. Larsen, R. Unger, and G. G. White. 2006. Assessing societal impacts when planning restoration of large alluvial rivers: A case study of the Sacramento River project, California. *Environmental Management* 37:862–79.

Halse, S. A., and T. Massenbauer. 2005. Incorporating research results into wetland management: Lessons from recovery catchments in saline landscapes. *Hydrobiologia* 552:33–44.

Hancock, P .J., and A. J. Boulton. 2005. The effects of an environmental flow release on water quality in the hyporheic zone of the Hunter River, Australia. *Hydrobiologia* 552:75–85.

Hillman, M., and G. J. Brierley. 2002. Information needs for environmental-flow allocation: A case study from the Lachlan River, New South Wales. *Annals of the Association of American Geographers* 92:617–30.

Hillman, M., and G. J. Brierley. 2005. A critical review of catchment-scale stream rehabilitation programmes. *Progress in Physical Geography* 29:50–76.

Hobbs, R. 1997. Future landscapes and the future of landscape ecology. *Landscape and Urban Planning* 37:1–9.

Imperial, M. 2005. Using collaboration as a governance strategy: Lessons from six watershed management programs. *Administration & Society* 37:281–320.

Jenkinson, R. G., K. A. Barnas, J. H. Braatne, E. S. Bernhardt, M. A. Palmer, and J. D. Allan. 2006. Stream restoration databases and case studies: A guide to information resources and their utility in advancing the science and practice of restoration. *Restoration Ecology* 14:177–86.

Kondolf, G. M. 1998. Lessons learned from river restoration projects in California. *Aquatic Conservation: Marine and Freshwater Ecosystems* 8:39–52.

Merritt, D. M., and E. E. Wohl. 2002. Processes governing hydrochory along rivers: Hydraulics, hydrology, and dispersal phenology. *Ecological Applications* 12:1071–87.

Nilsson, C., J. E. Pizzuto, G. E. Moglen, M. A. Palmer, E. H. Stanley, N. E. Bockstael, and L. C. Thompson. 2003. Ecological forecasting and the urbanization of stream ecosystems: Challenges for economists, hydrologists, geomorphologists, and ecologists. *Ecosystems* 6:659–74.

Palmer, M. A., E. S. Bernhardt, J. D. Allan, P. S. Lake, G. Alexander, S. Brooks, J. Carr, S. Clayton, C. Dahm, J. Follstad Shah, D. J. Galat, S. Gloss, P. Goodwin, D. H. Hart, B. Hassett, R. Jenkinson, G. M. Kondolf, R. Lave, J. L. Meyer, T. K. O'Donnell, L. Pagano, P. Srivastava, and E. Sudduth. 2005. Standards for ecologically successful river restoration. *Journal of Applied Ecology* 42:208–17.

Piégay, H., P. Dupont, and J. A. Faby. 2002. Questions of water resources management: Feedback of the French implemented plans SAGE and SDAGE (1992–1999). *Water Policy* 4:239–62.

Richter, B. D., A. T. Warner, J. L. Meyer, and K. Lutz. 2006. A collaborative and adaptive process for developing environmental flow recommendations. *River Research and Applications* 22:297–318.

Rogers, K., and H. Biggs. 1999. Integrating indicators, endpoints and value systems in strategic management of the rivers of the Kruger National Park. *Freshwater Biology* 41:439–51.

Roux, A. L. 1982. *Cartographie polythématique appliquée à la gestion écologique des eaux: Étude d'un hydrosystème fluvial, le haut-Rhône français*. Lyon, France: Published report CNRS.

Sime, P. 2005. St. Lucie Estuary and Indian River Lagoon conceptual ecological model. *Wetlands* 25:898–907.

Tress, B., G. Tress, and G. Fry. 2005. Researchers' experiences, positive and negative, in integrative landscape projects. *Environmental Management* 36:792–807.

Wohl, E., P. L Angermeier, B. Bledsoe, G. M. Kondolf, L. MacDonnell, D. M. Merritt, M. A. Palmer, N. L. Poff, and D. Tarboton. 2005. River restoration. *Water Resources Research* 41:W10301.

Woolmore, C., and M. D. Sanders. 2005. Project river recovery strategic plan 2006–2012. *Project River Recovery Report,* 04/05. Twizel, New Zealand: Department of Conservation.

Wu, J. G., and R. Hobbs. 2002. Key issues and research priorities in landscape ecology: An idiosyncratic synthesis. *Landscape Ecology* 17:355–65.

PART II

An Integrative Scientific Perspective to Guide the Process of River Repair

> While our insights and methods for analyzing the parts of the complex web of relations have become increasingly sophisticated and our tools for measurement incredibly precise, our conceptual framework for integrating the growing mass of information is still lagging behind.
>
> Rapport et al. 1998, xi

Part II addresses two primary themes: crossdisciplinary scientific considerations that underpin the process of river repair (chapters 4, 5, and 6), and concerns for the measurement of river condition and the analysis of societal relations to river systems (chapters 7 and 8). A whole-of-system approach to the process of river repair is framed in terms of ecosystem integrity, merging abiotic concerns that shape the biophysical template of river systems (i.e., physical integrity) with biotic concerns that fashion biodiversity and functional relationships (ecological integrity). The biophysical template builds upon a geomorphic platform that frames habitat availability. Ecological concerns fashion the viability and use of habitat. A key focus throughout this section is the need to apply scientific understanding in a system-by-system manner, making careful use of generalized theoretical understanding.

Chapter 4 analyzes the spatial organization of river systems, considering cross-scalar approaches to the analysis of river systems in terms of nested hierarchical arrangements. Notions of environmental gradients are framed in relation to patch dynamic perspectives. A catchment focus for river rehabilitation initiatives works with longitudinal, lateral, and vertical linkages in river systems. Subsequent chapters focus on analysis of landscape and ecosystem dynamics, placing trajectories of change in an evolutionary context (chapter 5) and controls upon aquatic ecosystem functionality (chapter 6).

Chapter 5 frames notions of river behavior and river change in relation to the evolutionary trajectory of river systems, placing impacts of human disturbance in context of natural variability. These principles are combined to develop an approach to place-based conceptual modeling of river systems, and a case study is presented.

Chapter 6 explores approaches to modeling elements of ecosystem integrity in a crossdisciplinary manner, and addresses three key themes: the importance of integrating structural and functional relationships, consideration of cross-scalar spatial and temporal relationships, and concerns for biophysical connectivity. Examples of the application of crossdisciplinary science examine two components of ecosystem functionality, namely organic matter processing and hyporheic zone processes.

Collectively, these various insights can be used to determine what is achievable in rehabilitation terms for any given catchment. Chapter 7 links this integrative scientific platform with which to approach targeted management to assessment of river condition, placing emphasis on concerns for ecosystem integrity and striving to develop approaches that "give a voice to the environment" (alternatively termed "thinking like an ecosystem"). A geomorphic template provides a biophysical platform with which to merge abiotic and biotic considerations in a manner that is appropriate to the specific system under investigation. Principles that should be incorporated within condition assessment procedures are outlined, building upon themes outlined in chapters 4, 5, and 6 (i.e., system-specific applications, concern for system dynamics, incorporation of elements of ecosystem functionality).

Ultimately, the process of river management must frame scientific perspectives within a social context. Chapter 8 relates ecosystem values to social values through place-based approaches to the analysis of river health. This chapter uses the idea of "connectivity" as an integrating theme, relating biophysical notions of landscape and ecosystem connectivity (or disconnectivity) to social relationships to place. These associations are then explored by examining the use of integrative river science in managing the process of river repair in differing parts of the world in part III.

Reference

Rapport, D. J., P. R. Epstein, R. Levins, R. Costanza, and C. Gaudet. 1998. *Ecosystem health*. Malden, MA: Blackwell.

Chapter 4

The Spatial Organization of River Systems

CAROLA CULLUM, GARY J. BRIERLEY, AND MARTIN THOMS

> Although the river and hill-side waste do not resemble each other at first sight, they are only extreme members of a continuous series, and when this generalization is appreciated, one may fairly extend the "river" all over its basin and up to its very divides. Ordinarily treated, the river is like the veins of a leaf, broadly viewed it is like the entire leaf.
>
> Davis 1899, 495

Complex geomorphic, hydrologic, and biotic feedbacks operating over multiple spatial and temporal scales generate remarkable diversity in riverine ecosystems (Poole 2002). Although spatial and temporal contingencies ensure that the biophysical makeup of every system is unique (Hynes 1975; Townsend 1996), many spatial relationships and patterns are repeated from system to system (Benda et al. 2004). The ability to unravel local influences from systematic relationships provides a critical basis for informed decision-making in river management.

The spatial arrangement of fluvial systems has been characterized in terms of broad-scale patterns. For example, Schumm (1977) described a general downstream transition from a zone of sediment production in the headwater region, through a zone dominated by sediment transport to an accumulation zone along the lower river reaches. More recently, the "continuum" perspective has been elaborated, incorporating notions of hierarchy and the effect of scale on patterns in all four dimensions (longitudinal, lateral, vertical, and temporal). For example, the "fluvial hydrosystem" proposed by Petts and Amoros (1996) seeks to describe and explain continua of forms and processes in both time and space, including lateral and vertical patterns alongside the longitudinal patterns described by Schumm, among others. Additionally, the biological implications of the physical gradient have been articulated, notably by Vannote et al. (1980) in their "River Continuum Concept."

Others have emphasized deviations from these patterns, claiming that local controls on the composition and spatial configuration of system elements are more important than position in the network (e.g., Thorp et al. 2006; Montgomery 1999; Junk et al. 1989). While the flows of water and sediment through the network are still seen as critical influences on the composition and spatial organization of the mosaic, new themes have been introduced by those adopting this "hierarchical patch dynamics" paradigm (Wu and Loucks 1995), including (Wiens 2002):

- The importance of scale, related to hierarchical levels of organization.
- A focus on both hierarchical constraints and emergent features. While higher levels of

organization constrain possibilities at the lower levels they contain ("top-down" control), some features emerge from processes occurring at lower levels ("bottom-up" controls).

- The role of patch boundaries as controls on system connectivity.
- Recognition that patch characteristics such as shape and size can influence the character and behavior both of an individual patch and its neighbors.
- The role played by organisms in shaping the mosaic. Organisms can not only change patch character and behavior, but can also influence the nature, intensity, frequency, and direction of fluxes between patches.
- Appreciation of the dynamic nature of systems, which are continually evolving and responding to episodic natural disturbances and human interventions. Responses may be cyclical, nonlinear and/or lagged, such that disturbance effects may overlap and interact with each other.
- Moving away from equilibrium and cyclical notions of landscape dynamics, as nonlinear relationships, stochasticity, and the possibility of alternative states are recognized.

The contrast between the continuum and patch dynamic approaches may be conceived as the dichotomy between Newtonian and Darwinian approaches to natural science, as described by Harte (2002, 29): "Physicists seek simplicity in universal laws. Ecologists revel in complex interdependencies. A sustainable future for our planet will probably require a look at life from both sides." In river science, Newtonian principles of simplification, ideal systems, and predictive understanding seemingly clash with Darwinian principles of complexity, contingency, and interdependence (Newman et al. 2006).

Optimal approaches to the spatial analysis of river systems need to encompass both the gradients formed by connectivities between fluxes of water and sediment across the landscape and also the discontinuities arising from local heterogeneity. Spatial structure (e.g., the arrangement of patches and tributary junctions) and function (e.g., connectivity and flow regimes) need to be considered at multiple scales. The types of elements that can be found at lower levels of organization (observed at finer scales) are constrained by boundary conditions imposed by the characteristics of higher levels of organization (observed at coarser scales). Furthermore, it must be recognized that each system is unique, set within its own context and history.

The relationship between a physical environment and the biota it supports is complex and varies widely between systems. Nevertheless, conceiving the physical structure of a river basin as a template upon which ecological dynamics are played out (Southwood 1977), geomorphological domains can be used to identify ecologically meaningful units. Identifying, describing, and mapping these domains at scales appropriate to a given system reduces the confusion in dealing with complex natural systems, providing:

- A spatial framework for organizing, interpreting, and applying data and knowledge across disciplines and between stakeholders
- A basis for generalizations across relatively homogenous spatial units
- A way of stratifying samples for assessment and monitoring
- A basis for comparing different sites in terms of their diversity, ecological integrity, potential for recovery, potential response to human intervention, natural disturbance, or environmental change

We begin this chapter by reviewing contemporary frameworks for the spatial analysis of river systems. A synthetic approach is then used to present a practical method for developing and mapping a nested hierarchical classification for the geomorphic domains of any given river system. Following a discussion of the biotic implications of such a classification, we conclude by exploring the potential management applications of this approach.

Perspectives on the Spatial Organization of River Systems

This section reviews four perspectives on the spatial analysis of landscapes:

1. Gradients within the four dimensions of river systems
2. Fractal relationships in network geometry
3. Discontinuities that create spatial heterogeneity
4. Hierarchical patch dynamics

The first two perspectives are both grounded in a network view of river systems, emphasizing idealized patterns and system-level controls. Patch-based approaches that focus on discontinuities and local controls are then discussed and placed in the context of hierarchy theory.

Gradients within the Four Dimensions of River Systems

Catchment-scale, longitudinal patterns in river structure have been recognized for decades and form the basis for much theoretical fluvial geomorphology. The patterns can largely be explained in terms of the downstream flow of water, mainly in channels, that erodes sediment and transports it across the landscape. Rivers can be partitioned into zones of sediment erosion (production), transport, and deposition (accumulation), each characterized by different geomorphological attributes and behavior (Schumm 1977).

In an idealized network, the stream power available to erode and transport sediment changes downstream in a predictable fashion (Church 2002). Stream power is a function of bed gradient and discharge. While channel slope generally decreases downstream, discharge increases as the drained area enlarges. The two factors interact so that after an initial sharp downstream increase, stream power slowly decreases toward the river mouth, sorting the grain size of bed material. Smaller grains are carried farther downstream, while larger boulders and cobbles are stored upstream. With increasing area drained, larger amounts of sediment are introduced to the channel and transported or stored within it (figure 4.1).

Together with the longitudinal distribution of stream power, downstream changes in slope and valley confinement influence river morphology and the types of erosional and depositional landforms that are found within the channel and floodplains. Thus river character and behavior differ between zones that are dominated by erosional processes (where bedrock-confined rivers dominate), transfer zones (in partly confined valley settings where floodplains are discontinuous), and accumulation zones (where alluvial streams are dominant) (table 4.1).

Physical differentiation of sediment source, transfer, and accumulation zones has been paralleled by ecological differentiation of upstream reaches from lowland rivers (Ilies and Botosaneanu 1963) and associated biological gradients that underpin the River Continuum Concept (RCC) (Vannote et al.1980). The RCC suggests that in headwaters, where much of the channel is shaded by riparian vegetation, there is comparatively little instream vegetation and respiration exceeds photosynthesis. Macroinvertebrate communities are dominated by shredders and collectors. Downstream, where a wider channel is typically less shaded, more instream vegetation is found, supporting communities of grazing macroinvertebrates and photosynthesis exceeds respiration (figure 4.1; table 4.1).

Physical gradients in river systems occur over multiple scales and dimensions. For example, the "fluvial hydrosystem" proposed by Petts and Amoros (1996) seeks to describe and explain patterns

TABLE 4.1

A Simplified Synthesis of Longitudinal Patterns in River Systems (based on Brierley and Fryirs 2005; Church 2002; Illies and Botosaneanu 1963; Knighton 1998; Minshall et al. 1983; Petts and Amoros 1996, Schumm 1977; Vannote et al. 1980). Note that levels are relative within a given system.

	Headwaters	Mid reaches	Lower reaches
Network characteristics			
Relief	Dissected, high-elevation landscape	Moderately dissected, medium-elevation landscape	Relatively uniform (low-relief) lower-elevation landscape
Area drained	Low	Medium	High
Stream order	High number and total length of streams but trunk stream has low order	Moderate number of streams—trunk stream has medium order	Low number and length of streams but trunk stream has high order
Bifurcation ratio	High	Medium	Low
Morphology			
Hillslope/channel gradient	Steeper	Medium	Shallow
Lateral confinement and floodplain characteristics	Narrow, often relatively straight confined valley with no floodplains or small, reworked floodplain pockets	Variable valley shape, typically with discontinuous floodplain pockets	Wide, continuous floodplains with deep alluvial soils
Channel morphology	Bedrock confined = narrow, relatively shallow (low width-depth ratio). Irregular forms prominent.	Mix of bedrock-forced and alluvial = medium width and depth with higher width-depth ratio. Banks are readily reworked (commonly composite sediments) with erosion prominent.	Predominantly alluvial channels, with high width and depth and high width-depth ratio. Increasingly symmetrical and asymmetrical forms. Banks commonly comprise cohesive sediments. Many lateral accretion features adjacent to banks.
Channel adjustment on valley floor	Largely an imposed condition	High-energy forms, with braid channel shifting, avulsion, and floodplain reworking common	Increasing prominence of progressive channel adjustment (lateral migration and downstream translation of bends)
Geomorphic units	Forced morphology dominated by bedrock eroded/sculpted and coarse boulder/wood features. Log/boulder jams, waterfalls, steps, cascades, rapids and pool-riffle, sequences are common	High diversity of bedload-dominated features. Downstream gradation from mid-channel to increasingly bank-attached forms. Floodplains commonly characterized by high-energy units, with levee and floodchannels prominent	Alluvial depositional forms dominate the channel (e.g., point/lateral bars, benches, and so on). Clear proximal-distal relationships on floodplains, with levees, crevasse splays, and sand sheets adjacent to the channel, cut-off and avulsion channels within the meander belt, and wetlands and backswamps in the distal zone
Bed substrate	Bedrock/boulders/cobbles	Variable (bedrock, cobbles, gravels)	Fine gravels, sands, mud

TABLE 4.1

(Continued)

	Headwaters	Mid reaches	Lower reaches
Large woody debris	Wood commonly spans the channel and is relatively difficult to move, exerting a large effect on flow intensity and direction	Logs are increasingly mobile as they scale less than channel width. Accumulate at areas of flow convergence to create log jams, commonly at bar heads. Deposited debris can trap sediment, seeds, and so on, and provide refugia for biota	Often restricted to bank-attached wood. Smaller amounts, less significant for biota
Flow-sediment interactions			
Surface processes	Sediment production zone, characterized by erosion in channel and on slopes. Bed-load dominated river	Sediment transport zone. Typically mixed load rivers	Sediment accumulation zone. Generally suspended load rivers
Sediment storage	Infrequent	Temporary	Longer-term, extensive storage instream and on floodplain
Flow conditions	Highly variable turbulent flow. Flash floods infrequently rework valley floor	Recurrent geomorphically effective flows	Persistent, more consistent flow—extensive flooding, characterized by events that are slow to rise and fall—low velocity, depositional flows
Stream power	Initially low, increasing sharply downstream	High	Low, decreasing toward outlet
Biophysical connectivity			
Dominant form	Predominantly longitudinal fluxes	Strong longitudinal and lateral fluxes, with significant prospect for floodplain reworking	Strong vertical fluxes and recurrent but episodic lateral connectivity between channel and floodplain
Hillslope/channel	Strongly coupled, sediment and nutrients flow from hillslopes to channel	Irregular (disconnected by floodplain pockets)	Low—materials supplied from low-slope hillslopes may be stored for extended periods at valley margins
Channel/floodplain	Limited by valley confinement	Irregular (disconnected in floodplain pockets)	High—ongoing exchange of surface and/or subsurface water, sediment, and nutrients
Tributaries/main channel	Strongly coupled	Irregular—tributaries may be trapped behind floodplain pockets	Generally coupled, although tributaries may be trapped behind levees
Channel/hyporheic zone	Limited/none (small/no hyporheic zone)	Confined to gravelly substrates (gravel bars, pool-riffles, and so on)—no connectivity when bedrock substrate dominates	High surface-subsurface exchange of water and nutrients is promoted by permeable alluvial materials

TABLE 4.1

(Continued)

	Headwaters	Mid reaches	Lower reaches
Biological characteristics			
Physical heterogeneity and habitat diversity	Moderate, given limited range of substrate types and geomorphic units, with high temporal variability given variability in flow	High diversity of substrate types and river forms, with regular reworking	Moderate, with increasingly smooth channel dominated by suspended load features that are less frequently reworked than upstream. Extensive floodplain features are (or were) common (e.g., wetlands)
Macroinvertebrate species richness and diversity/type	Low diversity, dominated by shredders and collectors	Moderate diversity, characterized by collectors and grazers	High diversity, dominated by filter feeders and burrowers
Role of riparian vegetation	High percentage of shading and high organic input, but limited instream vegetation. Results in low UV penetration and lower water temperatures, with little daily variation	Medium percentage of shading and moderate organic input, with increased presence of instream vegetation. Medium UV penetration and water temperatures, with high daily range	Low percentage of shading and organic input (often cleared for agriculture/urban development). Often high instream vegetation (commonly exotics). High UV penetration and water temperatures, with relatively low daily range
Organic content of water and substrate	Low, coarse, little retained	Medium—high detrital storage	High, fine, nutrients retained
Dissolved oxygen instream	High	Medium	Oversaturated during day, low by end of night
Production	Allochthonus: inputs from riparian vegetation and hillslopes	Autochthonous: instream vegetation	Autochthonous: instream vegetation and planktonic algae
Turbidity	Low	Variable	High
Fish zone (European)	Trout/Salmonid	Grayling/Barbel	Bream
Human impacts			
Land use	Forestry	Agriculture	Agriculture/Urban development
Types of common flow regulation	Dams	Irrigation from channel and groundwater	Channelized

in all four dimensions and at many scales (see also Ward 1989 and chapter 6). The *longitudinal* dimension includes upstream-downstream, tributary-trunk stream linkages, the *lateral* dimension includes slope-channel and channel-floodplain linkages, the *vertical* dimension includes inundation levels and surface-subsurface interactions, and the *temporal* dimension includes magnitude, frequency, and timing of water and sediment movement, disturbance regimes, and succession patterns following disturbance (see Chapter 5).

In the fluvial hydrosystem, the spatial arrangement of both physical and biological elements is largely determined by fluxes of water transporting

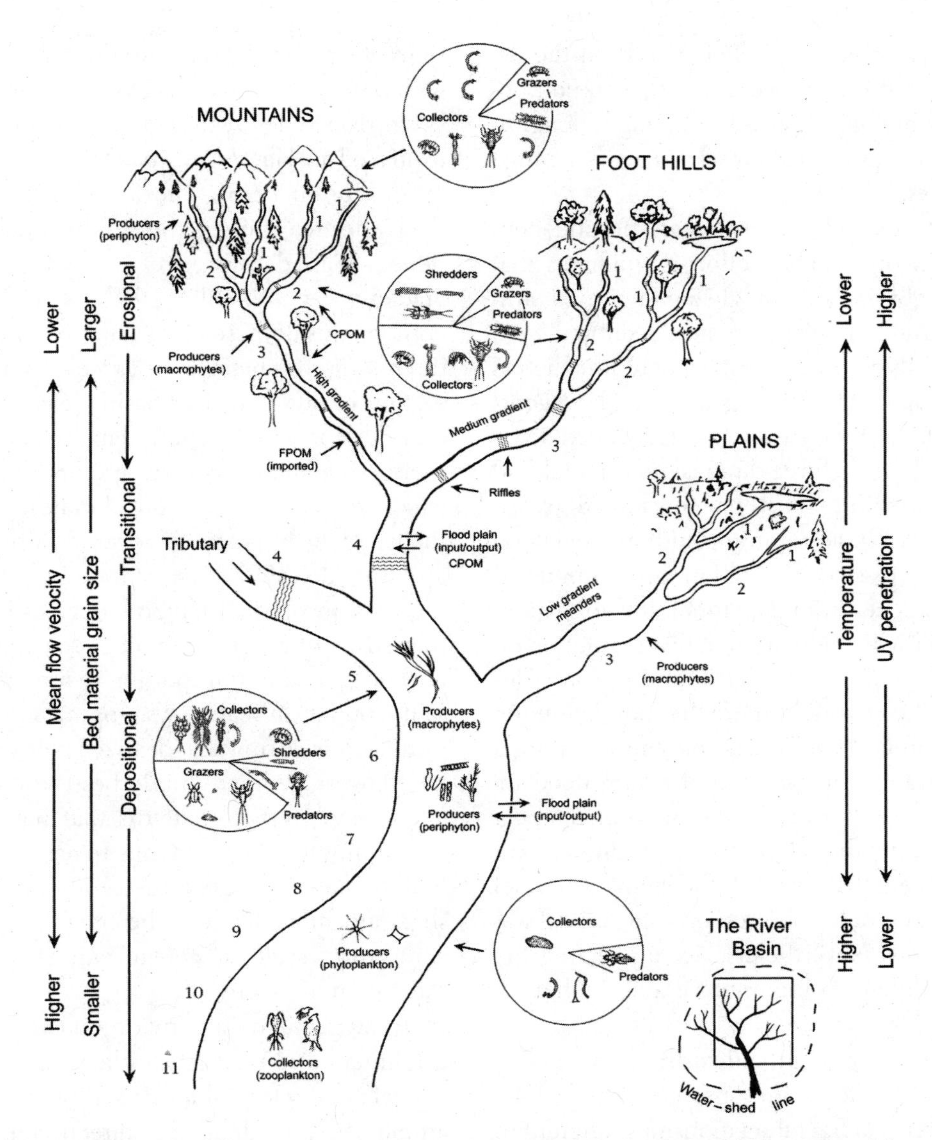

FIGURE 4.1. A revised view of the Riverine Continuum Concept (from Craig 2002; accessed from: http://www.benthos.org/digitallibrary/ and reproduced with permission from D. A. Craig, Department of Biological Sciences, University of Alberta). Longitudinal (upstream-downstream) and lateral (channel-floodplain) linkages expressed in this diagram that build upon Vannote et al. (1980) have been supplemented in recent years by enhanced appreciation of vertical (surface-subsurface) connectivity and notions of disconnectivity (including patch dynamics) in river systems (see text). Longitudinal linkages are also summarized in table 4.1.

sediment, nutrients, and organic material. Functional and genetic links between adjoining elements produce the clinal patterns conceptualized as continua by allowing cascading interactions and the transfer of matter and energy between landscape units (see chapter 6). Fundamentally, fluvial hydrosystems relate to the variability of the hydrological and biological processes that determine (a) the types of habitat patches present and (b) the strength, duration, and frequency of their connectivity. System integrity depends on the dynamic interactions of hydrological, geomorphological, and biological processes acting in longitudinal, lateral, and vertical dimensions over a range of time scales.

Gradients in each dimension are not independent of each other. The notion of gradients that exist in straight lines in a single direction is an artificial construct. In reality, water, sediment, and other materials do not move in straight lines across a flat landscape. In translating theoretical gradients to a particular system, local topography needs to be incorporated into a three-dimensional model and temporal variability must be considered (*sensu* Poole et al. 2006). For example, although a lateral gradient may exist in vegetation communities at different distances from a channel, this may interact with a vertical gradient, such that vegetation on high islands is similar to that found at some distance from the channel, where the inundation frequency is similar. The vegetation pattern is not a response to direction per se, but to the flood regime and flow paths over three-dimensional topography. Similarly, the lateral extent of the parafluvial zone (i.e., that part of the active channel without surface water) is a response to vertical water movements. Nevertheless, deconstructing the system into different dimensions is a useful heuristic exercise.

Idealizing the system still further, connectivities and the gradients they produce can be described in terms of fractal relationships inherent in the geometry of river systems.

Fractal Relationships in Network Geometry

Fractal relationships are seen in many systems where information, stresses, or materials are distributed through a network, including the Internet, animal brains, and cardiovascular systems, trees, and river networks (Dodds and Rothman 2000). Allometric relationships describe how the dimensions of different parts of the structure scale or grow with respect to each other (Mandelbrot 1982).

In river systems, these relationships have been expressed in Hack's law and Horton's laws (Hack 1957; Horton 1945). These laws relate stream order to the area drained, the number of streams (and hence stream density), stream length, and hillslope gradient (e.g., Strahler 1957). The exponential components of these ratios differ between systems, describing the shape and degree of landscape dissection of different river basins (e.g., Stankiewicz and de Wit 2005). These parameters reflect the weathering and erosion regime in the catchment, which depends mainly on climate and geology, mediated by vegetation cover. Rodriguez-Iturbe and Rinaldo (1997) suggest that river systems self-organize in optimal patterns that result from the propensity of water/sediment to maximize the energy dissipated in transporting sediment and water to the outlet. In reality, systems are suboptimal, as local factors disrupt the idealized patterns.

Network geometry underlies the longitudinal, lateral, and vertical patterns that have been observed in many river systems. In areas with consistent climate and geology, geometrical relationships are more likely to be fractal, holding true across many scales and organizational levels within the system hierarchy.

Basin geometry has important consequences at catchment and reach scales both for geomorphological and biological forms and processes. For example, the area drained is directly related to discharge (Leopold et al. 1964). The stream power available to erode or transport sediment is related to discharge and slope (Knighton 1999). The peakedness of hydrographs and the time taken for floods to reach their peak are related to basin shape (e.g., Gregory and Walling 1973). The implications of network geometry on the spatial configuration of geomorphic forms and the biota they support has been explored by Benda et al. (2004), with particular reference to confluences. Stream (drainage) density has many implications for both physical and biological processes. For example, in landscapes with relatively permeable surfaces and hence low stream density, water falling on hillslopes tends to infiltrate vertically, with relatively little (or very slow) lateral surface movement of water and sediment. In arid and semi-arid climates, this water is likely to be transpired or evaporated, with little being stored or transported below the surface. Under such conditions, there is little lateral movement of water or sediment into channels. By contrast, in areas where the surface is relatively impermeable, higher levels of runoff generate a higher stream density and more lateral surface movement of water and sediment. Differences in stream density also involve differences in the proportion of an area that is occupied by riverine vegetation, the distance between streams, and the size and gradient of hillslope patches—all factors with implications for the composition and dynamics of vegetation and faunal communities.

In reality, fractal relationships in network geometry can fluctuate quite widely within a given system and can vary depending on the resolution at which the system is observed (Dodds and Rothman 2000). Critics of clinal and fractal models of river systems point out that in many systems the expected patterns cannot actually be observed at some or all scales. Alternative conceptual models focus on patch heterogeneity rather than on predictable patterns in spatial organization.

Discontinuities That Create Spatial Heterogeneity

Notions of environmental gradients and network geometries refer to idealized systems that are never perfectly replicated in nature. In reality, many rivers appear to break all the rules, defying attempts to predict the spatial organization of different elements in two ways. First, the longitudinal sequence of geomorphological landforms and biological assemblages may not correspond to that predicted by classical theories. Second, many rivers are characterized by abrupt transitions in hydraulic characteristics, landforms, and biological attributes, rather than exhibiting the gradual changes suggested by the notion of a continuum (Townsend 1989; Montgomery 1999; Poole 2002).

While the RCC assumes that the spatial arrangement of the elements within a river network forms a continuum, new perspectives have emerged that emphasize discontinuity and patchiness. The Serial Discontinuity Concept (Ward and Stanford 1983) addressed the effects of dams, which "reset" the longitudinal continuum, highlighting abrupt transitions between river segments with dissimilar physical structure (e.g., canyon to floodplain, lake to stream) (Poole 2002). The Flood Pulse Concept (Junk et al. 1989) proposes that in large floodplain rivers, the existence, productivity, and interactions of major biota are determined by the episodic flood regime. In these circumstances, the system is shaped more by lateral exchanges of water, sediment, and nutrients than by upstream processes. Thus the character and behavior of large floodplain rivers do not form part of a longitudinal continuum in either time or space. Disruptions to the idealized patterns of river geometry and biological response are also emphasized in the Riverine Ecosystem Synthesis (Thorp et al. 2006). These notions concur with Poole (2002) and Montgomery (1999) in suggesting that local or regional hydrologic and geomorphic conditions are more important to ecosystem structure and function than simple location along a longitudinal dimension of the river network. Adjacent functional process zones may be similar or genetically linked, but they can be functionally disconnected. Indeed, pairs of adjacent zones can be less similar than pairs separated by greater distances, and functional process zones can also be replicated within a stream (e.g., Rice et al. 2001; Davey and Lapointe 2007). Their placement within and among stream systems, while not random, is not fixed or precisely predictable within a network.

Deviations from the continuum are often associated with regional or local variations in climate, geology, riparian conditions, tributaries, lithology, or geomorphology (e.g., Minshall et al. 1983) or with human interventions that disrupt the flow/sediment and/or disturbance regime, such as dams, channelization, changes to land cover, or fire suppression (e.g., Ward 1998). Many discontinuities and deviations reflect breaks in network connectivity. Strong clinal patterns depend on strong connections between system elements. These connections often vary in strength over time, being strongly influenced by natural disturbance and human intervention (Montgomery 1999). Boundaries play an important role in determining the pattern, rate, and intensity of fluxes and interactions between landscape elements (Wiens et al. 1985; Cadenasso et al. 2003). Boundary shape (especially perimeter/area ratio), tortuosity, and sharpness all influence the permeability of boundaries and hence the efficiency with which they absorb, amplify, transmit, reflect, or transform matter, energy, or information (Strayer et al. 2003).

Whereas continuum perspectives generally assume pristine conditions in which the system attains a dynamic equilibrium, it is now widely accepted that notions of stability and equilibrium are scale-dependent (Turner et al. 1993), that human intervention has irrevocably altered natural processes in most river systems, and that many processes involve stochasticity, such that outcomes cannot be predicted with certainty. Indeed, rivers may be better conceptualized as complex adaptive systems, characterized by a multitude of interactions and feedbacks, nonlinear relationships, and responses induced by the crossing of thresholds (Levin 1992). Outcomes are path dependent and are strongly influenced by the history of a specific location, such that deterministic models can never describe the full range of variability present in any river system.

Furthermore, many of the patterns observed in river systems change when viewed at different scales in space or time. Different processes appear to control the composition and configuration of system elements viewed at different scales (see Turner et al. 2001). In order to deal with these issues of scale, complexity, and variability, many river scientists now adopt the paradigm of hierarchical patch dynamics when describing or explaining spatial and temporal relationships (Wu and Loucks 1995).

Hierarchical Patch Dynamics

Hierarchy theory is used widely in ecology and earth sciences, providing a useful paradigm that addresses the complexity of natural systems and allows the integration of knowledge from diverse disciplines (Allen and Starr 1982; O'Neill et al. 1986). Within hierarchy theory, systems are conceived as elements arranged at different levels of organization (holons). Elements in higher levels of the hierarchy constrain the character and behavior of elements at lower levels. In spatial terms, this means that only certain types of smaller patches can be contained within larger patches of a particular type. The behavior of smaller patches is also limited by the type and state of the patch they are contained within. Elements at lower organizational levels may also influence the structure and function of higher levels, such that the composition and dynamics of small patches can explain the characteristics and behavior of larger patches at higher organizational levels. Thus both "top-down" and "bottom-up" controls are involved in shaping the system.

Applying hierarchy theory to ecosystems, Wu and Loucks (1995) developed the paradigm of "hierarchical patch dynamics." They suggested that the temporal and spatial heterogeneity of ecosystems could be described in terms of observable mosaics of patches generated by the dominance of different processes in different areas of a landscape and at different scales. Each patch can be considered as a "process domain," characterized by dominant processes that interact to produce the observed pattern (see Montgomery [1999] for application of these ideas to geomorphological processes in river systems). If the nature and rate of the dominant process(es) is(are) similar across the whole patch, then the patch will appear homogenous. Where the same patterns are seen across several scales (i.e., the pattern is fractal), this suggests that the same processes dominate across these scales, indicating the limits at which cross-scale extrapolations are valid (Wiens et al. 1995).

Each level of organization within a given ecosystem is associated with a characteristic spatial and temporal scale. These scales are determined by the rates of the processes that produce the observed patterns. Higher levels are characterized by patterns evident at coarse spatial and temporal scales, produced by processes that operate at relatively slow rates. Faster processes characterize lower levels of organization and are responsible for the patterns observed at finer scales (Turner et al. 2001).

The paradigm of hierarchical patch dynamics suggests that the spatio-temporal distribution of dominant landscape processes can be revealed through the analysis of spatial and temporal landscape patterns and connectivities (Wu and Loucks 1995). However, the link between an observed pattern and its presumed causal process regime is difficult to demonstrate conclusively. Pattern feeds back to process at many scales (Turner et al. 2001), such that differing initial conditions and histories may result in different patterns in areas that currently experience the same process regime. Conversely, the same pattern may be the result of several possible process regimes (e.g., Schumm 1991). Furthermore, different processes may operate on the same patch at different times. For example, ephemeral rivers exhibit very different flow-sediment regimes during wet and dry seasons.

As complex adaptive systems, ecosystems are characterized by nonlinear relations, threshold effects, historical dependency, and multiple possible states (Scheffer et al. 2001). Thus, even if patch processes can be identified, outcomes are often unpredictable. Additionally, local conditions and disturbance history may interact to produce unique outcomes that are not predictable by general theories.

Hierarchical patch dynamics thus offers a paradigm in which the higher levels of organisation impose structural and functional constraints on lower levels, while local heterogeneity can emerge from fine-scale interactions within and between patches. Whereas the continuum perspective tends to focus on landscape-scale connectivities, hierarchical patch dynamics invites consideration of linkages between system elements at different scales. At scales characteristic of a hierarchical level of organization, linkages and interactions between system elements are most numerous and intense (Allen and Starr 1982). Cross-scalar links are also seen, notably in the ways in which lower levels are contextually constrained by higher levels.

Building on these ideas, we now consider how river systems might be analyzed in terms of spatially nested hierarchies of geomorphic forms and processes, integrating concepts from all the perspectives considered above.

An Integrated Perspective: Analyzing River Systems as Spatially Nested Hierarchies

Despite local influences and discontinuities, the general patterns of spatial organization suggested by theories based on continua and fractal basin geometry are found in many river systems. Reconciling the perspectives of continuity and discontinuity involves a compromise between the general explanatory and predictive power delivered by continuum paradigms such as the fluvial hydrosystem and the improved local accuracy gained from the patch perspective. Such a compromise is needed in order to construct a process-based classification that identifies ecologically meaningful units and links observable physical attributes with likely future behavior. A framework for such a classification needs to:

- Be framed within hierarchy theory, recognizing that observed patterns are scale dependent, that processes at lower hierarchical levels (characterized by finer scales) are constrained within boundary conditions set by higher levels, and that some features may emerge from dynamics at lower levels in the hierarchy (Dollar et al. 2007).
- Be grounded in theory, recognizing the processes that are responsible for the spatial organization of elements within a given river system.
- Classify process domains, inferred from discontinuities in observed geomorphological patterns, while remaining cognizant of the perils associated with such inference (Schumm 1991).
- Recognize that system-specific characteristics are shaped by the local setting and history, rather than seeking to apply standard categories across all systems.
- Efficiently partition ecologically relevant system variability, using a minimal set of variables derived from data that are readily available or cost-effective to collect. Variable selection must be guided by theory, utilizing unifying concepts such as network geometry and stream power, which elegantly encapsulate a wide range of attributes.
- Accept the uncertainties associated with human interpretation of landscape forming processes from observed features and with the stochasticity of natural processes operating in complex systems with many interactions. While precise predictions may be impossible, likely responses to various impacts may be foresighted.

Recognition of the hierarchical controls imposed on small-scale (and short-term) physical features and processes in rivers by larger-scale (and longer-term) factors has led to the development of nested hierarchical models of physical organization (Frissell et al. 1986; Naiman et al. 1992; van Niekerk et al. 1995; Petts and Amoros 1996; Poole 2002; Rogers and O'Keefe 2003; Townsend 1996; Brierley and Fryirs 2005). Such models can be applied to any river system. Landforms associated with different geomorphological processes are used to identify and locate process domains at system-specific scales (figure 4.2).

Established theory is used to infer the dominant controls on forms and processes observed at each hierarchical level. For example, in the River Styles framework developed by Brierley and Fryirs (2005), tectonic influences combine with lithologic and climatic controls at a *regional* level, determining boundary conditions of relief, slope, valley width, substrate, flow, and vegetation cover (among other factors). *Catchment*-scale boundary conditions are then set mainly by geology and climate, through their influence on the patterns and rates of erosion, which determine drainage patterns, stream density, and the degree of landscape dissection. Morphometric parameters such as the elongation or relief ratio of a catchment, drainage pattern, and stream order can be used to analyze the topographic imprint upon biophysical interactions. Relief, slope, and valley morphology (width and shape) control the spatial limits within which rivers can adjust, the potential energy available within a landscape, and the pathways through which this energy is dissipated. Catchment-scale geological and climate controls interact with vegetative cover and land use to determine the amount of water and sediment available to the system as well as the nature, magnitude, frequency, and geomorphic effectiveness of disturbance events such as fire or flood.

Within a catchment, some areas may have quite different physical attributes, associated, for example, with differences in geology or land use. *Landscape units* are areas of similar topography that contain a characteristic pattern of landforms, produced by a relatively homogenous set of processes. These units are similar to the functional process zones of the riverine ecosystem synthesis (Thorp et al. 2006) or Montgomery's process domains (1999). Key factors used to identify landscape units include measures of relief, elevation, topography, geology, and network position (e.g., upland versus lowland settings). As landscape units are a function of slope, valley confinement, and lithology, they not only determine the calibre and volume of sediment made available to a reach, but they also impose major constraints on the distribution of flow energy that mobilizes sediments and shapes river morphology.

At the *reach* scale, the fluvial system is divided into its distinct components (channel, aquifer, hillslope, and floodplain), which can be considered as separate but interconnected systems. *Reaches* are defined as "sections of river along which boundary conditions are sufficiently uniform (i.e., there is no change in the imposed flow or sediment load) such that the river maintains a near consistent structure" (Kellerhals et al. 1976). Reach boundaries are demarcated by discernible changes to river character and behavior, such that individual reaches can be defined in terms of particular patterns of river planform, channel, and/or floodplain landforms and bed material texture. Downstream transitions between reaches are generally coincident with a change in valley setting and/or slope. Reach boundaries do not necessarily coincide with tributary confluences, as assumed by many computer models of stream networks.

Erosional and depositional landforms along river courses are referred to as *geomorphic units* (Brierley 1996). Characteristic assemblages of these channel and floodplain features are found at the reach scale. The availability of material and the potential for it to be reworked in any given reach determines the distribution of geomorphic units, and hence river structure. Some rivers comprise forms sculpted into bedrock (e.g., cascades, falls, pools), while others comprise channel and floodplain forms that reflect sediment accumulation in short- or long-term depositional environments (e.g., mid-channel versus floodplain forms). Each landform has a distinct form-process association. Analysis of geomorphic unit morphology, bounding surface, and sedimentological associations, along with interpretation of its distribution and genetic associations with adjacent features, provides a basis to interpret formative processes and behavior. Given the specific set of flow (energy) and sediment conditions under which each geomorphic unit is formed and reworked, they are often found in characteristic locations along river courses. Adjacent geomorphic units are commonly genetically linked, such as pool-riffle sequences, point bars with chute channels, and levee-floodchannel assemblages.

Hydraulic units are spatially distinct patches of relatively homogeneous surface flow and substrate character (Newson and Newson 2000). These can range from fast-flowing variants over coarse substrates to standing water environments on fine substrates. Flow-substrate interactions vary with flow stage. Several hydraulic units may comprise a single geomorphic unit. For example, distinct zones or patches may be evident within individual riffles, characterized by differing substrate, the height and spacing of roughness elements, flow depth, flow velocity, and hydraulic parameters such as the Froude and Reynolds numbers. Flow-sediment interactions should also be considered for floodplain and hillslope compartments.

This approach recognizes both hierarchical contextual constraints on the composition of system elements and the network constraints that determine their spatial arrangement. A nested hierarchical classification can be developed for a specific system and used to inform management decisions (see chapter 5). Scientists can also use the classification to explore relationships between form and process in analogous locations.

Challenges in Determining Scales and Patch Boundaries

Challenges involved in developing river classification schemes include identifying the appropriate scale of analysis for each level of organization

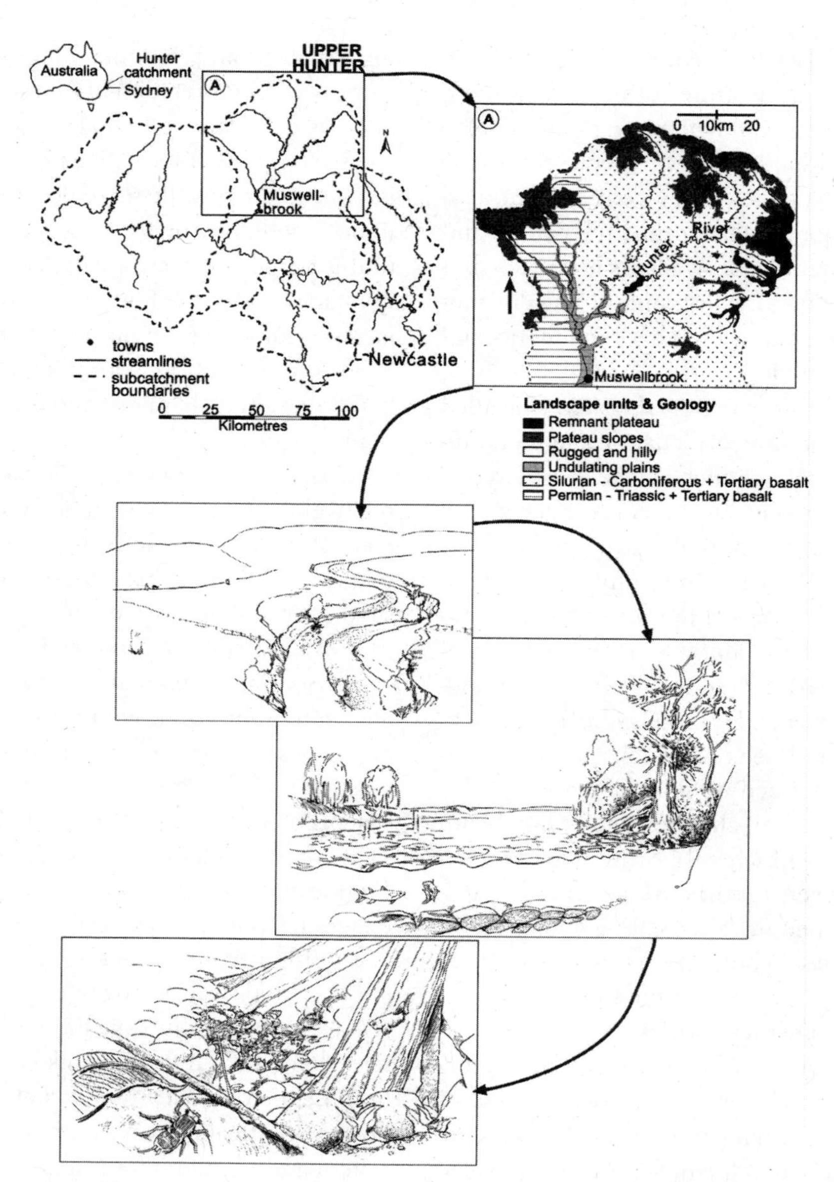

FIGURE 4.2. A conceptualization of the nested hierarchical arrangement of river attributes for a reach of the Upper Hunter River, NSW, Australia. The nested hierarchy follows those scales used within the River Styles framework (Brierley and Fryirs 2005). Primary subcatchment boundaries of the Hunter Catchment are shown in the upper left map. Nested within the catchment are areas of relatively homogenous topography that are called landscape units (inset A shows the distribution of these features for the Upper Hunter subcatchment). Landscape units are defined in terms of their geology, topography, and position within a catchment. Their configuration and connectivity determines the imposed boundary conditions within which rivers form and operate. They also influence the flux boundary condition that fashion river character and behavior of rivers by regulating the flow and sediment regimes and associated vegetation composition. The reach-scale morphology of the alluvial section of the Upper Hunter River at Muswellbrook is shown in the oblique image in mid-figure. This meandering gravel-bed river comprises compound point bar, bench, floodplain, and terrace geomorphic units. Each geomorphic unit has a distinct form-process association. Interpretation of the assemblage

within a hierarchy and delineating patches and their boundaries at each chosen scale. Scales associated with each hierarchical level are relative, such that in a large system (such as the Amazon), the absolute scales characterizing each level are much larger than those applicable to smaller networks. Thus the scale appropriate for the analysis of a particular organizational level differs between systems (Dollar et al. 2007). In general, landscape complexity within any ecoregion increases as size increases (Schumm 1991). For example, while a small subcatchment may lie within one climatic region, form on one lithologic unit, and be subjected to one type of land use, a larger catchment may span climatic, lithologic, and land-use boundaries, and is thus more complex, comprising several landscape units.

Perhaps inevitably, the larger the scale of analysis, the greater the level of generality of forms and processes involved. Large-scale attributes are delineated using large-scale characteristics such as relief and valley slope, and necessarily include considerable variation in small-scale characteristics such as flow type and substrate. Different scalar units in the nested hierarchy are commonly not discrete physical entities; rather, they are part of a continuum in which the dimensions of units at each scale overlap significantly (Naiman et al. 1992; Ward 1989). The challenge for riverine ecologists is to match the scales of their observations and experiments to the characteristic scales of the phenomena that they investigate (Cooper et al. 1998).

Selecting the appropriate scales for investigating pattern and process is important for the study and management of landscapes. In theory, appropriate scales are determined by identifying those scales which display maximum spatial variation in attributes associated with each hierarchical level. However, in most cases, neither the relevant levels of organization nor the attributes associated with them are known. The problem may be approached by predetermining the organizational levels that are to be investigated, somewhat arbitrarily selecting attributes that are likely to characterize different catchments, landscape units, reaches, and so on within a particular setting. Alternatively, a wide range of attributes may be analyzed simultaneously, searching for clusters of attributes with high spatial variance at a limited number of scales. This approach allows both the organizational levels and their attributes to emerge from the analysis. A third approach combines elements of both extremes (Parsons et al. 2004). For practical management purposes, the first approach is preferred, using the organizational levels and attributes described in an established method of classification (e.g., the River Styles framework; Brierley and Fryirs 2005).

Viewing patterns and processes in a multiscalar manner is critical in interdisciplinary settings because river scientists generally impose scales of observation commensurate with their disciplinary experience of the system in question (Thoms and Parsons 2003). Individual disciplines view patterns and processes according to their own paradigms and theories, resulting in mismatches of scales of observation between disciplines (see chapter 3). Patch identity clearly varies with scale. However, different disciplines may also define patches of interest in different ways at the same scale. There are no definitive patch types that are relevant across all disciplines or all projects, since each discipline (or project) tends to focus on different sets of variables and processes. Nevertheless, many physical and biotic processes are strongly aligned at similar scales

---

of geomorphic units which make up any given reach guides insight into river behavior. A pool within this reach is indicative of the geomorphic unit scale. This image shows the direct connection to fish habitat associations. Vegetation lines the channel, with localized woody debris. Finally, the lower image conveys hydraulic unit scale interactions on the stream bed, linking substrate/wood and flow hydraulic considerations to organic matter storage/processing and macroinvertebrate habitat. Hydraulic units are areas of homogenous flow and substrate characteristics that are nested within geomorphic units. Figure drawn by Dan Keating.

within a given system, such that it is possible to delineate geomorphological patches that have strong biological relevance.

Biotic Implications of the Spatial Arrangement of Geomorphic Process Domains

The physical structure of a river (i.e., substrate conditions, channel shape and size, assemblage of geomorphic units, floodplain attributes, and so on) provides a template for analysis of biotic interactions and associations at all scales (Southwood 1977; Newson and Newson 2000). While the physical template does not completely determine biological composition and dynamics, it does limit possibilities.

The physical characteristics of a geomorphic process domain constrain the types of species it can support. The spatial distribution of these domains therefore has clear consequences for the distribution of biota. For example, instream habitat "space" is defined by the heterogeneity and spatial arrangements of geomorphic and hydraulic units (Brierley and Fryirs 2005; Thomson et al. 2001) and the flow conditions (the quantity, timing, frequency, and variability of flows) that maintain and modify river structure (Clarke et al. 2003; Newson and Newson 2000). Hydrologic forces that shape physical, chemical, and biological interactions in river systems have been described as "the maestro that orchestrates pattern and process" (Walker et al. 1995). Geomorphically effective flows modify the type and pattern of geomorphic and hydraulic units, potentially altering the availability, pattern, and connectivity of aquatic habitats. Changes to the geomorphic structure of a river can modify and fragment the physical template, diminishing its capacity to support ecological systems, and hence its ecological integrity and/or condition (Baron et al. 2002; Karr 1995; Naiman et al. 2002; chapter 7).

Differing assemblages of landforms present a range of habitats that may be important at different life stages for different organisms. For example, different geomorphic units act as feeding (runs), resting or holding/refuge (pools, backwater areas), and spawning (gravel bars) sites, such that the reach-scale assemblage of geomorphic units influences the composition of fish assemblages (e.g., Aadland 1993; Wesche 1985). During the course of a day, fish may use pools, riffles, and adjacent units for different activities (e.g., Montgomery et al. 1995; Bisson and Montgomery, 1996; Hawkins et al. 1993). Changes to the arrangement of patches along a river's course may alter the ecological dynamics of the system, even if the relative proportion of patch types remains the same (Fisher et al. 1998; Poole 2002). In many instances, especially along lowland rivers, floodplain inundation is just as important as hydraulic interactions that shape instream patch dynamics (Thoms 2003). Many fish, bird, and amphibian species have adapted their life cycles to the periodicity of flooding and the reliability of floodplain inundation (e.g., Power et al. 1995; Beechie et al. 2001; Amoros et al. 1987).

Building on the relationship between habitat availability and species distribution, Poff (1997) developed a theory of "environmental filters," suggesting that the potential distribution and abundance of different species could be predicted on the basis of their physiological ability to "pass through" the abiotic filters present in different patches. While it is certainly the case that the availability of suitable habitat limits the species that can survive in a particular location, ecological factors such as the distribution of gene pools, species interactions, and the local history of natural and human disturbance also play important roles in determining the precise composition of individual communities (e.g., Jacobson et al. 2001).

The relationship between a physical template and the biota it supports is not one way. Biotic communities shape the physical template as well as being shaped by it as the system self-organizes over time (O'Neill et al. 1986). For example, the nature and extent of vegetative cover alters erosion rates. In turn, vegetative cover is affected by herbivore densities and disturbances such as fire or flood. Animals and plants may also physically engineer their environment (Jones et al. 1994). Beaver dams, for example, may dramatically alter flow/sediment regimes, as do caddis fly nets on a much smaller scale (Cardinale et al. 2002; Westbrook et al. 2006).

Even though the relationship is two way, it is clear that geomorphic hierarchies and process domains are important influences on the distribution and abundance of biota and that species diversity is linked to geomorphic diversity. However, this does not necessarily imply that biological hierarchies and domains are coincident with geomorphological hierarchies, either in space or time. Furthermore, while geomorphological hierarchies and domains are defined fairly consistently among authors, biological hierarchies and domains vary enormously, usually being defined in terms of a species or phenomenon of interest (Levin 1992; O'Neill 2001). Dollar et al. (2007) suggest that each ecological phenomenon studied needs to be positioned within a framework encompassing biological, hydrological, and geomorphological hierarchies. Depending on the subject of study, these hierarchies may be aligned at different scales.

Whilst these issues are being debated, it seems sensible to use a geographical analysis of the composition and configuration of geomorphic domains as a proxy for ecological domains that encompass both biotic and abiotic elements and processes since:

- Relationships between geomorphological form and process appear to be less complex and are better understood than relationships between biological structure and dynamics. It is therefore much easier to identify physical domains by interpreting the landforms present in a given system than it is to identify biological domains.
- There are strong two-way relationships between physical and biological domains in any landscape.

Management Implications

Spatial thinking underpins our efforts to conserve, rehabilitate, and manage river landscapes. In order to manage ecosystems we need to understand the processes that generate and maintain them. We also need to identify what is happening where and the likely trajectory of change following management intervention at a particular location. Only with such understanding can we assess the likely impact of human interventions and create sustainable solutions capable of delivering ecosystem services that are vital to human survival and development. Acquiring this knowledge is a daunting task, given the complexity of ecological systems, where vast numbers of interactions operate across a multitude of scales, often in a stochastic, unpredictable manner. The nonuniform, heterogeneous, and multiscaled nature of river landscapes is particularly daunting to those policy-makers and decision-makers who seek uniformity and simplicity. Indeed, a first step in the quest for sustainable approaches to ecosystem management is explicit recognition of spatial diversity and the nonequilibrium nature of many ecosystem interactions (Kay et al. 1999). As noted in chapter 14, a greater awareness of limitations in our understanding is a critical first step in management endeavors to cope with uncertainty.

Awareness of the spatial organization of river systems is a key issue in the management of river futures. Increased availability and use of remotely sensed and real time data has brought about a revolution in spatial awareness, with increased appreciation of big-picture relationships within an increasingly connected world. However, endeavors to (re)generate vibrant aquatic ecosystems must relate directly to the specific place under consideration (see chapter 8), framing river courses in their landscape context. Ecosystem management entails concerns for a multitude of interactions and feedbacks. Adopting a whole-of-system approach requires acceptance of the fact that phenomena cannot be treated in isolation within a single disciplinary context. While a physical/geomorphological template represents a compromise, it provides a valuable starting point on which to ground insights into ecosystem interactions and their management. Management actions should not be framed around static spatial considerations at any given scale. Each application should consider at least three adjacent hierarchical levels, placing due regard upon higher scale attributes to provide context and lower scales attributes to appraise emergent features and/or properties (Forman 1995; Wu and Loucks 1995). Due regard must also be given to dimensions and rates of biophysical interactions/fluxes that determine landscape/ecosystem connectivity (*sensu* Kondolf et al. 2006). Hence, spatial considerations must be intimately linked to knowledge of ecosystem dynamics across varying timescales (see chapters 5 and 6). However, when used effectively, nested hierarchical frameworks are an elegant tool with which to organize information, providing a logical basis for "whole of landscape" or "whole of ecosystem" management applications.

Rehabilitation can be conducted at a wide variety of scales, but in practice all activities should be approached with a spatially explicit landscape perspective, in order to ensure the suitability of flows, interactions, and exchanges with contiguous ecosystems (SER 2004). The catchment scale is the most appropriate scale at which to plan such endeavors. However, ecoregion scale considerations must be integrated within catchment management plans, as many key processes operate outside the physical domain of landscapes and catchment relationships. For example, demographic processes such as dispersal, along with evolutionary forces of gene flow, genetic drift, and natural selection, are important not only in structuring viable populations, but also in maintaining them (Wishart and Davies 2003).

Ecoregions seemingly present the most logical basis for conservation programs (e.g., management of rare/endangered species, wild and scenic rivers, and so on), as this scale of inquiry most effectively encapsulates process-based insights into biophysical attributes, relationships, and linkages. However, smaller-scale analyses of catchment biophysical interactions and connectivity are required to capture the local variability needed to frame predictions of likely future river adjustments in relation to offsite and/or lagged responses to disturbances. This approach takes into account the underlying causes of change, rather than merely addressing the symptoms, thereby ensuring that rehabilitation strategies are sustainable and work with nature (see chapter 5). Within the catchment context, even finer-scale, site-based analysis must address key elements of ecosystem functionality for a given reach, framing local considerations in light of broader-scale pressures and recovery prospects. These spatial considerations provide a basis for prioritization of management activities in efforts to develop proactive strategies that strive to maximize prospects for success.

Spatial and scalar matters are also critical in appraising the transferability of insight from one management application to another, predicting locations that are likely to respond to environmental change, disturbance events, and management treatments in similar ways. While catchment-specific management strategies strive to maintain unique or distinctive attributes of river courses, we also need to know how readily findings from treatments can be applied elsewhere. Ideally, biomonitoring programs should be designed to maximize our potential to learn from analogous systems and management practices (see Ralph and Poole 2003; Bernhardt et al. 2005; Palmer et al. 2005; Wohl et al. 2005).

Regrettably, the most meaningful scales and boundaries within which to consider spatial organization and process linkages in river systems seldom coincide effectively with scales of governance structures (i.e., management and institutional arrangements). In this light, even greater regard must be given to the specific articulation and documentation of scalar and connectivity issues in reporting management goals and lessons learned for any given system (see chapter 15).

As landscape configuration influences the pattern and rate of process linkages in any catchment, changes to the biophysical template in any compartment may alter longitudinal, lateral, and vertical linkages of biophysical processes. Ultimately, concerns for spatial distributions, whether viewed in terms of process zones, ecological ranges/assemblages, or biotic/biophysical interactions, must be framed in their temporal context, as highlighted in the following chapter.

Conclusion

This chapter has highlighted the need to combine static hierarchical frameworks of landscape organization that differentiate scalar components of any given system with frameworks that consider process dynamics, linkages, and connectivity through network/gradient analysis. Understanding the role of spatial organization and complexity in ecological systems remains challenging to scientists, but is an even greater challenge to managers, for whom such endeavors must be practically focused and achievable.

ACKNOWLEDGMENTS

We thank Geoffrey Poole for particularly insightful comments in his review of an earlier version of this chapter. Kirstie A. Fryirs, Helen Farmer, and Kevin Rogers are also thanked for supportive and constructive comments.

REFERENCES

Aadland, L. P. 1993. Stream habitat types: Their fish assemblages and relationship to flow. *North American Journal of Fisheries Management* 13:790–806.

Allen, T., and T. Starr. 1982. *Hierarchy*. Chicago, IL: University of Chicago Press.

Amoros, C., A. L. Roux, and J. L. Reygrobellet. 1987. A method for applied ecological studies of fluvial hydrosystems. *Regulated Rivers: Research and Management* 1:17–36.

Baron, J. S., N. L. Poff, P. L. Angermeier, C. N. Dahm, P. H. Gleick, R. B. Jackson, C. A. Johnston, B. D. Richter, and A. D. Steinman. 2002. Meeting ecological and societal needs for fresh water. *Ecological Applications* 12:1247–60.

Beechie, T. J., B. D. Collins, and G. R. Pess. 2001. Holocene and recent geomorphic processes, land use, and salmonid habitat in two North Puget Sound river basins. In *Geomorphic processes and riverine habitat*, ed. J. M. Dorava, D. R. Montgomery, B. B. Palcsak, and F. A. Fitzpatrick, 37–54. Washington, DC: American Geophysical Union.

Benda, L., N. L. Poff, D. Miller, T. Dunne, G. Reeves, G. Pess, and M. Pollock. 2004. The network dynamics hypothesis: How channel networks structure riverine habitats. *Bioscience*, 54:413–27.

Bernhardt, E. S., M. A. Palmer, J. D. Allan, G. Alexander, K. Barnas, S. Brooks, J. Carr, S. Clayton, C. Dahm, J. Follstad-Shah, D. Galat, S. Gloss, P. Goodwin, D. Hart, B. Hassett, R. Jenkinson, S. Katz, G. M. Kondolf, P. S. Lake, R. Lave, J. L. Meyer, T. K. O'Donnell, L. Pagano, B. Powell, and E. Sudduth. 2005. Synthesizing U.S. river restoration efforts. *Science* 308:636–37.

Bisson, P. A., and D. R. Montgomery. 1996. Valley segments, stream reaches and channel units. In *Methods in stream ecology*, ed. F. R. Hauer and G. A. Lamberti, 23–52. San Diego, CA: Academic Press.

Brierley, G. J. 1996. Channel morphology and element assemblages: A constructivist approach to facies modelling. In *Advances in fluvial dynamics and stratigraphy*, ed. P. Carling and M. Dawson, 263–98., Chichester, UK: Wiley InterScience.

Brierley, G. J., and K. A. Fryirs. 2005. *Geomorphology and river management: Applications of the River Styles framework*. Malden, MA: Blackwell.

Cadenasso, M. L., S. T. A. Pickett, K. C. Weathers, S. S. Bell, T. L. Benning, M. M. Carreiro, and T. E. Dawson. 2003. An interdisciplinary and synthetic approach to ecological boundaries. *Bioscience* 53:717–22.

Cardinale, B. J., M. A. Palmer, and S. L. Collins. 2002. Species diversity enhances ecosystem functioning through interspecific facilitation. *Nature* 415:426–29.

Church, M. 2002. Geomorphic thresholds in riverine landscapes. *Freshwater Biology* 47:541–57.

Clarke, S. J., L. Bruce-Burgess, and G. Wharton. 2003. Linking form and function: Towards an eco-hydromorphic approach to sustainable river restoration. *Aquatic Conservation: Marine and Freshwater Ecosystems* 13:439–50.

Cooper, S. D., S. Diehl, K. Kratz, and O. Sarnelle. 1998.

Implications of scale for patterns and process in stream ecology. *Australian Journal of Ecology* 23:27–40.

Craig, D. A. 2002. A new view of the River Continuum Concept. *Bulletin North American Benthological Society* 19:380–81.

Davey, C., and M. Lapointe. 2007. Sedimentary links and the spatial organization of Atlantic salmon (*Salmo salar*) spawning habitat in a Canadian Shield river. *Geomorphology* 83:82–96.

Davis, W. M. 1899. The geographic cycle. *Geographical Journal* 14:481–504.

Dodds, P. S., and D. H. Rothman. 2000. Scaling, universality and geomorphology. *Annual Review of Earth and Planetary Sciences* 28:571–610.

Dollar, E. S., C. S. James, K. H. Rogers, and M. C. Thoms. 2007. A framework for interdisciplinary understanding of rivers as ecosystems. *Geomorphology* 89:147–62.

Fisher, S. G., N. B. Grimm, and R. Gomez. 1998. Hierarchy, spatial configuration and nutrient cycling in a desert stream. *Australian Journal of Ecology* 23:41–52.

Forman, R. T. 1995. *Land mosaics: The ecology of landscapes and regions*. Cambridge, UK: Cambridge University Press.

Frissell, C. A., W. J. Liss, W. J. Warren, and M. D. Hurley. 1986. A hierarchical framework for stream habitat classification: Viewing streams in a watershed context. *Environmental Management* 10:199–214.

Gregory, K. J., and D. E. Walling. 1973. *Drainage basin form and processes*. Hoboken, NJ: Wiley.

Hack, J. T. 1957. Studies of longitudinal stream profiles in Virginia and Maryland. *US Geological Survey Professional Paper*, 45–97.

Harte, J. 2002. Toward a synthesis of the Newtonian and Darwinian worldviews. *Physics Today* 55:29–43.

Hawkins, C. P., J. L. Kershner, P. A. Bisson, M. D. Bryant, L. M. Decker, S. V. Gregory, D. A. McCullough, C. K. Overton, G. H. Reeves, R. J. Steedman, and M. K. Young. 1993. A hierarchical approach to classifying stream habitat features. *Fisheries* 18:3–12.

Horton, R. E. 1945. Erosional development of streams and their drainage basins, hydrophysical approach to quantitative morphology. *Bulletin of the Geological Society of America* 56:275–310.

Hynes, H. 1975. The stream and its valley. *Verhandlungen der Internationalen Vereinigung für Theoretische und Angewandte Limnologie* 19:1–15.

Ilies, J., and L. Botosaneanu. 1963. Problèmes et méthodes de la classification de la zonation écologique des eaux courantes, considérées surtout du point de vue faunistique. *Mitteilungen der Internationale Vereinigung für Theoretische und Angewandte Limnologie* 12:1–57.

Jacobson, R. B., M. S. Laustrup, and M D. Chapman. 2001. Fluvial processes and passive rehabilitation of the Lisbon Bottom side-channel chute, Lower Missouri River. In *Geomorphic processes and riverine habitat*, ed. J. M. Dorava, D. R. Montgomery, B. B. Palcsak, and F. A. Fitzpatrick, 199–216. Washington, DC: American Geophysical Union.

Jones, C. G., J. H. Lawton, and M. Shachak. 1994. Organisms as ecosystem engineers. *Oikos* 69: 373–86.

Junk, W. J., P. B. Bayley, and R. E. Sparks. 1989. The flood pulse concept in river-floodplain systems. *Canadian Special Publication of Fisheries and Aquatic Sciences* 106:110–27.

Karr, J. R. 1995. Ecological integrity and ecological health are not the same. In *Engineering within ecological constraints*, ed. P. Schulze, 1–15. Washington, DC: National Academy of Engineering, National Academy Press.

Kay, J. J., H. A. Regier, M. Boyle, and G. Francis. 1999. An ecosystem approach for sustainability: Addressing the challenge of complexity. *Futures* 31:721–42.

Kellerhals, R., M. Church, and D. I. Bray. 1976. Classification and analysis of river processes. *Journal of the Hydraulics Division, American Society of Civil Engineers* 102:813–29.

Knighton, D. 1998. *Fluvial forms and processes: A new perspective*. London, UK: Arnold.

Knighton, D. 1999. Downstream variation in stream power. *Geomorphology* 29:293–306.

Kondolf, G. M., A. J. Boulton, S. J. O'Daniel, G. C. Poole, F. J. Rahel, E. H. Stanley, E. Wohl, Å. Bång, J. Carlstrom, C. Cristoni, H. Huber, S. Koljonen, P. Louhi, and K. Nakamura. 2006. Process-based ecological river restoration: Visualizing three-dimensional connectivity and dynamic vectors to recover lost linkages. *Ecology and Society* 11(2): Article 5. http://www.ecologyandsociety.org/vol11/iss2/art5/ (accessed December 16, 2007).

Leopold, L. B., M. G. Wolman, and J. P. Miller. 1964. *Fluvial processes in geomorphology*. San Francisco, CA: W.H. Freeman and Co.

Levin, S. A. 1992. The problem of pattern and scale in ecology. *Ecology* 73:1943–67.

Mandelbrot, B. B. 1982. *The fractal geometry of nature*. San Francisco, CA: W.H. Freeman and Co.

Minshall, G. W., R. C. Petersen, K. W. Cummins, T. L. Bott, J. R. Sedell, C. E. Cushing, and R. L. Vannote. 1983. Interbiome comparison of stream ecosystem dynamics. *Ecological Monographs* 53:1–25.

Montgomery, D. R. 1999. Process domains and the river continuum. *Journal of the American Water Resources Association* 35:397–410.

Montgomery, D. R., J. M. Buffington, N. P. Peterson, D. Schuett-Hames, and T. P. Quinn. 1995. Stream-bed scour, egg burial depths, and the influence of salmonid spawning on bed surface mobility and embryo survival.

Canadian Journal of Fisheries and Aquatic Sciences 53:1061–70.

Naiman, R. J., S. E. Bunn, C. Nilsson, G. E. Petts, G. Pinay, and L. C. Thompson. 2002. Legitimizing fluvial ecosystems as users of water: An overview. *Environmental Management* 30:455–67.

Naiman, R. J., D. G. Lonzarich, T. J. Beechie, and S. C. Ralph. 1992. General principles of classification and the assessment of conservation in rivers. In *River conservation and management*, ed. P. J. Boon, P. Calow, and G. E. Petts, 93–123. Washington, DC: American Geophysical Union.

Newman, B. D., B. P. Wilcox, S. R. Archer, D. D. Breshears, C. N. Dahm, C. J. Duffy, N. G. McDowell, F. M. Phillips, B. R. Scanlon, and E. R. Vivoni. 2006. Ecohydrology of water-limited environments: A scientific vision. *Water Resources Research* 42:W06302.

Newson, M. D., and C. L. Newson. 2000. Geomorphology, ecology and river channel habitat: Mesoscale approaches to basin-scale challenges. *Progress in Physical Geography* 24:195–217.

O'Neill, R. V. 2001. Is it time to bury the ecosystem concept? (With full military honors of course!). *Ecology* 82:3275–84.

O'Neill, R. V., D. L. DeAngelis, J. B. Waide, and T. F. H. Allen. 1986. *A hierarchical concept of ecosystems*. Princeton, NJ: Princeton University Press.

Palmer, M. A., E. S. Bernhardt, J. D. Allan, P. S. Lake, G. Alexander, S. Brooks, J. Carr, S. Clayton, C. N. Dahm, J. Follstad Shah, D. L. Galat, S. G. Loss, P. Goodwin, D. D. Hart, B. Hassett, R. Jenkinson, G. M. Kondolf, R. Lave, J. L. Meyer, and T. K. O'Donnell, L. Pagano, P. Srivastava, and E. Sudduth. 2005. Standards for ecologically successful river restoration. *Journal of Applied Ecology* 42:208–17.

Parsons, M., M. C. Thoms, and R. H. Norris. 2004. Using hierarchy to select scales of measurement in multiscale studies of stream macroinvertebrate assemblages. *Journal of the North American Benthological Society* 23:157–70.

Petts, G. E., and C. Amoros. 1996. *Fluvial hydrosystems*. London: Chapman & Hall.

Poff, N. L. 1997. Landscape filters and species traits: Towards mechanistic understanding and prediction in stream ecology. *Journal of the North American Benthological Society* 16:391–409.

Poole, G. C. 2002. Fluvial landscape ecology: Addressing uniqueness within the river discontinuum. *Freshwater Biology* 47:641–60.

Poole, G. C., J. A. Stanford, S. W. Running, and C. A. Frissell. 2006. Multi-scale geomorphic drivers of groundwater flow path dynamics: Subsurface hydrologic dynamics and hyporheic habitat diversity. *Journal of the North American Benthological Society* 25:288–303.

Power, M. E., A. Sun, G. Parker, W. E. Dietrich, and J. T. Wootton. 1995. Hydraulic food-chain models. *BioScience* 45:159–67.

Ralph, S. C., and G. C. Poole. 2003. Putting monitoring first: Designing accountable ecosystem restoration and management plans. In *Restoration of Puget Sound rivers*, ed. D. R. Montgomery, S. Bolton, D. B. Booth, and L. Wall, 226–47. Seattle, WA: University of Washington Press.

Rice, S. P., M. T. Greenwood, and C. B. Joyce. 2001. Tributaries, sediment sources, and the longitudinal organisation of macroinvertebrate fauna along river systems. *Canadian Journal of Fisheries and Aquatic Sciences* 58:824–40.

Rodriguez-Iturbe, I., and A. Rinaldo. 1997. *Fractal river basins: Chance and self-organisation*. Cambridge, UK: Cambridge University Press.

Rogers, K. H., and J. O'Keefe. 2003. River heterogeneity: Ecosystem structure, function and management. In *The Kruger experience: Ecology and management of savanna heterogeneity*, ed. J. T. Du Toit, K. Rogers, and H. C. Biggs, 189–218. Washington, DC: Island Press.

Scheffer, M., S. Carpenter, J. A. Foley, C. Folke, and B. Walker. 2001. Catastrophic shifts in ecosystems. *Nature* 413:591–96.

Schumm, S. A. 1977. *The fluvial system*. New York, NY: Wiley.

Schumm, S. A. 1991. *To interpret the Earth: Ten ways to be wrong*. Cambridge, UK: Cambridge University Press.

SER (Society for Ecological Restoration Science and Policy Working Group). 2004. *The SER primer on ecological restoration*. www.ser.org (accessed March 30, 2006).

Southwood, T. R. E. 1977. Habitat, the templet for ecological strategies? *The Journal of Animal Ecology* 46:336–65.

Stankiewicz, J., and M. J. de Wit. 2005. Fractal river networks of Southern Africa. *South African Journal of Geology* 108:333–44.

Strahler, A. N. 1957. Quantitative analysis of watershed geomorphology. *American Geophysical Union Transactions* 38:913–20.

Strayer, D. L., M. E. Power, W. F. Fagan, S. T. A. Pickett, and J. Belnap. 2003. A classification of ecological boundaries. *Bioscience* 53:723–29.

Thoms, M. C. 2003. Floodplain-river ecosystems: Lateral connections and the implications of human interference. *Geomorphology* 56:335–49.

Thoms, M. C., and M. Parsons. 2003. Identifying spatial and temporal patterns in the hydrological character of the Condamine-Balonne River, Australia, using multivariate statistics. *River Research and Applications* 19:443–57.

Thomson, J. R., M. P. Taylor, K. A. Fryirs, and G. J. Brierley. 2001. A geomorphic framework for river characterization and habitat assessment. *Aquatic Conservation: Marine and Freshwater Ecosystems* 11:373–89.

Thorp, J. H., M. C. Thoms, and M. D. Delong. 2006. The riverine ecosystem synthesis: Biocomplexity in river networks across space and time. *River Research and Applications* 22:123–47.

Townsend, C. R. 1989. The patch dynamics concept of stream community ecology. *Journal of the North American Benthological Society* 8:36–50.

Townsend, C. R. 1996. Concepts in river ecology: Pattern and process in the catchment hierarchy. *Archiv für Hydrobiologie* Supplement 113:3–21.

Turner, M. G., R. H. Gardner, and R. V. O'Neill. 2001. *Landscape ecology in theory and practice: Pattern and process*. New York, NY: Springer.

Turner, M. G., W. H. Romme, R. H. Gardner, R. V. O'Neill, and T. K. Kratz. 1993. A revised concept of landscape equilibrium: Disturbance and stability on scaled landscapes. *Landscape Ecology* 8:213–27.

van Niekerk, A. W., G. L. Heritage, and B. P. Moon. 1995. River classification for management: The geomorphology of the Sabie River in the Eastern Transvaal. *South African Geographical Journal* 77:68–76.

Vannote, R. L., G. W. Minshall, K. W. Cummins, J. R. Sedell, and C. E. Cushing. 1980. River continuum concept. *Canadian Journal of Fisheries and Aquatic Sciences* 37:130–37.

Walker, K. F., F. Sheldon, and J. T. Puckridge. 1995. A perspective on dryland river ecosystems. *Regulated Rivers: Research & Management* 11:85–104.

Ward, J. V. 1989. The four-dimensional nature of lotic ecosystems. *Journal of the North American Benthological Society* 8:2–8.

Ward, J. V. 1998. Riverine landscapes: Biodiversity patterns, disturbance regimes, and aquatic conservation. *Biological Conservation* 83:269–78.

Ward, J. V., and J. A. Stanford. 1983. The serial discontinuity concept of lotic ecosystems. In *Dynamics of lotic ecosystems*, ed. T.D. Fontaine and S.M. Bartell, 29–42. Ann Arbor, MI: Ann Arbor Scientific Publishers.

Wesche, T. A. 1985. Stream channel modifications and reclamation structures to enhance fish habitat. In *The restoration of rivers and streams: Theories and experience*, ed. J. A. Gore, 103–63. Ann Arbor, MI:.Ann Arbor Science.

Westbrook, C. J., D. J. Cooper, and B. W. Baker. 2006. Beaver dams and overbank floods influence groundwater–surface water interactions of a Rocky mountain riparian area. *Water Resources Research* 42:W06404.

Wiens, J. 2002. Riverine landscapes: Taking landscape ecology into the water. *Freshwater Biology* 47: 501–15.

Wiens, J. A., C. S. Crawford, and J. R. Gosz. 1985. Boundary dynamics: A conceptual framework for studying landscape ecosystems. *Oikos* 45:421–27.

Wiens, J. A., T. O. Crist, K. A. With, and B. T. Milne. 1995. Fractal patterns of insect movement in microlandscape mosaics. *Ecology* 76:663–66.

Wishart, M. J., and B. R. Davies. 2003. Beyond catchment considerations in the conservation of lotic biodiversity. *Aquatic Conservation: Marine and Freshwater Ecosystems* 13:429–37.

Wohl, E. E., P. L. Angermeier, B. Bledsoe, G. M. Kondolf, L. MacDonnell, D. M. Merritt, M. A. Palmer, N. L. Poff, and D. Tarboton. 2005. River restoration. *Water Resources Research* 41:W10301.

Wu, J. G., and O. L. Loucks. 1995. From balance of nature to hierarchical patch dynamics: A paradigm shift in ecology. *Quarterly Review of Biology* 70:439–66.

Chapter 5

Working with Change: The Importance of Evolutionary Perspectives in Framing the Trajectory of River Adjustment

Gary J. Brierley, Kirstie A. Fryirs, Andrew Boulton, and Carola Cullum

Nothing is permanent but change.

Heraclitus

Research in "pristine" environments provides an intriguing sense of natural river function and evolution (e.g., Collins and Montgomery 2001; Brooks and Brierley 2002). We typically fail to appreciate just how profoundly rivers have been altered by human activities. For example, Brooks et al. (2003) document a 700 percent increase in channel capacity and a 150-fold increase in the rate of lateral channel migration within a few decades of clearance of riparian vegetation and removal of wood from a river in southeastern Australia. Long-term evolutionary insights are required to interpret river responses to human disturbance relative to natural variability. Most rivers have been fundamentally altered from those that existed prior to human disturbance.

River systems can be characterized as shifting mosaics of patch dynamic relationships (see chapter 4), for which disturbance is a fundamental requirement for the maintenance of ecosystem integrity. Indeed, change is a natural, vital component of aquatic ecosystem functioning. However, changes to the geomorphic structure of a river can modify and fragment the physical template, severely diminishing its capacity to support ecological systems. For example, substantial declines in the integrity of ecosystems have been associated with habitat change, fragmentation, and loss (Bunn and Arthington 2002; Dudgeon et al. 2006; Postel and Richter 2003).

Analyses of river dynamics have often been framed in *equilibrium* terms that describe fluctuations around a characteristic state. Increasingly, variation around a mean state is considered to provide a misleading representation of the evolutionary pathway of a system because it reflects the relatively short timeframes and small spatial scales to which these ideas have been applied (e.g., Wu and Loucks 1995; Perry 2002; Phillips 1992, 2003). Recognition of the nonlinear nature of river adjustment over time is a key element of contemporary approaches to river analysis. Like all natural systems, rivers are subjected to indeterminacy, whereby many interrelated variables change simultaneously, limiting our capacity to accurately predict future trends. Effective approaches to the rehabilitation of riparian ecosystems strive to mimic their functionality as mobile habitat mosaics characterized by variability and unpredictability (Hughes et al. 2005; see chapters 14 and 15). Prospects for success in ecological terms are maximized when management interventions work with recovery mechanisms (Gore 1985; Gore and Shields 1995).

Understanding system-specific evolution and trajectory of change provides a critical basis to explain contemporary river condition and determine what is biophysically achievable in rehabilitation efforts. Reconstruction of the past provides a means to test models of change that, if verified, can be used to predict likely future river behavior. Consideration of prospective trajectories of change provides a fundamental platform with which to guide management programs. Determination of emergent properties entails analysis of the inherent sensitivity/resilience of the system, the influence of historical factors, and the range of pressures, stressors, and disturbance events to which the system is likely to be subjected. Particular attention should be given to identification of potential threshold conditions that may induce shifts in the state of the system.

In this chapter, we build upon the River Styles framework (Brierley and Fryirs 2005) to highlight the application of geomorphic notions of river adjustment and evolution as tools to promote healthier river futures. Several themes are addressed. First, contemporary river dynamics are placed in their evolutionary context, differentiating river behavior from river change. Second, scales and forms of river adjustment are considered. Biotic relationships to abiotic controls are then briefly reviewed. These principles are then used to analyze prospects for river recovery, documenting various examples of evolutionary trajectory. Finally, we demonstrate the use of a place-based conceptual model to integrate our understanding of catchment-scale river responses to human disturbance and to guide management actions.

Framing Contemporary River Dynamics in Their Evolutionary Context

Approaches to river management that strive to work with nature build upon insights into the types and rates of river adjustment that are expected for any given system. Various concepts have been developed to assist in this process (see Kondolf and Piégay 2003). Key themes that are addressed in this section are the differentiation of river behavior from river change, analysis of timeframes of river adjustment, and appraisal of the importance of river history.

River Behavior versus Change

The nature and rate of river adjustment vary from system to system and over differing timeframes. Meaningful differentiation can be made between river behavior and river change. River *behavior* reflects ongoing geomorphic and ecological adjustments over timeframes in which flux boundary conditions (i.e., flow/sediment regimes and vegetation interactions) remain relatively uniform, such that a reach retains a characteristic set of process-form relationships (figure 5.1; Brierley and Fryirs 2005). River *change* (or metamorphosis) occurs when a reach experiences a wholesale shift in form-process relationships (Schumm 1969). This may be induced by natural or human disturbance. Sometimes, rivers are subjected to dramatic, near-instantaneous changes in response to catastrophic events. Alternatively, if the system lies close to a threshold condition, relatively small events can reconfigure the system to an alternate state (Scheffer and Carpenter 2003; Scheffer et al. 2001). This may occur when intrinsic (e.g., bed slope) or extrinsic (e.g., climate) factors breach threshold conditions. Alternatively, change may occur from progressive, incremental adjustments (e.g., land clearance).

Some systems are inherently more sensitive to physical and biological disturbance than others. In abiotic terms, the *capacity for adjustment* is a measure of the potential range and extent of geomorphic adjustments that can occur along a river (Brierley and Fryirs 2005). This can range from an imposed condition (natural or human-induced) to

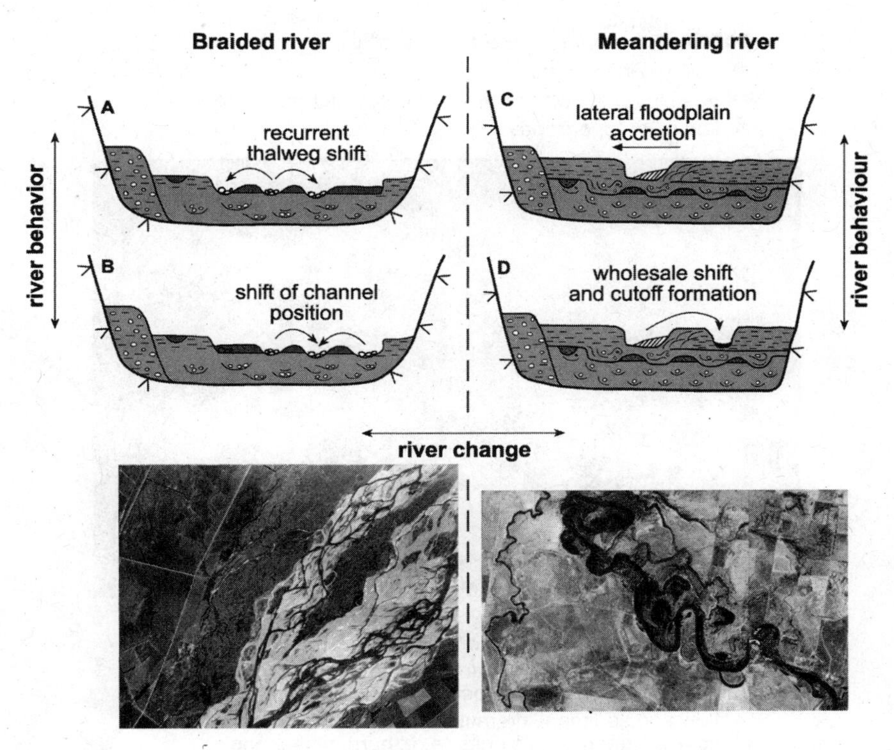

FIGURE 5.1. River behavior and river change. River *behavior* reflects adjustments that occur within the natural capacity for adjustment of a river such as those depicted for the braided river in sections A and B and for the meandering river in sections C and D. River *change* reflects a wholesale shift in river type as depicted by the transition from a braided to a meandering river.

a freely adjusting condition (figure 5.2). An example of a naturally resilient situation is a bedrock-confined gorge where river forms and processes are geologically controlled, whereas regulated rivers downstream from dams or channelized rivers in urban settings are human-forced situations. While a gorge or an urban channel is relatively insensitive to change, a freely migrating meandering river, or the active channel zone of a braided river, may be very sensitive to change.

In biotic terms, ecological resilience is measured as the degree of disturbance that can be absorbed by a system before a threshold is crossed, metamorphosis occurs, and the system adopts an alternative state (Gunderson 2000). The likelihood of a shift in characteristic state increases when humans reduce the inherent ecological resilience of ecosystems, making them more vulnerable to changes that previously could be absorbed by the system (Folke et al. 2004).

Timeframes of River Adjustment

Biophysical processes in river systems operate across a wide range of timescales. Over long periods, *geologic influences* such as tectonic and base-level controls drive landscape evolution, shaping relief, slope, valley confinement, and associated patterns of aggradation and degradation along

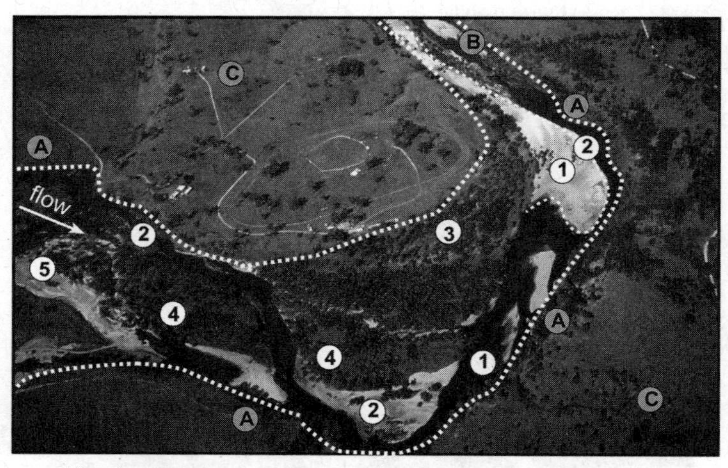

Figure 5.2. Imposed and flux boundary conditions for a partly confined valley with bedrock-controlled discontinuous floodplain River Style (Clarence River, northern NSW).

a river (Schumm and Lichty 1965). These are *imposed boundary conditions* (Brierley and Fryirs 2005). In contrast, *climatic influences* set the *flux boundary conditions* within which rivers operate. Interactions of flow, sediment, and vegetation determine the behavioral regime of a reach. Climate changes may alter the flux boundary conditions, potentially creating a new river type with a different capacity for adjustment (Brierley and Fryirs 2005). For example, the planform of many European rivers was transformed from braided to meandering as systems adjusted to postglacial conditions (e.g., Kozarski and Rotnicki 1977).

Over shorter time frames, rivers are subjected to a suite of disturbance events, such as floods, droughts, and fire. The nature, timing, and magnitude of these disturbances vary in a system-specific manner. *Pulse* disturbances are episodic events of low frequency, high magnitude, and limited duration (e.g., seasonal floods) whose effects tend to be localized (Lake 2000). Offsite impacts are minimal and brief, such as the reorganization of bed materials within a reach or the temporary displacement of river biota after a spate. During a *press* disturbance (e.g., the construction of a dam), there is a sustained shift in flux boundary conditions as the system adjusts to a more permanent shift in input conditions. Such events and their responses occur over larger areas than pulsed events, although responses are not spatially uniform and tend to be more permanent (Brunsden and Thornes 1979; Schumm 1979). In *ramp* disturbance events, the strength of the disturbance steadily increases over time and space (Lake 2000). Examples of ramp disturbances include drought, increasing sedimentation of a stream bed following vegetation clearance, or the incremental spread of an exotic organism. While it is important to understand system responses to individual disturbance events, it is the cumulative effect that determines ecosystem health and the trajectory of river change (chapter 7).

History Matters

Aquatic ecosystem relationships are not merely a product of contemporary forms and processes. History matters, as rivers are influenced to various degrees by past events that have shaped their evolutionary pathway. This imprint from the past, also referred to as path dependence, inheritance, antecedence, system memory, or historical contingency, preconditions system responses to contemporary disturbance events (e.g., Trofimov and Phillips 1992). In many instances, contemporary river adjustments reflect responses to offsite disturbances. Landscape connectivity exerts a fundamental control upon the efficiency and lag time with which impacts of disturbance events are conveyed through landscapes (see chapters 4 and 8).

Scales and Forms of Geomorphic Adjustment

Forms of geomorphic adjustment can be differentiated into bed (substrate) organization, channel geometry, and channel-floodplain relationships that determine the planform of a river (i.e., its configuration in plan view) (figure 5.3). Inevitably, mutual adjustments occur among these attributes. Types of geomorphic adjustment for rivers in differing valley settings are summarized in figure 5.4.

Substrate Organization and Surface-Subsurface Flow Interactions

Adjustments to alluvial bedforms reflect short-term alterations to hydraulic conditions and/or sediment flux resulting from adjustments to flow geometry, the distribution of flow energy across the channel bed, and the availability and calibre of sediment. These flow-sediment interactions influence benthic habitat diversity, providing a template for the analysis of ecohydraulic and patch dynamic relationships (e.g., Downes et al. 2000; Thomson et al. 2001).

Channel Geometry

Channel geometry records the balance of erosional and depositional processes along a reach, reflecting flow energy and alignment, channel position on the valley floor, sediment availability, and distribution of forcing elements such as bedrock outcrops, instream/bank vegetation, and woody debris. Channel size and morphology influence, and in turn are influenced by, the range of within-channel landforms or geomorphic units. Specific combinations of bedrock-sculpted units and depositional features (bank-attached and mid-channel) can be determined (Brierley and Fryirs 2005). Geomorphic units interact with local flow regimes to influence riverine ecological processes from sites of local leaf-pack accumulation and breakdown (Hoover et al. 2006) to reach-scale surface-groundwater exchange (Elliott and Brooks 1997; Boulton 2007) and invertebrate and fish distribution (Downes et al. 2000; Pusey et al. 2000). They also mediate the differential responses by river plants and animals to disturbances such as drought (Boulton 2003).

FIGURE 5.3. Scales of river adjustment used to assess the physical integrity of a reach. This is separated into (a) fine-scale adjustments to substrate and bedform organization, (b) adjustments to channel geometry and the assemblage of geomorphic units within a channel, and (c) coarse-scale adjustments to channel-floodplain interactions and planform.

Channel-Floodplain Relationships

When flows go overbank, floodplains are formed and reworked. Interpretation of the form-process associations of floodplain geomorphic units along a reach, and their history of formation and reworking, reveals long-term adjustments to channel planform (Richards et al. 2002; Brierley and Fryirs 2005). These channel adjustments should not be considered independently of the surrounding slopes. Vegetation change on hillslopes affects the hydrological partitioning of rainfall into evapotranspiration, infiltration, and runoff as well as altering patterns of soil formation and erosion. Alterations to flow paths or initiation of new channels affect the transport and distribution of water, sediment, soluble nutrients, and propagules. Faunal communities may also respond to vegetation change. For

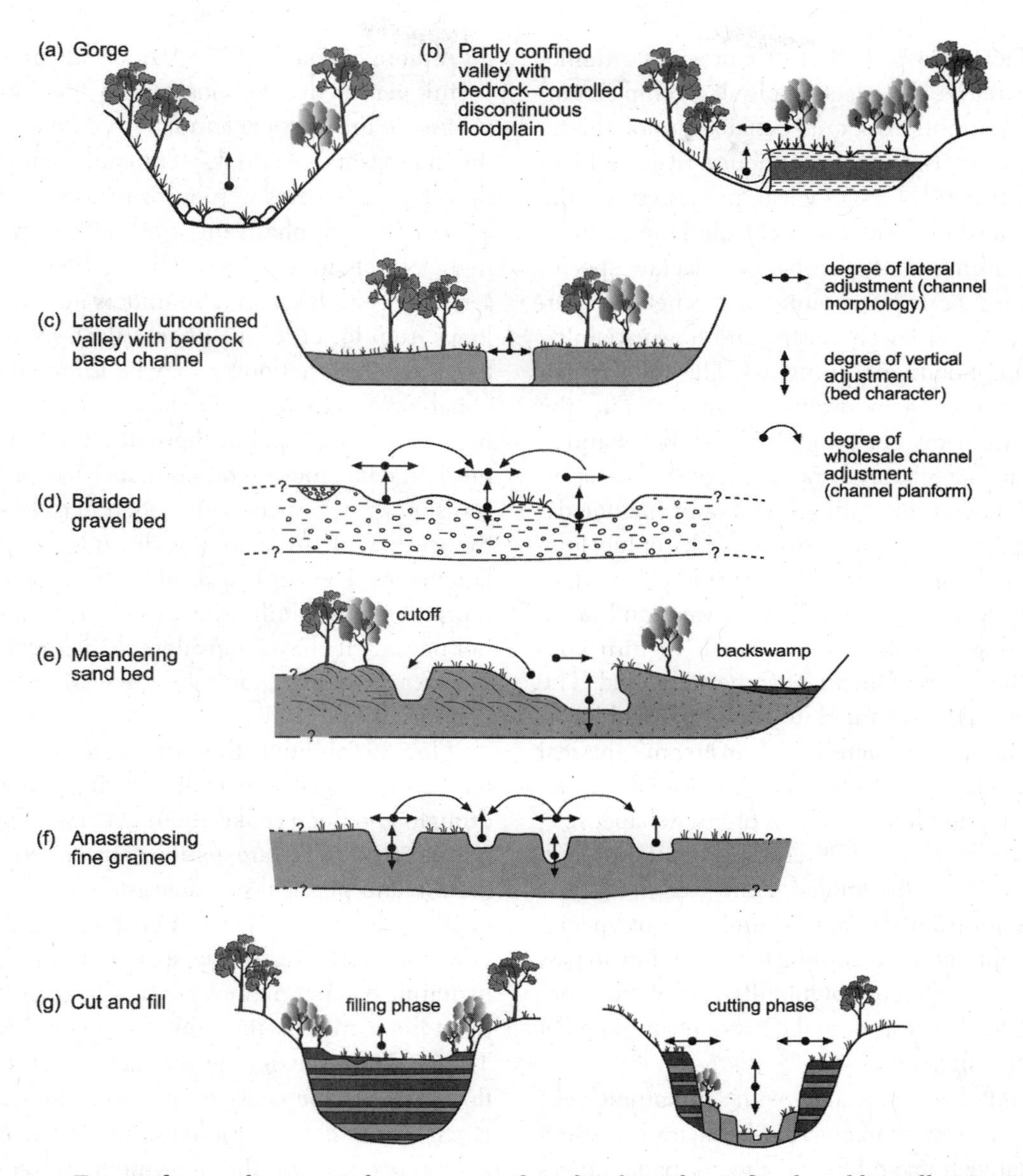

FIGURE 5.4. Forms of river adjustment along a range of confined, partly confined, and laterally unconfined river types. Forms of adjustment are categorized into vertical, lateral, and wholesale types. Reproduced from Brierley and Fryirs (2005) with permission from Blackwell Publishing, Oxford, UK.

example, herbivory pressure affects vegetation structure, thereby exerting a secondary influence upon hydrological partitioning and disturbance regimes (e.g., magnitude/frequency of fire or flood).

Linkages between Abiotic and Biotic Adjustments along Rivers

The geomorphic template of a river system provides the three-dimensional habitat for a vast range of interdependent forms of life across multiple scales (chapter 4). The concept of environmental filters developed by Poff (1997) provides a framework to analyze the effects of physical templates on species distributions. Four spatial scales of abiotic factors are considered to influence riverine biota and, by inference, biological processes. At the basin/watershed scale, historical, climatic, geological, and human land-use patterns dictate species pools (biogeography, evolutionary genetics), thermal regimes, water chemistry, and flow variability (including floods and droughts). The valley/reach scale imposes geomorphic constraints in the channel and on the floodplain. In the channel, sediment size, channel morphology, and riparian vegetation controls influence organic matter dynamics, secondary productivity, and trophic linkages. On floodplains and hillslopes, landforms and topography influence the flow of water and nutrients through the landscape, which in turn constrains the forms of life that can be supported. The next two scales—channel unit and microhabitat—determine the availability of instream physical habitat, substrate stability, and flow variability, affecting riverine biota in terms of body shape, size, and use of oxygen and other resources (Gordon et al. 2004). Using this model, Poff (1997) proposed that the distribution and abundance of species could be predicted according to their ability to pass through the different abiotic filters, explicitly considering the environmental constraints imposed at different scales.

Habitat loss associated with human-induced changes to river character and behavior has eliminated native flora and fauna, altered spatial ranges and interactions, and facilitated the incursion of exotic species (Dudgeon et al. 2006). Simplification of river courses has reduced the geomorphic complexity of channels and altered channel-floodplain connectivity, reducing the diversity of habitat and refugia in rivers (Sheldon et al. 2002; Allan 2004). Changes to the physical template may alter community composition of inhabitants and affect ecosystem processes such as photosynthesis, respiration, or nutrient spiraling (Allan 2004).

Although Poff's (1997) framework provides a useful perspective for analysis of the biotic responses to the physical template at different scales, the framework overlooks temporal factors (e.g., time lags in species' responses to physical disturbance). The emphasis on constraints imposed by higher levels in an organizational hierarchy (observed at coarser scales) also undervalues feedback loops from lower levels (finer scales). Disruptions to ecosystem functioning may be imposed by loss of habitat or refugia (at any stage in the life cycle of organisms), chemical or thermal limitations (e.g., nutrient deficiencies, contaminants, loss of shade), invasion by new species, degradation or loss of native species, or any factor that disrupts food web relationships. Hence, two reaches of the same geomorphic type within the same catchment will not necessarily have equivalent viable habitat and biotic assemblages, despite physical similarities (Thomson et al. 2004).

Flow variability is the primary agent of disturbance in rivers, especially during periods of drought and flood (Lake 2000). As biological communities in rivers are partly constrained by hydraulic and physical characteristics such as water depth, flow velocity, bed material size, and turbulence, the assemblage and spatial arrangement of hydraulic patches may exert a significant influence upon the biotic composition of a stream (Dyer and Thoms 2006). Hydraulic interactions at differing flow stages shape ecosystem relationships such as patch dynamics, nutrient spiraling, hyporheic zone processes, and the movement and storage of organic matter. As noted by Walker et al. (1995), the hydrologic forces that shape physical, chemical, and biological interactions in river systems can be viewed as "the maestro that orchestrates pattern and process."

Changes to the natural pattern of hydrologic variation and disturbance alter habitat dynamics, creating new conditions to which native biota may be poorly adapted (Poff et al. 1997; Richter et al. 1997; Sparks 1995). Flow-induced variability such as pulse events may affect seed dispersal and seedling establishment, animal migration, reproduction, groundwater–surface water exchange, and bed material organization (Richards et al. 2002). Ramp disturbances such as droughts that cause longer-term changes in flow regime may alter geomorphic river structure and function, exerting secondary impacts upon riparian vegetation and groundwater inputs, with major consequences for biotic interactions (Lake 2000; Boulton 2003).

Manipulation of flow probably represents the greatest single human impact on river ecosystems (Postel and Richter 2003). Bunn and Arthington (2002) review four major reasons why flow modifications devastate river biota and ecosystems. First, because river flows—and particularly floods—shape the physical habitats of rivers and their floodplains, changes to flow affect the distribution and abundance of plants and animals. Loss of a particular component of the flow regime may completely eliminate certain habitats and their dependent species. Second, survival and reproductive strategies of many aquatic species are keyed to natural flow conditions. If the flow conditions for a species to successfully complete its life cycle no longer exist, the species will inevitably disappear. Third, many species require adequate water depth at critical times of the year to provide refugia during drought and/or to facilitate their movements along the river or onto the floodplain. Flow alterations that reduce hydrological connectivity and inhibit these movements may prevent them from reaching crucial survival, feeding, and breeding sites. Fourth, altered flow conditions often favor exotic species over native ones. For example, a regulated flow regime with less variability may give exotic fish a competitive advantage over native species that are better adapted to unpredictable flooding. As native species decline, rapidly dispersing exotic species may quickly invade the vacated niches. It is increasingly recognized that our capacity to improve river health is dependent upon re-instigation of ecohydraulic relationships that mimic natural variability in flow regimes (e.g. Poff et al. 1997; Naiman et al. 2002). In framing these determinations, management of flow variability must be framed in relation to the heterogeneity of river structure with which flow interacts. However, management strategies will not succeed unless colonizers are available and minimal biophysical requirements are met for all stages of life cycles. For example, biodiversity loss may occur if fragments or patches are not linked within an appropriate range for a given species. Hence, fragmentation and connectivity exert a significant influence upon the adaptive capacity of ecosystems and their ability to respond to pressures such as climate change. Proximity to nearest colonizers limits the capacity for ecosystems to recover, presenting a serious constraint on the so-called field of dreams hypothesis (Hilderbrand et al. 2005).

Although abiotic controls upon habitat availability and measures of ecosystem functionality are relatively self-evident, it is incorrect to see this as a one-way set of interactions. In many instances, biota influence the form and function of the physical template. For example, the extent and nature of in-channel vegetation profoundly affect the formation of islands and bars. Alternatively, changes to hillslope vegetation may affect the magnitude and frequency of lateral flows of water and sediment, altering the physical template both locally and downstream. Modifications to habitat availability on hillslopes can have knock-on effects, as changes to the patterns and intensity of herbivory affect vegetation structure, generating further biophysical adjustments. "Ecosystem engineers" from beavers to caddis flies also shape attributes of river morphology (see Moore 2006).

However, just as some geomorphic processes operate independently of biological factors, biotic functions that operate independently of the geomorphic template also drive ecosystem relationships. For example, dispersal or migration zones in various climatic regions influence the occurrence of a given species. Food web processes and species-species interactions, such as predator-prey relationships, may dictate the occurrence or otherwise of a given organism in any given system. Similarly, evolutionary adjustments to the gene pool may fashion biotic interactions independently of landscape considerations (Wishart and Davies 2003). In urban streams, recovery may be constrained by poor water quality or altered organic matter dynamics even when near-natural geomorphic features are restored (Paul and Meyer 2001).

Integrating these abiotic and biotic considerations is critical in developing an understanding of key controls upon ecosystem functionality with which to guide targeted river management actions (see chapter 6). The effectiveness of these actions will be fashioned, in large part, by the extent to which they appreciate (and work with) the evolutionary trajectory of a river system, as outlined in the following section.

Conceptualizing River Evolution and Recovery as a Basis for Management Planning and Action

Determination of what is biophysically achievable for a given ecosystem is a key consideration in any river conservation or rehabilitation initiative. Such proactive endeavors strive to maximize ecological and social benefits while minimizing costs, working within a fifty- to one hundred-year management timeframe (chapter 2). Evolutionary analyses are required to relate the contemporary condition of the system (chapter 7), and its trajectory of change, to assessment of recovery potential.

The potential for river recovery following disturbance reflects a river's inherent sensitivity to change and the severity of impacts to which the system is (or has been) subjected. Sometimes, there may be a nonlinear relationship between sensitivity to change and the capacity for recovery. For example, human-induced changes to river forms and processes in many parts of Australia are irreversible, as irregular flow variability and limited sediment availability constrain prospects for geomorphic river recovery (Brooks and Brierley 2004; Fryirs and Brierley 2001). In contrast, the greater availability of sediment and more consistent flow of rivers in New Zealand present greater prospects for recovery (Fryirs et al. 2007).

The river recovery diagram developed by Brierley and Fryirs (2005) provides a framework to appraise river responses to disturbance and prospects for recovery (figure 5.5). The vertical axis on the left conveys a *degradation* pathway, starting from an intact state at the top, with progressively more degraded conditions down this axis. The axes on the right represent the potential recovery pathways of a reach. The position when initial signs of recovery are noted is represented by the turning point. The *restoration* pathway reflects a system that shows signs of returning toward the intact state. These reaches have experienced *reversible* change from their intact condition (i.e., adjustments that have occurred are part of the *behavioral regime* for that type of river). The creation pathway reflects recovery toward a new, alternative condition that did not exist previously at the site. These reaches have experienced *irreversible change* so it is no longer realistic or expected that the degraded river will return toward the pre-disturbance condition over management timeframes. The pathway of adjustment will depend on disturbance events and management strategies instigated to enhance recovery.

In a few parts of the world, it is meaningful to frame these analyses in the context of a near-pristine condition, scarcely influenced by human disturbance. This pre-disturbance condition may itself be subject to significant natural variability (i.e., a behavioral regime), reflecting its own stage of evolutionary adjustment. For most rivers, however, it is more realistic to frame these analyses in relation to prevailing boundary conditions (i.e., flow, sediment, and vegetation interactions). In these instances, the state at the top of the degradation pathway reflects a period when catchment boundary conditions have reached the contemporary stage of development. Reconstruction of river evolution

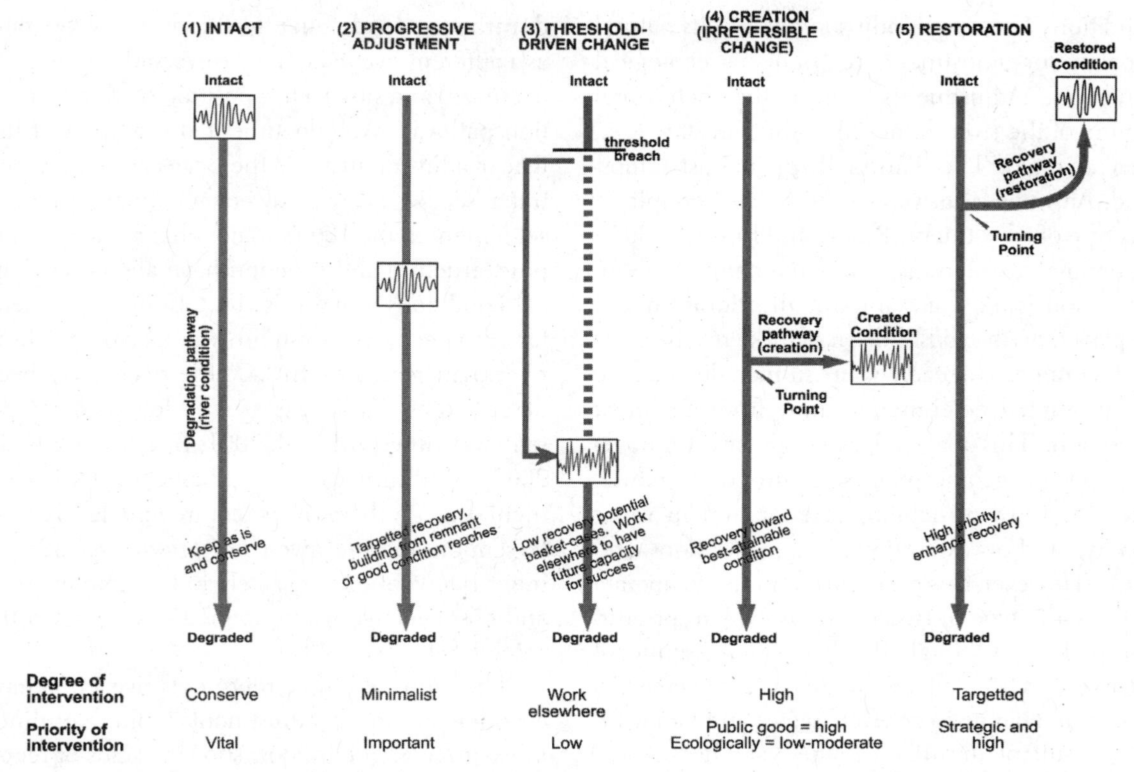

FIGURE 5.5. The river recovery diagram for a range of scenarios. Each reflects a type of degradation and recovery response and the degree of management intervention and priority placed on that intervention.

enables differing "evolutionary time-slices" to be placed on the recovery diagram, representing various stages of adjustment for reaches of the same river type (Fryirs and Brierley 2000).

Limiting factors and pressures determine the potential for that reach to recover. Hence, identifying limiting factors and pressures is crucial to any river conservation or rehabilitation program (Brierley and Fryirs 2005). Limiting factors are internal to the system and may include changes to sediment availability (e.g., passage of sediment slugs or sediment starvation), runoff relations, and vegetation cover. Pressures refer to factors that are external to the system, such as climate variability, human changes to landscape forms and processes, and a myriad of socioeconomic and cultural changes (e.g., population, land use, and so on). Analysis of limiting factors and pressures is a catchment-specific exercise. Each reach must be placed in its catchment context, interpreting lagged and offsite impacts in the conveyance of disturbance responses.

Examples of River Trajectories

Five examples of evolutionary trajectories of rivers affected by human disturbance, and interpretations of recovery prospects, are conceptualized in figure 5.5. Example 1 is an intact river. The reach sits at the top of the degradation pathway, as it has not experienced human-induced deterioration in condition. The river is adjusting within its natural capacity for adjustment, retaining its ecological condition. Adjustments reflect the behavioral regime of the river, rather than shifts in state (i.e., river change). The Thurra River in East Gippsland, Australia, is a well-documented example of this type of adjustment (Brooks and Brierley 2002). Under these conditions, *minimal* rehabilitation intervention is required, and a high priority should be placed on the *conservation* of these reaches.

Example 2 is a reach where human disturbance has prompted deterioration away from an intact condition. This may reflect progressive homogenization of geomorphic structure and, consequently, a loss of habitat, deterioration in water quality, and/or extinction of aquatic flora and fauna. However, this example continues to operate as the same type of river that was evident prior to disturbance. Although the behavioral regime of the river has been altered, there has not been an irreversible change in river character and behavior. As a result of disturbance, the river has moved down the degradation pathway. The River Murray in Australia exemplifies this type of adjustment. Since European settlement in the early 1800s, the biophysical condition of the river has progressively deteriorated. The loss of native fauna such as the Murray Cod reflects declining water quality, altered flow regime, loss of riparian vegetation (e.g., the dieback of River Red gums), and a simplification of geomorphic structure (Benson 2004). In terms of rehabilitation, these examples require *minimalist* intervention strategies that target recovery by building out from remnant reaches of river that are in good condition. In ecosystem terms, targeting the recovery of endangered keystone species (e.g., Murray Cod) is required to improve ecosystem condition. Rivers displaying this progressive form of adjustment are considered *important priorities* in river management plans.

The third example represents rivers that have experienced threshold-induced adjustments in river character and behavior in response to human disturbance. Consequently, the river now operates as a different river type (i.e., irreversible change has occurred) and sits at a low position on the degradation pathway. Well-documented examples of this type of adjustment are in the Bega catchment, Australia (see Bega case study below). Since European settlement in the 1860s, major changes to geomorphic structure and function have altered ecological conditions along this river. Limited prospects for river recovery constrain what is achievable in river management terms. These rivers have been described as "basket-cases" with low recovery potential (Rutherfurd et al. 2001; Brierley and Fryirs 2005). Within a catchment-based river management strategy, these rivers require *high* levels of intervention and are given *low priority* as management interventions are likely to be more efficient and cost-effective when applied elsewhere in the catchment.

The fourth example represents rivers that have experienced similar adjustments to those outlined in example 3, but are now showing signs of recovery. Given the poor condition of the reach, it sat low on the degradation pathway, limiting prospects for recovery along a restoration pathway. The trajectory of change is along the creation pathway, whereby the biophysical characteristics of the river differ from those that have occurred in the recent (i.e., 10^3 years) evolution of the reach. Although a change in river type has occurred and the behavioral regime of the river is now improving, recovery prospects are limited to the best-attainable structure and function for this new type of river. Examples of this type of recovery occur along urban and regulated rivers. For example, Fairfield Council in Sydney, Australia, is rehabilitating a former chain-of-ponds river that has been channelized and severely degraded into a meandering pool and riffle configuration (Ruddell 2005). The ultimate aim is to reinstate some natural geomorphic and ecological function to the river, improving its condition. These situations often require costly, *high* intervention river rehabilitation strategies, with *low-moderate* prospects in terms of ecosystem condition.

Example 5 reflects the recovery pathway that may be experienced by systems shown in example 2. This may occur naturally or be enhanced by human intervention. The trajectory of change is along a restoration pathway, recovering some of the former biophysical structures and functions. In these instances, river change has not occurred, and improvements in the behavioral regime are evident. These self-restoration tendencies can be enhanced by management actions. For example, strategic installation of wood structures and replanting of riparian vegetation enhanced habitat availability and improved ecological conditions along the Never Never River, Australia (Cohen et al. 1998). This is a good example of managers working with the natural character and behavior of a river within a recovery enhancement framework. For this strategy to be successful, the initial reach should be in relatively good condition (i.e., the reach sits high on the degradation pathway) such that natural processes can be enhanced rather than reinstated. Within a catchment-based river management strategy, these rivers require *minimal* and *targeted* intervention, as they represent *strategic* and/or *high recovery potential* priorities.

These various examples demonstrate the range of system responses to human disturbance and variable prospects for geoecological recovery. Even within a relatively homogenous region, adjacent systems are unlikely to respond to external stimuli in a directly equivalent manner. Nonsynchronous timing, pattern, and rates of biophysical responses to differing forms of disturbance reflect the character, configuration, connectivity, and evolutionary trajectory of each river system. These considerations, along with appraisal of pressures and limiting factors, and lagged and offsite responses, require catchment-specific information with which to predict likely future river character and behavior.

Place-Based Conceptual Modeling

Place-based conceptual models provide a practical basis for management purposes in both scientific and social terms. In scientific terms, these models can be used to test hypotheses and reformulate our knowledge. In social terms, there is much greater prospect for uptake of scientific guidance if it relates to a given place and/or community (see chapters 2 and 8). Various themes that must be integrated in the development of place-based conceptual models with which to scope river futures are outlined in figure 5.6.

Using these principles, a place-based conceptual model of aquatic ecosystem functionality for Bega catchment, Australia, is outlined below.

Geomorphic Changes and River Recovery in Bega Catchment

The history of river change, contemporary river condition, and river recovery potential in Bega catchment have been thoroughly documented (Brierley et al. 1999; Brierley and Fryirs 2005; Brooks and Brierley 1997; Fryirs 2002, 2003; Fryirs and Brierley 1998, 2001, 2005). Human-induced changes to river character and behavior since European settlement have caused irreversible geomorphic changes (figure 5.7). However, in recent decades, geomorphic river recovery has commenced (figure 5.8). The potential for restoration is limited to the reinstatement of swamps within the incised channel networks of the upper catchment. River recovery is constrained by several factors, such as the lack of riparian vegetation cover and the modified sediment budget of the catchment (Fryirs and Brierley 2001). Upstream reaches are so starved of sediment that it will take thousands of years to refill incised channels at the base of the escarpment. In contrast, the lower Bega River has been oversupplied with sand, much of which is now trapped by willows and other exotic

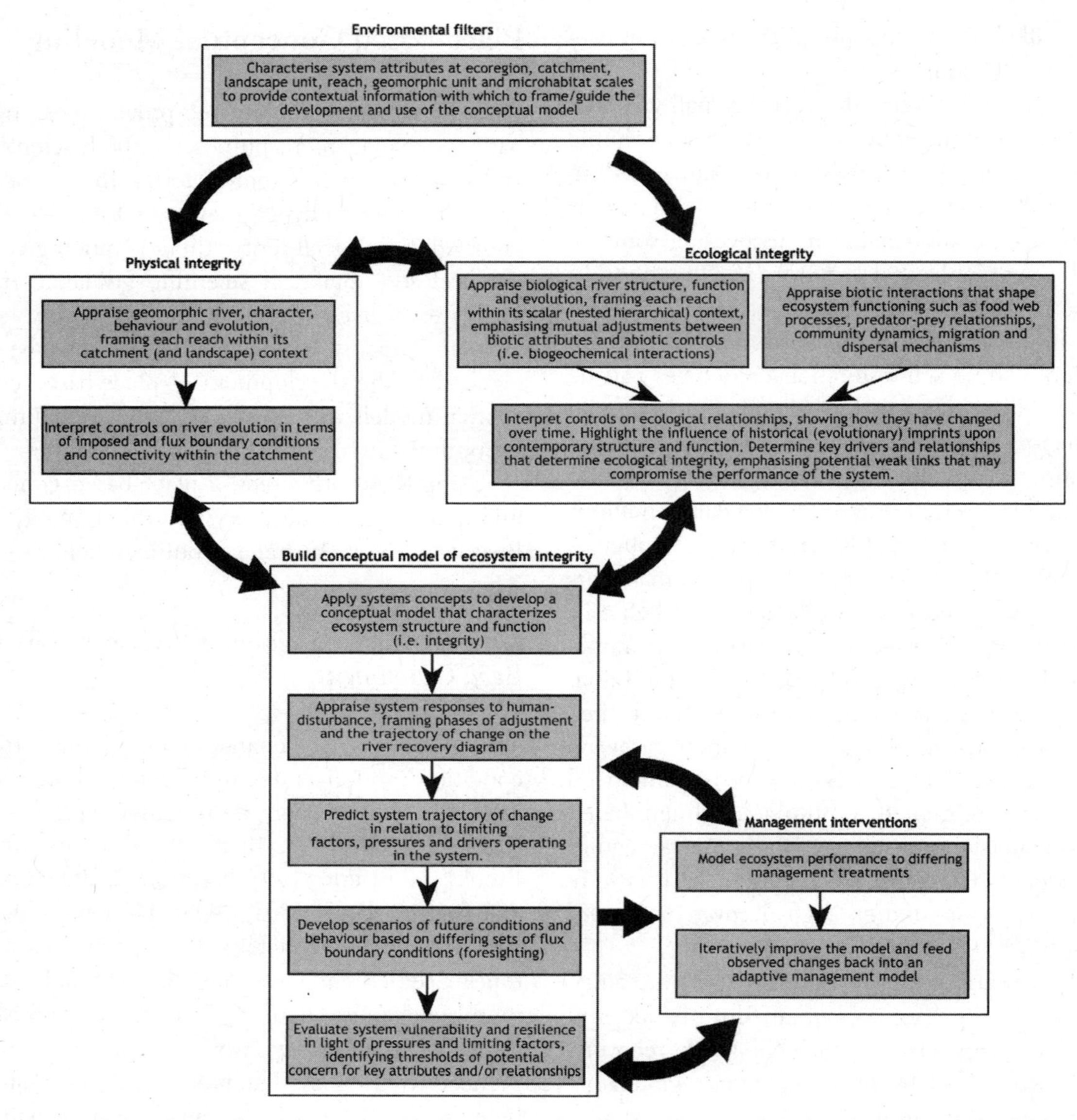

FIGURE 5.6. An approach to construction of conceptual models of aquatic ecosystem condition.

vegetation in a sand slug. Most reaches are now adjusting toward a created condition.

Biotic Responses to Abiotic Changes in Bega Catchment

Changes to river morphology have directly affected habitat availability in Bega catchment, impacting upon its ecological condition and functioning. Reaches in good geomorphic condition sustain richer native biodiversity of macrophytes and invertebrates than reaches in poor geomorphic condition (Chessman et al. 2006). Inferred ecological responses to geomorphic change are summarized in figure 5.8. Since European settlement, the channel network has become *longitudinally* connected. Discontinuous watercourses that

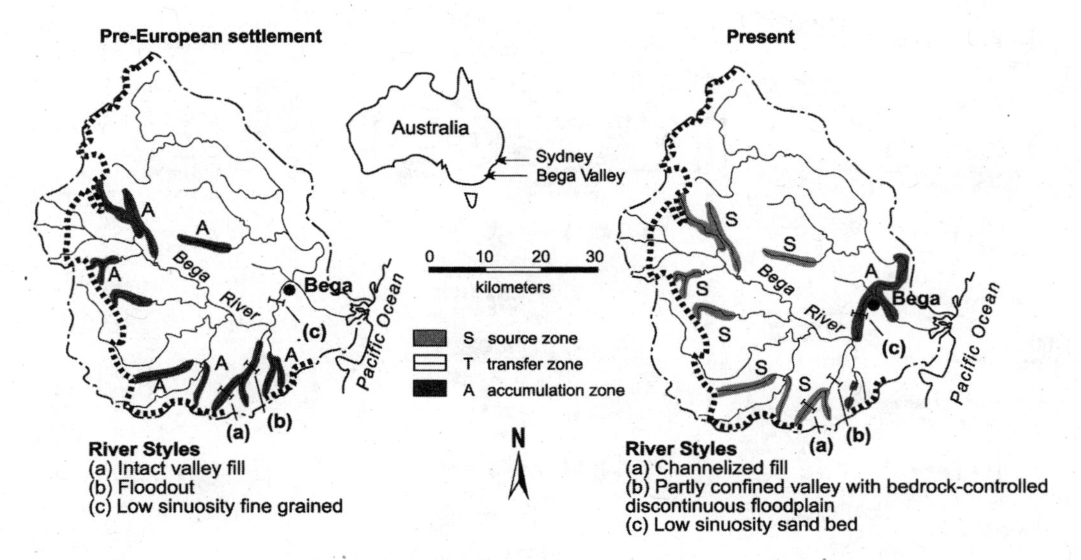

FIGURE 5.7. Pre- and post-European settlement changes to the distribution of sediment source, transfer and accumulation zones in Bega catchment.

characterized base of escarpment and mid-catchment locations prior to European settlement have been transformed into continuous yet ephemeral channels. The enlarged channel capacity along the lowland plain ensures that higher flows are now retained within its banks. Enhanced longitudinal connectivity has increased the rate and efficiency of downstream transfer of flow, sediment, organic matter, and other nutrients, decreasing their retention within the system. In most rivers, this increased "leakiness" deprives the upper reaches of essential carbon and nutrients, reducing biomass and diversity of instream plants and animals (Wolfenden et al. 2005). Given the extensive deposition of sand sheets, a large percentage of flow is now subsurface. The mixed load river of the pre-disturbance era has been transformed into a bedload-dominated system.

Contemporary channel character throughout Bega catchment is more homogenous than that before European settlement. Reduced structural complexity and heterogeneity limits the diversity of riverine habitats, as noted for other rivers subjected to sand slugs (e.g., Downes et al. 2006). Uniform, sand-sized materials now line the bed and banks. Habitats have been smothered, reducing their biodiversity. The functional role of pools as refugia has diminished. Many of the biotic attributes of swampy, discontinuous watercourses have been lost. Riparian vegetation associations have been transformed.

Changes to channel form have also altered *lateral* exchanges of water, sediment, organic matter, and nutrients between the channel and its floodplain. Floodplains are no longer inundated at the base of the escarpment because of the deeply incised channels. In contrast, extensive swathes of coarse sand have been deposited atop the formerly fine-grained floodplain of the lowland plain. This has partially infilled backswamps, reducing the capacity of the lowland plain to act as a nutrient sink or to provide habitat and food for aquatic fauna. Exotic vegetation incursion along the lowland plain has fundamentally altered the composition, seasonality, and rates of breakdown of leaf litter and organic matter. In mid-catchment locations

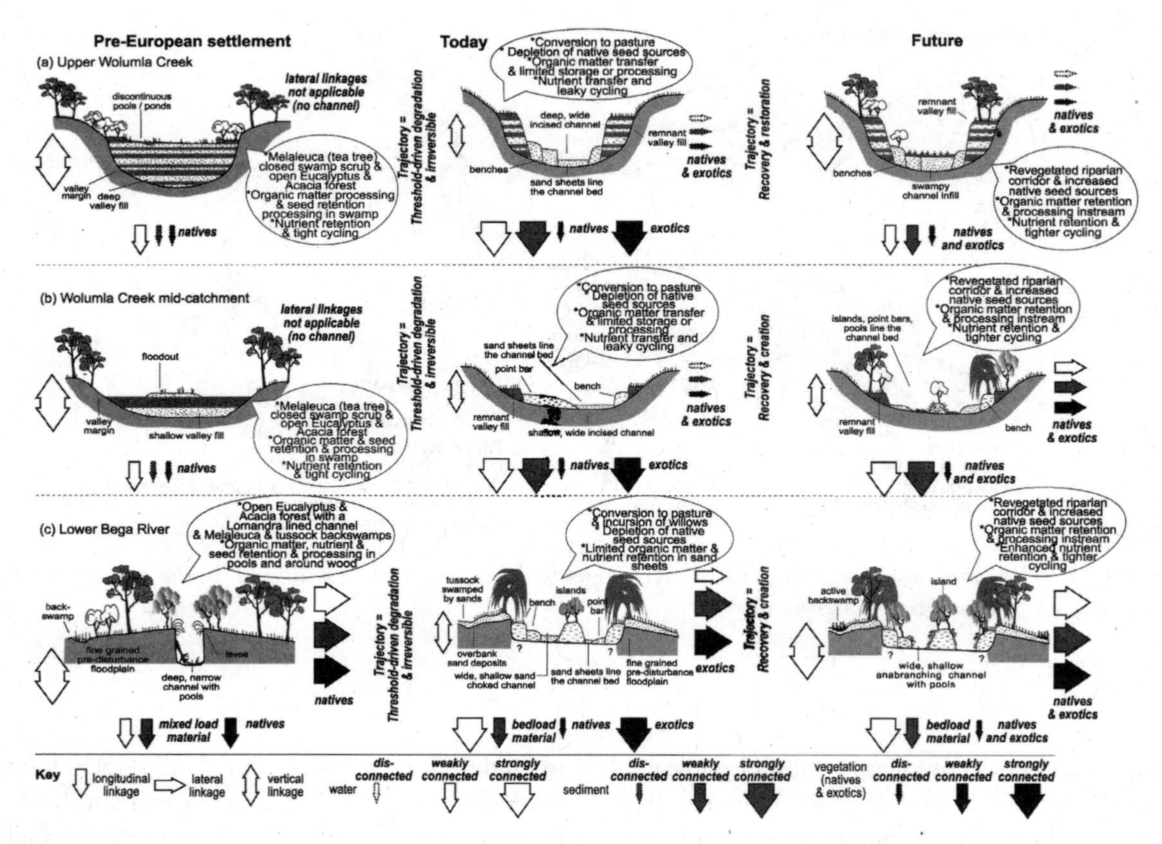

FIGURE 5.8. Conceptual model showing changes to geo-ecological relationships in upper, middle and lowland reaches of Bega catchment.

the conversion to pasture has modified channel-floodplain connectivity, such that there is little if any organic matter input and shading is nonexistent. These modifications have significantly reduced the value of edge habitats to native plants and animals.

The development of a continuous network of channels, associated changes to channel-floodplain relationships, and the highly degraded riparian vegetation cover have altered *vertical* water exchange and depth of the water table throughout Bega catchment. While the water table has been lowered in the base of escarpment setting, where swamp deposits have been deeply incised, the water table has risen in lowland areas of the catchment, where extensive bed aggradation has occurred. These changes, along with the homogeneity of bed material size (contrasting with the fine-grained swamps, bedrock pools, and discontinuous sand bars of the pre-disturbance channel) have modified the hydrologic regime and, potentially, the function of hyporheic zone processes along rivers (chapter 6, Boulton 2007).

Management Applications in Bega Catchment

Given the limiting factors operating in the system, the potential for rivers in Bega Catchment to attain structures that equate to their pre-European settlement condition is virtually zero over management timeframes of fifty to one hundred years. The functioning of aquatic ecosystems has been transformed by changes to geomorphic river structure and processes, constraining the capacity for ecological recovery. This understanding of historical and contemporary processes has been used to predict realistic targets for river rehabilitation, using a conservation-first prioritization framework to protect remnant swamps and develop rehabilitation strategies that extend from areas in good ecological condition (Fryirs and Brierley 2005). Other things being equal (e.g. there is no deterioration in water quality), the prospects for ecological recovery depend on improvements to abiotic condition.

Conclusion

Three primary implications are drawn from this chapter:

1. *Appreciate the range of physical and biological behavior of any given system.* The character, rate, and permanence of river changes have enormous implications for river management. System dynamics, sensitivity to change, and proximity to threshold conditions are critical considerations in interpretations of river condition and trajectories of change, and related assessment of the potential for geo-ecological recovery. Understanding contemporary biophysical conditions, framed in the context of historical channel change, underpins effective river rehabilitation programs. In a sense, just as contemporary river management frameworks recognize the importance of "space to move" initiatives (Piégay et al. 2005), we also need to integrate "time to adjust" perspectives into planning and decision-making. This requires greater appreciation of evolutionary trajectories and lagged responses operating within any given system, unraveling the impacts of human disturbance from natural variability.

2. *Develop an effective understanding of place-based, system-specific behavior.* Appreciating the range of system behavior, variability of system response to disturbance events, and the changing nature of key drivers and/or relationships that fashion ecosystem condition presents a critical knowledge base with which to frame management actions. Efforts to enhance ecological resilience must build upon a coherent, system-specific understanding of controls upon aquatic ecosystem condition. This encompasses an appreciation of the changing nature of system adjustments over time, appraising ecohydraulic relationships atop a geomorphic template across multiple scales. These insights provide the foundations for conceptual models with which to predict likely future scenarios.

3. *Use evolutionary insights to develop proactive management strategies.* Piecemeal, reactive strategies do not provide efficient and cost-effective ways to achieve rehabilitation success. Costs for repair and maintenance are excessive. The key to resilience lies in placing short-term dynamics in relation to longer-term adjustments, shaping determinations of what is natural or expected for any given system. Management programs that aim to enhance recovery of aquatic ecosystems maximize the likelihood of achieving the greatest benefits for the least cost. Understanding causes of system adjustment is required to determine the level of intervention that is needed in working toward specified goals. In some instances, intervention may not be required, as noted by Wheeler (1998, 299): "our real task isn't to manage, but to stop intervening, stand aside, and let the miracle of life assert itself."

Elsewhere, proactive plans may be required to target strategic initiatives that tackle the most serious issues first (see chapter 15). Proactive plans frame target conditions for individual reaches within their catchment context, recognizing the changing character, pattern, and linkages of ecosystem forms and processes over time.

Acknowledgments

We thank Angela Arthington and Hervé Piégay for constructive comments that improved this chapter. Gary J. Brierley, Kirstie A. Fryirs, and Andrew Boulton also thank the Australian Research Council for financial support.

References

Allan, J. D. 2004. Landscapes and riverscapes: The influence of land use on stream ecosystems. *Annual Review of Ecology, Evolution and Systematics* 35:257–84.

Benson, L. 2004. *The science behind "The Living Murray Initiative"—Part 2*. Deniliquin, Australia: Murray Irrigation Limited. www.murrayirrigation.com.au/download/3217000.pdf (accessed December 16, 2007).

Boulton, A. J. 2003. Parallels and contrasts in the effects of drought on stream macroinvertebrate assemblages. *Freshwater Biology* 48:1173–85.

Boulton, A. J. 2007. Hyporheic rehabilitation in rivers: Restoring vertical connectivity. *Freshwater Biology* 52:632–50.

Brierley, G. J., T. Cohen, K. Fryirs, and A. Brooks. 1999. Post-European changes to the fluvial geomorphology of Bega Catchment, Australia: Implications for river ecology. *Freshwater Biology* 41:839–48.

Brierley, G. J., and K. A. Fryirs. 2005. *Geomorphology and river management: Applications of the River Styles framework*. Oxford, UK: Blackwell.

Brooks, A. P., and G. J. Brierley. 1997. Geomorphic responses of lower Bega River to catchment disturbance, 1851–1926. *Geomorphology* 18:291–304.

Brooks, A. P., and G. J. Brierley. 2002. Mediated equilibrium: The influence of riparian vegetation and wood on the long-term evolution and behaviour of a near pristine river. *Earth Surface Processes and Landforms* 27:343–67.

Brooks, A. P., and G. J. Brierley. 2004. Framing realistic river rehabilitation programs in light of altered sediment transfer relationships: Lessons from East Gippsland, Australia. *Geomorphology* 58:107–23.

Brooks, A. P., G. J. Brierley, and R. G. Millar. 2003. The long-term control of vegetation and woody debris on channel and floodplain evolution: Insights from a paired catchment study in southeastern Australia. *Geomorphology* 51:7–29.

Brunsden, D., and J. B. Thornes. 1979. Landscape sensitivity and change. *Transactions of the Institute of British Geographers* NS4:463–84.

Bunn, S. E., and A. H. Arthington. 2002. Basic principles and ecological consequences of altered flow regimes for aquatic biodiversity. *Environmental Management* 30:492–507.

Chessman, B. C., K. A. Fryirs, and G. J. Brierley. 2006. Linking geomorphic character, behaviour and condition to fluvial biodiversity: Implications for river rehabilitation. *Aquatic Conservation: Marine and Freshwater Research* 16:267–88.

Cohen, T., I. Reinfelds, and G. J. Brierley. 1998. River styles in Bellinger-Kalang catchment. *Macquarie Research Limited Report for NSW Department of Land and Water Conservation*. 119 pp.

Collins, B. D., and D. R. Montgomery. 2001. Importance of archival and process studies to characterizing pre-settlement riverine geomorphic processes and habitat in the Puget Lowland. In *Geomorphic processes and riverine habitat*, ed. J. M. Dorava, D. R. Montgomery, B. B. Palcsak, and F. A. Fitzpatrick, 227–43. Washington, DC: American Geophysical Union.

Downes, B. J., J. S. Hindell, and N. R. Bond. 2000. What's in a site? Variation in invertebrate density and diversity in a spatially replicated experiment. *Austral Ecology* 25:128–39.

Downes, B. J., P. S. Lake, A. Glaister, and N. R. Bond. 2006. Effects of sand sedimentation on the macroinvertebrate fauna of lowland streams: are the effects consistent? *Freshwater Biology* 51:144–60.

Dudgeon, D., A. H. Arthington, M. O. Gessner, Z. Kawabata, D. Knowler, C. Lévêque, R. J. Naiman, A. H. Prieur-Richard, D. Soto, and M. L. J. Stiassny. 2006. Freshwater biodiversity: Importance, threats, status, and conservation challenges. *Biological Reviews* 81:163–82.

Dyer, F. J., and M. C. Thoms. 2006. Managing river flows for hydraulic diversity: An example of an upland regulated gravel-bed river. *River Research and Applications* 22:257–67.

Elliott, A. H., and N. H. Brooks. 1997. Transfer of nonabsorbing solutes to a streambed with bed forms: Theory. *Water Resources Research* 33: 23–136.

Folke C., S. R. Carpenter, B. Walker, M. Scheffer, T. Elmqvist, L. Gunderson, and C. S. Holling. 2004. Regime shifts, resilience and biodiversity in ecosystem management. *Annual Review in Ecology, Evolution and Systematics* 35:557–81.

Fryirs, K. 2002. Antecedent landscape controls on river character, behaviour and evolution at the base of the escarpment in Bega catchment, South Coast, New South Wales, Australia. *Zeitshrift für Geomorphologie* 46:475–504.

Fryirs, K. 2003. Guiding principles of assessing the geomorphic condition of rivers: Application of a framework in Bega catchment, South Coast, NSW, Australia. *Catena* 53:17–52.

Fryirs, K., and G. J. Brierley. 1998. The character and age structure of valley fills in upper Wolumla Creek catchment, South Coast, New South Wales, Australia. *Earth Surface Processes and Landforms* 23:271–87.

Fryirs, K., and G. J. Brierley. 2000. A geomorphic approach for identification of river recovery potential. *Physical Geography* 21:244–77.

Fryirs, K., and G. J. Brierley. 2001. Variability in sediment delivery and storage along river courses in Bega catchment, NSW, Australia: Implications for geomorphic river recovery. *Geomorphology* 38:237–65.

Fryirs, K. A., and G. J. Brierley. 2005. *Practical applications of the River Styles framework as a tool for catchment-wide river management: A case study from Bega Catchment, NSW, Australia*. E-book published on www.riverstyles.com (accessed December 1, 2007).

Fryirs, K., G. J. Brierley, N. J. Preston, and M. Kasai. 2007. Buffers, barriers and blankets: The (dis)connectivity of catchment-scale sediment cascades. *Catena* 70:49–67.

Gordon, N. D., T. A. McMahon, B. L. Finlayson, C. J. Gippel, and R. J. Nathan. 2004. *Stream hydrology: An introduction for ecologists*. Chichester, UK: Wiley.

Gore, J. A., ed. 1985. *The restoration of rivers and streams: Theories and experience*. Boston, MA: Butterworth Publishers.

Gore, J., and F. Shields Jr. 1995. Can large rivers be restored? *BioScience* 45:142–52.

Gunderson, L. 2000. Ecological resilience—In theory and application. *Annual Review in Ecology, Evolution and Systematics*. 31:425–39.

Hilderbrand, R. H., A. C. Watts, and A. M. Randle. 2005. The myths of restoration ecology. *Ecology and Society* 10 (1): Article 19. http://www.ecologyandsociety.org/vol10/iss1/art19/ (accessed December 16, 2007).

Hoover, T. M., J. S. Richardson, and N. Yonemitsu. 2006. Flow-substrate interactions create and mediate leaf litter resource patches in streams. *Freshwater Biology* 51:435–47.

Hughes, F. M. R., A. Colson, and J. O.Mountford. 2005. Restoring riparian ecosystems: The challenge of accommodating variability and designing restoration trajectories. *Ecology and Society* 10 (2): Article 12.

Kondolf, G. M., and H. Piégay, eds. 2003. *Tools in fluvial geomorphology*. Chichester, UK: Wiley.

Kozarski, S., and K. Rotnicki. 1977. Valley floors and changes of river channel patterns in the north Polish Plain during the late Wurm and Holocene. *Quaestiones Geographicae* 4:51–93.

Lake, P. S. 2000. Disturbance, patchiness, and diversity in streams. *Journal of the North American Benthological Society* 19:573–92.

Moore, J. W. 2006. Animal ecosystem engineers in streams, *Bioscience* 56:237–46

Naiman, R. J., S. E. Bunn, C. Nilsson, G. E. Petts, G. Pinay, and L. C. Thompson. 2002. Legitimizing fluvial ecosystems as users of water: An overview. *Environmental Management* 30:455–67.

Paul, M. J., and J. L. Meyer. 2001. Streams in the urban landscape. *Annual Review of Ecology and Systematics* 32:333–65.

Perry, G. L. W. 2002. Landscapes, space and equilibrium: Shifting viewpoints. *Progress in Physical Geography* 26:339–59.

Phillips, J. D. 1992. Nonlinear dynamical systems in geomorphology: Revolution or evolution? *Geomorphology* 5:219–29.

Phillips, J. D. 2003. Sources of nonlinearity and complexity in geomorphic systems. *Progress in Physical Geography* 27:1–23.

Piégay, H., S. E. Darby, E. Mosselman, and N. Surian. 2005. A review of techniques available for delimiting the erodible river corridor: A sustainable approach to managing bank erosion. *River Research and Applications* 21:773–89.

Poff, N. L. 1997. Landscape filters and species traits: Towards mechanistic understanding and prediction in stream ecology. *Journal of the North American Benthological Society* 16:391–409.

Poff, N. L., J. D. Allan, M. B. Bain, J. R. Karr, K. L. Prestegaard, B. D. Richter, R. E. Sparks, and J. C. Stromberg. 1997. The natural flow regime: A paradigm for river conservation and management. *BioScience* 47:769–84.

Postel, S., and B. D. Richter. 2003. *Rivers for life: Managing water for people and nature*. Washington, DC: Island Press.

Pusey, B. J., M. J. Kennard, and A. H. Arthington. 2000. Discharge variability and the development of predictive models relating stream fish assemblage structure to habitat in north-eastern Australia. *Ecology of Freshwater Fishes* 9:30–50.

Richards, K., J. Brasington, and F.Hughes. 2002. Geomorphic dynamics of floodplains: Ecological implications and a potential modeling strategy. *Freshwater Biology* 47:559–79.

Richter, B. D., J. V. Baumgartner, R. Wigington, and D. P. Braun. 1997. How much water does a river need? *Freshwater Biology* 37:231–49.

Ruddell, N. 2005. Conversion of a dry retention basin to a wet basin to achieve water quality outcomes and stormwater harvesting in Bonnyrigg. Stormwater Industry Association 2005 Regional Conference, Port Macquarie, NSW. *Sustainable Stormwater: You Are Responsible—Justify Your Decisions*. 20–21 April, 2005.

Rutherfurd, I., K. Jerie, and N. Marsh. 2001. Planning for stream rehabilitation: Some help in turning the tide. *Water: Journal of the Australian Water and Wastewater Association* 28:25–27.

Scheffer, M., and S. R. Carpenter. 2003. Catastrophic regime shifts in ecosystems: linking theory to observation. *Trends in Ecology and Evolution* 18:648–56.

Scheffer, M., S. R. Carpenter, J. A. Foley, C. Folke, and

B. Walker. 2001. Catastrophic shifts in ecosystems. *Nature* 413:591–96.

Schumm, S. A. 1969. River metamorphosis. *Journal of the Hydraulics Division, Proceedings of the American Society of Civil Engineers* 95:255–73.

Schumm, S. A. 1979. Geomorphic thresholds: The concept and its applications. *Transactions of the Institute of British Geographers* N54:485–515.

Schumm, S. A., and R. W. Lichty. 1965. Time, space and causality in geomorphology. *American Journal of Science* 263:110–19.

Sheldon, F., A. J. Boulton, and J. T. Puckridge. 2002. Conservation value of variable connectivity: Aquatic invertebrate assemblages of channel and floodplain habitats of a Central Australian Arid-Zone River, Cooper Creek. *Biological Conservation* 103:13–31.

Sparks, R. E. 1995. Need for ecosystem management of large rivers and their floodplains. *BioScience* 45:168–82.

Thomson, J., M. P. Taylor, and G. J. Brierley. 2004. Are River Styles ecologically meaningful? A test of the ecological significance of a geomorphic river characterization scheme. *Aquatic Conservation: Marine and Freshwater Ecosystems* 14:25–48.

Thomson, J., M. P. Taylor, K. A. Fryirs, and G. J. Brierley. 2001. A geomorphological framework for river characterisation and habitat assessment. *Aquatic Conservation: Marine and Freshwater Ecosystems* 11:373–89.

Trofimov, A. M., and J. D. Phillips. 1992. Theoretical and methodological premises of geomorphological forecasting. *Geomorphology* 5:203–11.

Walker, K. F., F. Sheldon, and J. T. Puckridge. 1995. A perspective on dryland river ecosystems. *Regulated Rivers: Research and Management* 11:85–104.

Wheeler, R. S. 1998. *The buffalo commons*. New York, NY: Tom Doherty Associates.

Wishart, M. J., and B. R. Davies. 2003. Beyond catchment considerations in the conservation of lotic biodiversity. *Aquatic Conservation: Marine and Freshwater Ecosystems* 13:429–37.

Wolfenden, B., S. Mika, A. Boulton, and D. Ryder. 2005. Assessing change in riverine organic matter dynamics in the Hunter River, NSW, over the last 200 years: Implications for stream restoration. In *Proceedings of the 4th Australian Stream Management Conference: Linking rivers to landscapes*, ed. I. Rutherfurd, I. Wiszniewski, M. J. Askey-Doran, and R. Glazik, 697–703. Hobart, Australia: Department of Primary Industries, Water and Environment.

Wu, J., and O. L. Loucks. 1995. From balance of nature to hierarchical patch dynamics: A paradigm shift in ecology. *The Quarterly Review of Biology* 70:439–66.

Chapter 6

Ecological Function in Rivers: Insights from Crossdisciplinary Science

Sarah Mika, Andrew Boulton, Darren Ryder, and Daniel Keating

> We know from science that nothing in the universe exists as an isolated or independent entity.
>
> Wheatley 2002

Because much ecological research in rivers applies theories developed elsewhere to a diverse array of habitats renowned for their spatial and temporal complexity, riverine ecology lacks a clear conceptual cohesiveness (Fisher 1997). Hence, the quest to identify, explain, and predict dominant ecological patterns and processes has led to the proposition of many conceptual models that also vary across spatial and temporal scales. These models range from the structure of river networks through to reach-scale models of flow regimes, patch dynamics, sediment organization, and stream hydraulics. Not surprisingly, the explicitness of these conceptual models to specific river types (e.g., headwaters, alluvial rivers, floodplain rivers) contributes significantly to the processes and linkages emphasized by the models.

Despite the obvious lack of cohesion in conceptual models of river function, three themes are common to all such models and these are fundamental to riverine ecology: (1) identifying interactions between structure and function; (2) understanding the processes driving the arrangement of structural components in space and time; and (3) identifying how specific habitats and processes are connected in space and time. Critical reviews of conceptual models of river function are given elsewhere (see Thorp et al. 2006). Our aim here is to discuss these three themes as they relate to understanding river function. First, we define ecosystem structure and function and discuss benefits of measuring ecological processes for managing river systems. Second, we look at how ecosystem structure and function interact through space and time. We use the different conceptual models of river function to highlight key structuring processes and introduce the roles of longitudinal, lateral, and vertical linkages in river function. Third, we define and discuss the concept of connectivity within riverine ecosystems. A holistic understanding of these three themes relies on a crossdisciplinary approach to river science. We present two case studies to illustrate how a crossdisciplinary approach to specific riverine processes and habitats incorporates both the concept of form-process interactions and connectivity among structural components of the riverine ecosystem. We finish by discussing the implications of crossdisciplinary approaches to river ecology for managing and rehabilitating river systems.

Interactions between Structure and Function

Ecosystems encompass hierarchical levels of biological organization. At the organism-level, structure refers to the composition, size, shape, and orientation of biological features, such as beak shape (Fisher et al. 2007). At a population level, structure refers to abundance, frequency-distribution of life stages (population dynamics), and genetic fitness and diversity (SER 2004). Unlike organism or population structure, community structure is spatially explicit, defined as all the organisms of all the species in a specified area (Fisher et al. 2007). Ecosystem structure is defined as the numbers and kinds of component parts in an ecosystem (Grimm 1995). It is at the ecosystem-level where abiotic and biotic structure are integrated (Fisher et al. 2007), although organisms, populations, and communities are influenced by and influence their abiotic environment. Thus, determining the ecological factors that influence the abundance and diversity of biota within a certain habitat means integrating biological knowledge of the biota (habitat requirements, genetic structure, population dynamics, predator-prey interactions), with knowledge of the historical and contemporary physical system (flow regime, geomorphic template, disturbances).

Historically, measures of biological structure within rivers such as the diversity of macroinvertebrate or fish assemblages, or physical structure such as the geomorphic template and hydrologic regime, have dominated the literature, as it is often assumed that if the pattern or structure of an ecosystem has been identified, then so too have the functions pivotal to ecosystem sustainability (Kentula 2000). This is not always the case (Ryder and Miller 2005). Biota are also classified by their functional role in the ecosystem, for example, as primary producers, herbivores, carnivores, detritivores, or as nitrogen-fixers (Boulton and Brock 1999; Ward et al. 2002). Recent studies of the relationships between ecosystem structure and function have centred on whether ecosystems with increasingly fewer species would be able to maintain functional properties and process rates comparable with unaffected systems (reviews in Loreau 2000). For example, local extinctions driven by environmental change can reduce the probability of particular species being present; these species may maintain fundamental ecosystem functions such as primary productivity that supports entire food webs (Vinebrooke et al. 2004). The difficulty in quantifying these relationships lies in the implicit knowledge needed of ecosystem- or landscape-level interactions and processes influencing the organisms or populations (Ehrenfeld 2000). This approach is somewhat more complicated in rivers because fluxes of water, transported sediments and nutrients, and organisms occur between different geomorphic features, resulting in a mosaic of interdependent habitats, each one suitable for different species and communities (Pedroli et al. 2002).

A functional perspective of river science provides a useful conceptual framework within which most aspects of ecology, geomorphology, and hydrology can be incorporated. At the ecosystem level of organization, structure and function affect each other as ecological processes shape ecosystem morphology, and landscape form influences ecosystem processes (Fisher et al. 2007). Thus, in considering the ecosystem-level functions of a river system, researchers must consider the location of the river (or study reach) in the landscape—its boundaries, its connections to adjoining ecosystems, and its inputs and losses of materials and energy from its physical surroundings (Ehrenfeld and Toth 1997). Quantifying fluxes of energy and materials into, out of, and within ecosystems is a hallmark of analyses of river function. For example, the energy of flowing water provides the physical template upon which the instream and riparian habitats are structured (Friedman et al. 1996). The mosaic of habitats provides sites where the fluxes of energy may subsidize the growth of organisms (e.g., low flows promoting benthic autotrophic production; Ryder 2004), or prevent their growth (e.g., high velocity flows scouring benthic autotrophs; Ryder et al. 2006). Understanding riverine function therefore depends on a comprehensive understanding of the complex interplay of fluvial geomorphology, hydrology, and instream and riparian biota.

Ecosystem-level processes such as transformations of matter or the fate of energy and matter can be ideal measures of the ecological condition of rivers because they provide an integrated response to a broad range of disturbances (Bunn et al. 1999). Measurements of ecosystem function in rivers are becoming more widespread in the scientific literature, with process-based ecological indicators being used to quantitatively assess biogeochemical cycles (McKee and Faulkner 2000), primary production (French-McCay et al. 2003), and organic matter breakdown (Gessner and Chauvet 2002). However, there are a number of difficulties with implementing this framework. Definitions of ecosystem function in the literature are vague and involve different processes. Processes that make up ecosystem function vary widely, including purely physical or biological phenomena (storm disturbance or diversity, respectively), those that integrate the activities of numerous organisms (nutrient cycling, energy flow), and processes that integrate complex interactions between abiotic and biotic forces (soil formation, water quality; Ehrenfeld 2000). The spatial and temporal scales over which these processes apply and are measured also vary widely and there are often multiple methods of quantifying them (see Ehrenfeld 2000).

Nevertheless, efforts to understand river function through crossdisciplinary research provide insight into ecosystem processes that are integral to the sustainability of rivers. Determining the rates and types of processes that occur in rivers relies heavily on specifying the spatial and temporal *scales* of each process as well as the scales at which these processes might *interact* (in turn, governed by connectivity). When viewing rivers from this dynamic and process-oriented perspective, "functional zones" are commonly used as spatial constructs to help understand how a river works and where the action takes place.

Interactions in Space and Time

This paradigm of river function and the spatial arrangement of functional zones has had a relatively long history in various disciplines. In geomorphology, for example, functional concepts of sediment transfer and deposition/erosion (e.g., Horton 1945) date back to early classifications of rivers into "bedrock controlled" and "alluvium controlled" sections to reflect the extent to which processes of erosion at a broad scale occur along the river continuum from upland streams to lowland rivers (reviewed in Schumm 1977). The way that river geomorphology might influence ecological processes at a range of scales (e.g., Parsons et al. 2003; Benda et al. 2004) explicitly acknowledges the functional significance of drainage patterns and catchment linkages (Newson 1994; chapter 4). These longitudinal patterns also relate to hydrological processes that vary along the river's length. Although attention typically focused on instream processes (Leopold et al. 1964), there was also early recognition of the way that groundwater hydrology influences surface flows and river function along its course (review in Freeze and Cherry 1979), and this has been incorporated into theories of hyporheic function at several spatial scales along a river (Boulton et al. 1998).

In the discipline of river ecology, early workers distinguished zones along rivers as not only supporting different biota (e.g., Illies 1961) but also being functionally different in the way that the biota responded to these conditions (Illies and Botosaneanu 1963; Ward et al. 2002). Disparate European literature supporting this view was drawn together nicely by Hynes (1970), who also recognized the crucial lateral links of streams with their catchments (Hynes 1975) and groundwater (Hynes 1983). However, the much-publicized River Continuum Concept (RCC; Vannote et al. 1980) is often credited for being the initial creative synthesis that related the processes of photosynthesis and respiration (expressed as a ratio to indicate their relative importance) to spatial location along the river continuum (Fisher 1997). These different sources of energy were hypothesized to have repercussions for expected proportions of functional feeding groups (e.g., shredders, collectors, grazers). This concept, while now accepted as oversimplistic (Boulton and Brock 1999), heralded a new way for aquatic ecologists to consider rivers in a more functional context. Over the last three decades, there have been numerous concepts and paradigms to help explain river ecosystem function, and these have been subject to frequent revision and resynthesis (e.g., Ward et al. 2002; Thorp et al. 2006).

If we consider function in a broad sense (e.g., photosynthesis, respiration), perhaps the catchment-scale "upland stream" versus "lowland river" approach suffices to define suitably broad functional zones. Efforts to further subdivide rivers into more specific, longitudinally ordered zones in a sequence have failed to correlate with functional processes (Townsend 1996). Although the RCC acknowledges that these zones are more of a continuum, the concept does not adequately recognize the functional significance of lateral (e.g., floodplain) or vertical (e.g., hyporheic zone) linkages in river ecosystems. Regardless of the applicability of these various models, several broad zones are commonly recognized at the river and reach scale (figure 6.1), providing useful terms of reference for researchers. To some degree, these zones can be considered functional in that one can speculate about the relative importance of processes such as photosynthesis versus respiration. For example, in the hyporheic zone, we would expect photosynthesis to be minimal compared with respiration whereas in the open water of an unshaded, limpid river, photosynthesis would exceed respiration during daylight hours.

This broad functional classification, while commonly espoused in textbooks, hides important information. First, defining the edges of such zones is difficult. Indeed, these edges may be "critical transition zones" in their own right (Wall et al. 2001), with specific functional properties relevant at finer scales. Second, discontinuities may be more common than we might expect from the doctrines of the RCC and other concepts (Poole 2002). For example, where tributaries join the main stream, biological hot spots are hypothesized to occur (Benda et al. 2004) and there is much evidence for the ubiquity of patches rather than continua in many stream ecosystems at a range of scales (Townsend 1989). These patches can be functionally discrete and support different rates of particular processes (e.g., leaf litter breakdown; Kobayashi and Kagaya 2004). Patch heterogeneity is likely related to biodiversity (i.e., more habitats lead to greater biodiversity), and some river survey methods (e.g., River Habitat Survey, UK) consider "functional habitats," defined by substrate and velocity, as the key link between river processes and river biodiversity (Harper and Everard 1998), although this hypothesis deserves further testing.

Many crucial processes in rivers occur at fine spatial scales within the sediments of the water's edge, the riparian zone, or the hyporheic zone (Boulton et al. 1998; Wall et al. 2001). Here, interstitial microclimates create gradients in redox potentials and rates of microbial activity to yield "functional zones" that may be less than several millimeters, yet these tiny patches differ greatly in their biogeochemical processes and products (Dahm et al. 1998). With increasing technology, our ability to sample functional processes at these microscales is improving dramatically. Putting this spatial definition and classification of functional zones in a crossdisciplinary context leads to serious practical and philosophical problems, especially at these fine scales. River managers, hydrologists, and geomorphologists are typically more familiar with broad scales and most data have been collected at these levels (Benda et al. 2002; chapter 3). Implicit

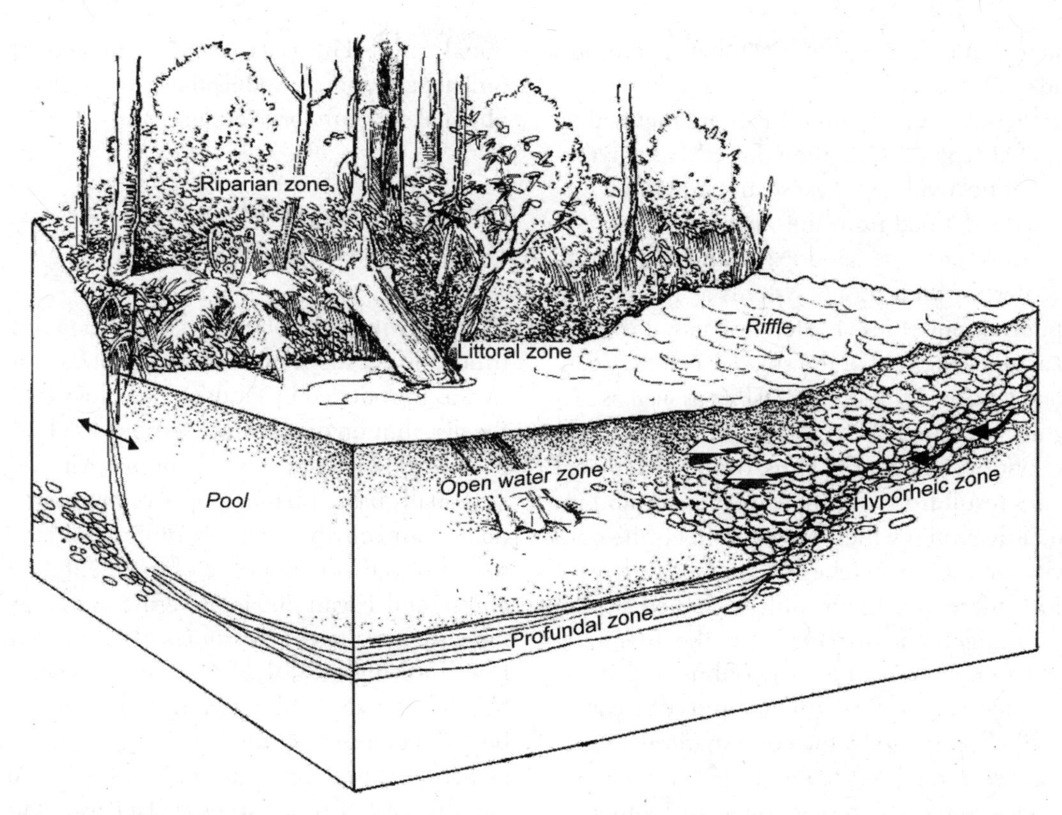

FIGURE 6.1. Rivers are often subdivided into pools (slow to zero flow) and riffles (faster flow, turbulent), and within these are habitats such as open water and profundal and littoral zones. Below, where water exchanges, is the hyporheic zone and laterally, the river interacts with bankside vegetation in the riparian zone.

is the hope that fine-scale patterns and processes somehow average out at the scale above in the hierarchies so popular in stream research (Frissell et al. 1986; Stanley and Boulton 2000). This may be a false assumption given the nonlinear nature of complex ecological systems (Burkett et al. 2005).

The "holy grail" of this crossdisciplinary approach to functional zones seems to be finding a common scale at which aquatic ecologists, fluvial geomorphologists, and hydrologists (at least) might work. This appears to have led to the proposal of functional process zones (FPZs) by Thorp et al. (2006; but see below). These are defined as hydrogeomorphic patches formed by catchment geomorphology and flow characteristics. FPZs result from shifts in hydrological and geomorphic conditions, distinguishing them from the more common but arbitrarily defined habitats such as reaches, segments, and sections (Frissell et al. 1986). Thus, physical hydrogeomorphic patches of direct interest to fluvial geomorphologists serve as a template for the ecological FPZs which are the research province of aquatic ecologists (Thorp et al. 2006). Of course, the consistent relationship of these FPZs to particular stream functions is yet to be established at a range of spatial and temporal scales, but perhaps this is a reasonable starting place where the disciplines of geomorphology, hydrology, and ecology might share common ground.

Interactions among these FPZs are dictated by their spatial arrangement, their physical and hydrological connectivity (see next section), and the way the patch is defined from the various perspectives of organisms, processes, and scale. As one moves to finer scales, by viewing ecosystems as nested, discontinuous hierarchies of patch mosaics, it is possible to analyze the role of the smaller patches (e.g., substrate types) within broader FPZs as well as factoring in temporal changes at different spatial scales (Thorp et al. 2006). Thus, overall ecological dynamics result from a composite of intra- and inter-patch dynamics which in turn reflects the connectivity among the patches. The rate and type of function, therefore, is not only governed by the spatial arrangement of patches in the hierarchy (defined from microscales of biofilms and interstices to megascales of catchments and riverscapes; Wiens 2002), but also by the combination of functional zones at different scales, their temporal activity cycles, and their connection to each other.

As yet, no conceptual model of river function provides a unified model of the interdependence of physical and biological structure and function across all river types. Furthermore, these conceptual models often lack clear definitions of the spatial and temporal scales at which they apply, and few data exist to test the validity of these models. However, the various conceptual models of river function provide ways of integrating ecological function with the physical structure of river systems and the temporal variability in hydrological and organic matter regimes. By emphasizing longitudinal, lateral, and vertical linkages within river systems, they integrate spatial and temporal drivers and constraints of ecological processes in rivers. Thus, they help synthesize structure-function relationships across the disciplines of biology, ecology, geomorphology, and hydrology. Currently accepted models of river function explicitly acknowledge the significance of connectivity across functional zones. This is a marked advance on the historical "zonation" models that were based on the physical structure of river channels.

Connectivity within Riverine Ecosystems

Connectivity is considered to be a fundamental attribute of all ecosystems. The concept of connectivity was introduced by landscape ecology to explain the distribution of species (Merriam 1984; Kondolf et al. 2006). Definitions of connectivity vary and are usually based on metapopulation dynamics (genetic connectivity) or the continuity of landscape structure (i.e., the opposite of fragmentation; Calabrese and Fagan 2004). Riverine ecosystems are largely controlled by hydrological connectivity, defined by Pringle (2001, 981) as the "water-mediated transfer of matter, energy, and organisms within or between elements of the hydrological cycle," such as between the channel and floodplain (Amoros and Roux 1988), or between surface waters and the alluvial aquifer (Gibert et al. 1997).

Variance in the vectors of hydrological connectivity underpins most ecosystem patterns and processes across multiple spatial and temporal scales (see Kondolf et al. 2006 and references within). Longitudinal hydrological linkages generally involve the downstream movement of water, sediment, nutrients, organic matter, aquatic macroinvertebrates, and seeds of aquatic and riparian plants. Hydrologic connectivity between river reaches also enables the upstream movement of some fish and invertebrate species. Although less widespread than downstream water-mediated transport (e.g., invertebrate drift), these upstream movements can have fundamental repercussions for upstream reaches, such as the nutrient input of migrating salmonids in headwater streams (Ben-David et al. 1998; Gresh et al. 2000).

Lateral connectivity between river channels and floodplains is highly significant for successional processes of floodplain vegetation, organic matter and nutrient dynamics, and the maintenance of the biophysical heterogeneity of floodplains (Ward et al. 1999). Flow variability and channel migration promote diverse arrays of lotic and lentic water bodies, vegetation structure, and concomitant fauna (Ward et al. 1999) with differing degrees of spatial and temporal connectivity to the main river channel and groundwater (Marmonier et al. 1992). Lateral linkages may be biological, as well as hydrological, such as the lateral energy transfer of aquatic insects to terrestrial predators, and terrestrial insects to aquatic predators such as fish (Fisher et al. 1998).

The third spatial dimension of hydrological connectivity in river systems is the vertical linkages between surface water and groundwater. Despite early recognition of the size of hyporheic zones in some alluvial rivers (e.g., Stanford and Gaufin 1974), and the importance of groundwater in solute transformations (Stream Solute Workshop 1990), these linkages have received less ecological attention than longitudinal or lateral connectivity (Stanford and Ward 1993). Yet, hydrological connectivity between surface and groundwater via the hyporheic zone (see case study 2) may be significant in large floodplain rivers. For example, the hyporheic zone extends laterally for several kilometers in the Flathead River in Montana, United States, (Stanford et al. 1994) and vertically for many meters in the Rhône River in France (Marmonier et al.1992). In a Sonoran Desert stream, the hyporheic zone can have a volume several times greater than that of the surface stream (Valett et al. 1990). Hydrological connectivity between surface and groundwater occurs across multiple spatial and temporal scales, influenced by flows and catchment geology and morphology. Shallow mixing zones occur at the scales of geomorphic units such as riffles, instream bars, and meander bends. Water may travel relatively quickly through these units, facilitating a high turnover of nutrients. Phreatic (deep) groundwater interacts less frequently with surface water, but it still provides habitat for specialized invertebrate fauna (stygofauna, Hancock et al. 2005). During dry periods, groundwaters maintain subsurface hydrological connectivity between pools via interstitial flow (e.g., Stanford and Ward 1993).

It is crucial to consider temporal dimensions of connectivity in river systems (Kondolf et al. 2006). Genetic connectivity in rivers is driven by season and lifecycle stages interacting with hydrological connectivity. Examples include seed dispersal, macroinvertebrate drift, and fish movement, all of which are negatively affected by disruptions to hydrological connectivity such as dams, weirs, excessive extraction and flow regulation (Bunn and Arthington 2002; Pringle 2003). Hydrological connectivity also varies over time, due to variability in precipitation, runoff, and flow regulation. In floodplain rivers, hydrological variability is a key structuring process as it controls the transfer of aquatic matter to floodplain habitats and terrestrial matter to instream habitats (see case study 1). It also is a key driver of sediment dynamics and the structure and composition of plant and animal communities (Ward et al. 1999). Due predominantly to the variability in discharge, connectivity among components of riverine ecosystems is dynamic in space and time. This is integral to healthy river function, as processes may need time in certain locations and to be in these locations at certain times.

Describing connectivity within rivers is difficult because such descriptions must include both spatial and temporal dimensions and often, connectivity may be low at one spatio-temporal scale, while simultaneously high at another. To illustrate, channelization, incision, and widening of rivers may disconnect surface water from groundwater at a catchment or regional scale. However, geomorphic diversity may have increased at the bar unit scale as a result of reworking of sediments within the incised, widened channel (Frothingham et al. 2002), leading to increased hydrological connectivity between surface water and shallow aquifers (i.e., hyporheic exchange) at these much finer spatial and temporal scales. Alternatively, habitats may be spatially connected (e.g., feeding habitat, spawning habitat, refuge habitat may be juxtaposed in space), but can be unavailable to biota when needed through temporal disconnection (e.g., artificially low discharge from regulation or extraction of flows preventing fish from migrating upstream to spawning habitats). Connectivity is often considered a positive ecological attribute, but this is not always the case (see chapter 8). Spatial and temporal disconnections can enhance genetic diversity by isolating populations, and prevent contaminants from moving through the ecosystem (e.g., Hancock 2002). Thus, *variability* in spatial and temporal connectivity is critical for maintaining maximum diversity of ecological structure and function (Kondolf et al. 2006).

Examples of Crossdisciplinary Research on Ecological Function

In this section we present case studies that exemplify the application of crossdisciplinary research in framing the conceptual understanding of two elements of ecological functionality, focusing on organic matter processing by streams and hyporheic zone processes.

Case Study 1: Stream Processing of Organic Matter

The concept of rivers as "riverscapes" was proposed by Wiens (2002) to emphasize the spatial connectivity between landscape ecology and aquatic ecology. Underlying this concept was a necessary integration of knowledge from terrestrial and aquatic scientists for the sustainable management of river systems. As riverine systems are governed by a unidirectional flow of water, they become effective agents in linking landscape elements, where changes in catchment land-use or disruptions to longitudinal linkages can have disproportionate impacts on river function. The supply of organic matter to rivers from catchment, riparian, and instream sources is pivotal for the maintenance of riverine ecosystem functions (Robertson et al. 1999), including photosynthesis, respiration, and nutrient retention. The sources, sinks, and transformations of organic matter in river systems, and the reliance of these processes on landscape connectivity, provide an ideal template to demonstrate the need for crossdisciplinary research in examining river function.

Hynes (1975) suggested that the catchment, rather than the stream, should be the basic unit of study because of the dependence of the stream on inputs of organic and inorganic material from the surrounding landscape. This was reinforced almost thirty years later by Gergel et al. (2002), who suggest landscape or catchment-based indicators of human impacts on riverine systems are required because of the diverse chemical, biological, hydrological, and geophysical components that must be assessed. The linkages among disciplines are evident in the complex interactions among inputs of organic matter from the terrestrial origins that dominate the energy dynamics of headwater streams. An example of this linkage is provided by Naiman et al. (1993), who demonstrate the benefits of riparian vegetation in promoting and maintaining the biodiversity of aquatic and terrestrial biota. However, these ecological benefits were necessarily placed in the context of human use of catchment resources, concluding that the sustainable management of rivers would require political and institutional cooperation as well as a sound scientific base.

As the concept of FPZs and their connectivity requires a spatial context, the management of river systems has increasingly involved ecologists and geographers through the subdiscipline of spatial ecology. The longitudinal and lateral connectivity to the floodplain created by flooding is recognized as a major driver of the structure and function of river systems, and has generated many models of how organic matter enters streams. Spatial ecology has informed the development of these models by treating rivers as a continuum, where changes to organic matter supply in the headwaters will have consequences downstream. Benke et al. (2000) used spatial information to construct floodplain inundation models for the unregulated Ogeechee River, producing long-term (decadal) predictions of riverine metabolism and energy supply based on flooding frequency. This model was applied at a finer spatial scale by Strayer et al. (2006), who used spatial ecology to link aquatic macroinvertebrate communities with the spatial distribution of geomorphic features and concomitant ecological habitats. They concluded that large-scale habitats (meter-scale) may be effective at explaining macroinvertebrate distributions in large rivers because they are integrative and describe habitat at the spatial scales of dominant controlling processes such as organic matter dynamics.

Organic matter in rivers is not homogeneous, but consists of an array of sources such as wood, leaves, water plants, algae, and bacteria. The provision of a variety of food sources promotes a diversity of aquatic consumers such as macroinvertebrates and fish. The distinction between allochthonous (*terrestrial* origin) and autochthonous (*aquatic* origin) organic matter is important because components of these materials (i.e., carbon and nutrients) are differentially available to biota and require processing before they are available for uptake. Processing involves a multitude of mechanisms including physical retention, abiotic leaching, abrasion, invertebrate grazing, and microbial decomposition, which release labile compounds available for stream biota (Boulton and Boon 1991). Each of these mechanisms regulating the sources, sinks, and transformations of stream organic matter requires discipline-specific techniques, from geomorphologists describing the physical template for organic matter retention to microbiologists documenting the microbial processing of organic fractions.

Case Study 2: Processes in the Hyporheic Zone

In many rivers, the hyporheic zone—the saturated sediments lying beneath the river channel—directly links surface water to permeable alluvial aquifers below the riparian zone (the parafluvial zone), and "true" groundwater (Boulton and Brock 1999; figure 6.1). The boundaries of the hyporheic and parafluvial zones are hard to identify as they fluctuate in response to variations in the depth and volume of water exchanged with the surface stream (White 1993), which are affected by surface water discharge and channel shape (Boulton 1993; Boulton et al. 2004). Gibert et al. (1990) proposed a "dynamic ecotone model" for the hyporheic zone, which emphasizes the *direction* of hydrologic exchange between the surface and subsurface as this governs many processes occurring in the hyporheic zone. Downwelling river water enters the sediments and introduces dissolved oxygen, nutrients such as DOC (dissolved organic carbon), organic matter, and small surface invertebrates into the hyporheic zone (Boulton et al. 2004). As the water percolates through the sediments, it is filtered physically as particulates are caught in the sediment matrix, and biogeochemically as solutes interact with the immense surface areas of microbial biofilms coating the sediment grains (Boulton et al. 1998). These biogeochemical transformations depend on the influent concentrations of oxygen and nutrients, and the porosity and local gradients in the chemical microenvironments of the sediment (reviews in Brunke and Gonser 1997; Jones and Mulholland 2000). Depending on its residence time in the hyporheic zone, water that reenters the surface stream (upwelling) may differ dramatically in its chemistry from that of surface water. Typically, upwelling water has lower concentrations of dissolved oxygen and higher concentrations of nutrients (such as nitrogen) in a chemically reduced form, which may be more available for uptake by surface biota (Coleman and Dahm 1990).

Hydrologic exchange between the surface stream and hyporheic zone is driven by the combination of geomorphic complexity and flow. Convectional exchange is caused by pressure differences arising from interactions between the topography of the streambed, surface flow, and the groundwater table (Boulton et al. 2004). As a stream flows over a bedform such as a crest, increased pressure drives water into the stream bed (Thibodeaux and Boyle 1987). An undulating streambed causes alternating areas of high and low pressure, promoting hydrological exchange of stream and subsurface water (Packman and Bencala 2000). A second mechanism driving hydrological exchange is through the interaction of stream flow and porewater. The velocity of stream flow is greater than the velocity of porewater, creating a gradient of exponentially decreasing flow velocities within the stream bed (Packman and Bencala 2000). Thus, streamflow near the bed surface "drags" porewater upward into the surface stream (advective exchange; Packman and Salehin 2003). Surface-subsurface exchange typically is greater in gravel-bed than sand-bed rivers because of the larger size of bedforms and the greater turbulence in near-surface streamflow in gravel-beds (Boulton et al. 2004).

The hyporheic zone is functionally significant for surface water processes in many rivers (reviews in Boulton et al. 1998; Dent et al. 2000). A key ecological role of the hyporheic zone is the alteration of water chemistry and production of nutrients that may be limiting in the surface stream (Coleman and Dahm 1990; Valett et al. 1994). Where surface water or groundwater is polluted, the hyporheic zone may form a barrier, inhibiting cross-contamination, and in some cases may remove contaminants through biochemical filtration (Harvey and Fuller 1998; Hancock 2002). Much of the decomposition and microbial processing of organic matter occurs in river sediments, and the hyporheic zone is a storage zone and processing site for this material (Marmonier et al. 1995). The hyporheic zone forms a potential refuge for surface stream invertebrates from flooding and drying (review in Boulton 2000) and polluted surface water (Jeffrey et al. 1986; Hancock 2002). Freely exchanging hyporheic zones are important sites for spawning and egg incubation of some fish in the northern hemisphere (e.g., salmonids, Vaux 1962; freshwater catfish, Boulton et al. 2004).

Human alterations to catchments and river channels have smoothed the bed profile of many rivers. Accelerated sedimentation has led to the loss of topographic relief of streambeds through infilling of pools (Schälchli 1992) and colmation as fine sediments percolate deep into the streambed (Hancock 2002). Together, these processes sever the surface stream from the hyporheic zone and groundwater aquifer (Boulton 2007). Colmated sediments are resistant to flushing unless the bed material is substantially reworked (Hancock and Boulton 2005). Severing linkages between the surface stream and hyporheic zone has adverse effects on both surface and subsurface biota and processes. Interstitial invertebrates and aerobic biofilms are starved of oxygen and nutrients from the surface stream. The capacity of the hyporheic zone to store and process organic matter is reduced. Upwelling diminishes and the chemical content of upwelling water is altered so that it no longer supplies the surface stream with transformed, biologically available nutrients. This has severe implications for the surface stream, particularly where surface water is nutrient-limited or where surface biota rely on actively exchanging hyporheic zones for flow refuges, spawning sites, and temperature buffering (Boulton 2000; Dent et al. 2000; Malcolm et al. 2003).

Repairing hyporheic zone linkages and processes has begun only recently (Boulton et al. 2004). Because sedimentation, siltation, and colmation have severed or reduced hydrological linkages, rehabilitation strategies commonly focus on using environmental flows to flush interstitial sediments (Hancock and Boulton 2005). Early results indicate very large flows that move bed material are needed and that the effects of flushing may be temporary as silt rapidly reenters the interstices (Boulton et al. 2004). Thus, long-term solutions for sedimentation, siltation, and colmation should focus on controlling silt inputs through improvements to land management and rehabilitation of riparian vegetation. A second rehabilitation strategy is to increase geomorphic complexity at the reach or sub-reach scale by installing wood "bedforms." Wood structures increase hydrologic exchange in the hyporheic zone and affect nutrient regeneration and export to the surface stream (Boulton 2007). However, the effectiveness of these structures may decrease with time between flood events as silt clogs the sediment matrix.

Understanding hyporheic zone structure and function clearly requires a crossdisciplinary approach. Hydrologic exchange drives most hyporheic processes and is determined by interdependent interactions between geomorphology and hydrology. Key ecological functions of the hyporheic zone such as the interstitial breakdown of organic matter and nutrient regeneration occur as a result of interactions among geomorphology, hydrology, biogeochemistry, physics, microbiology, and zoology. Because understanding hyporheic processes explicitly requires collaboration across individual scientific disciplines, hyporheic zone research provides an elegant example of how crossdisciplinary science is necessary to develop a comprehensive understanding of a riverine functional zone at the reach-scale. This provides a framework for understanding how structure and function interact at the reach-scale, how human activities have impacted on this structure and functioning, and perhaps, the mechanisms by which river managers can effectively protect and rehabilitate hyporheic zones to contribute toward benefiting the integrity of the riverine ecosystem.

Conclusion

Riverine function refers to *processes* that are influenced by a variety of physical, chemical, hydrological, and biotic factors. These factors influence each other. Combining ideas of riverine function with the recognition that broadly homogeneous patches recur predictably in river systems led to the concept of functional process zones (Thorp et al. 2006). Different processes occur at different locations in the catchment, and there is often spatial and temporal specificity for these processes, creating a patchy mosaic of functional habitats and critical transition zones (chapter 4). *How* FPZs are connected in space and time is critical to their function. There are different types of linkages, such as hydrologic and genetic connectivity, but in riverine landscapes, connectivity is predominantly driven by hydrology. Consequently, connectivity is also variable in space and time, and this variability affects the integrity of riverine ecosystems. Thus, the key question is: can spatially and temporally complex interactions among processes influenced by geological, geomorphological, hydrological, chemical, and biotic factors be effectively synthesized using a crossdisciplinary framework to answer specific questions about riverine function (e.g., organic matter dynamics and hyporheic processes) at scales and levels of complexity that are relevant to issues of river management?

Significant difficulties must be overcome for crossdisciplinary science to successfully advance our understanding of ecological function in rivers. The foremost issue is the multiplicity of spatial and temporal scales involved in feedbacks between biophysical structure and function. An approach taken from spatial ecology is to set the scale of investigation of ecological processes at the scale of the dominant controlling processes (Strayer et al. 2006), but this can be complicated by knowledge gaps (chapter 3), and the complexity and nonlinearity of biophysical processes (Frothingham et al. 2002). Geomorphology provides a hierarchical framework for this approach (chapter 4), both in setting the spatial and temporal context for ecological processes, and in determining potential trajectories of system change (chapter 5). Measuring ecosystem-level processes provides an integrated response to a broad range of disturbances by incorporating cumulative effects (direct, indirect, and synergistic) across multiple spatial and temporal scales. As teasing apart the ecological repercussions of specific disturbances often involves research at relatively fine spatial and temporal scales, it may prove difficult to integrate these insights with current geomorphological and hydrological knowledge.

Connectivity is a fundamental attribute of ecosystems (Kondolf et al. 2006), and its measurement is both spatially and temporally explicit. Temporal and spatial variability in the hydrological connectivity within river systems is crucial to maintaining maximum diversity of habitats, populations, communities, and ecological processes. The concept of connectivity explicitly integrates interrelationships between biophysical structure and function in space and time. Because connectivity is quantifiable (e.g., inundation frequencies of floodplain water bodies, gene flow through populations), it is useful for monitoring the health of river systems, and as a means of guiding and assessing rehabilitation activities within rivers.

A growing emphasis on riverine processes and their connections has necessarily led to multi- and interdisciplinary research to identify the key biophysical interactions that drive ecological processes. This approach is not without its difficulties, as researchers struggle to find common spatial and temporal scales among disciplines that have historically focused on significantly different scales, and to balance the increased complexity of broader-scale crossdisciplinary research with the information needs of river managers. The examples given here of research into organic matter dynamics and hyporheic zone processes demonstrate encouraging advancement toward a holistic understanding of riverine function facilitated by crossdisciplinary research teams. The next challenge is to build on this and develop crossdisciplinary river science to aid in the holistic rehabilitation of river systems.

Acknowledgments

Many of the concepts and ideas expressed here arose from a workshop in Muswellbrook in March, 2005, and we thank Gary J. Brierley and Kirstie A. Fryirs for organizing that event. We acknowledge the Australian Research Council and various industry partners for funding the work that led to the case studies discussed here. We also thank the editors, Barb Downes, and an anonymous referee for constructive comments on an earlier draft of this chapter.

References

Amoros, C., and A. L. Roux. 1988. Interaction between water bodies within the floodplain of large rivers: Function and development of connectivity. In *Connectivity in landscape ecology, Muenstersche Geographische Arbeiten* 29, ed. K.-F. Schreiber, 125–30. Paderborn, Germany: Schöningh.

Benda, L., N. L. Poff, D. Miller, T. Dunne, G. Reeves, G. Pess, and M. Pollock. 2004. The network dynamics hypothesis: How channel networks structure riverine habitats. *Bioscience* 54:413–27.

Benda, L. E., N. L. Poff, C. Tague, M. A. Palmer, J. Pizzuto, S. Cooper, E. Stanley, and G. Moglen. 2002. How to avoid train wrecks when using science in environmental problem solving. *BioScience* 52:1127–36.

Ben-David, M., T. A. Hanley, and D. M. Schell. 1998. Fertilization of terrestrial vegetation by spawning Pacific salmon: The role of flooding and predator activity. *Oikos* 83:47–55.

Benke, A. C., I. Chaubey, G. M. Ward, and E. L. Dunn. 2000. Flood pulse dynamics of an unregulated river floodplain in the Southeastern U.S. coastal plain. *Ecology* 81:2730–41.

Boulton, A. J. 1993. Stream ecology and surface-hyporheic exchange: implications, techniques and limitations. *Australian Journal of Marine and Freshwater Research* 44:553–64.

Boulton, A. J. 2000. The functional role of the hyporheos. *Internationale Vereinigung für Theoretische und Angewandte Limnologie Verhandlungen* 27:51–63.

Boulton, A. J. 2007. Hyporheic rehabilitation in rivers: Restoring vertical connectivity. *Freshwater Biology* 52: 632–50.

Boulton, A. J., and P. I. Boon. 1991. A review of methodology used to measure leaf litter decomposition in lotic environments: Time to turn over an old leaf? *Australian Journal of Marine and Freshwater Research* 42:1–43.

Boulton, A. J., and M. A. Brock. 1999. *Australian freshwater ecology: Processes and management*. Adelaide, Australia: Gleneagles Publishing.

Boulton, A. J., S. Findlay, P. Marmonier, E. H. Stanley, and H. M. Valett. 1998. The functional significance of the hyporheic zone in streams and rivers. *Annual Review of Ecology and Systematics* 29:59–81.

Boulton, A. J., S. J. Mika, D. S. Ryder, and B. J. Wolfenden. 2004. Raising the dead: Can we restore the health of subsurface aquatic ecosystems by recovering geomorphic complexity using conventional river rehabilitation techniques? In *Airs, waters, places: Transdisciplinary research in ecosystem health*, ed. G. Albrecht, 41–61. Callaghan, Australia: University of Newcastle.

Bunn, S. E., and A. H. Arthington. 2002. Basic principles and ecological consequences of altered flow regimes for aquatic biodiversity. *Environmental Management* 30:492–507.

Bunn, S. E., P. M. Davies, and T. D. Mosisch. 1999. Ecosystem measures of river health and their response to riparian and catchment degradation. *Freshwater Biology* 41:333–45.

Burkett, V. R., D. A. Wilcox, R. Stottlemyer, W. Barrow, D. Fagre, J. Baron, J. Price, J. L. Nielsen, C. D. Allen, D. L. Peterson, G. Ruggerone, and T. Doyle. 2005. Nonlinear dynamics in ecosystem response to climatic change: Case studies and policy implications. *Ecological Complexity* 2:357–94.

Brunke, M., and T. Gonser. 1997. The ecological significance of exchange processes between rivers and groundwater. *Freshwater Biology* 37:1–33.

Calabrese, J. M., and W. F. Fagan. 2004. A comparison-shopper's guide to connectivity metrics. *Frontiers in Ecology and the Environment* 10:529–36.

Coleman, R. L., and C. N. Dahm. 1990. Stream geomorphology: Effects on periphyton standing crop and primary production. *Journal of the North American Benthological Society* 9:293–302.

Dahm, C. N., N. B. Grimm, P. Marmonier, H. M. Valett, and P. Vervier. 1998. Nutrient dynamics at the interface between surface waters and groundwaters. *Freshwater Biology* 40:427–51.

Dent, C. L., J. D. Schade, N. B. Grimm, and S. G. Fisher. 2000. Subsurface influences on surface biology. In *Streams and ground waters*, ed. J. B. Jones and P. J. Mulholland, 381–402. San Diego, CA: Academic Press.

Ehrenfeld, J. G. 2000. Defining the limits of restoration: The need for realistic goals. *Restoration Ecology* 8:2–9.

Ehrenfeld, J. G., and L. A. Toth. 1997. Restoration ecology and the ecosystem perspective. *Restoration Ecology* 5:307–17.

Fisher, S. G. 1997. Creativity, idea generation, and the functional morphology of streams. *Journal of the North American Benthological Society* 16:305–18.

Fisher, S. G., N. B. Grimm, E. Martí, R. M. Holmes, and J. B. Jones. 1998. Material spiraling in stream corridors: A telescoping ecosystem model. *Ecosystems* 1:19–34.

Fisher, S. G., J. B. Heffernan, R. A. Sponseller, and J. R.Welter. 2007. Functional ecomorphology: Feedbacks between form and function in fluvial landscape ecosystems. *Geomorphology* 89:84–96.

Freeze, R. A., and J. A. Cherry. 1979. *Groundwater*. Englewood Cliffs, NJ: Prentice-Hall.

French-McCay, D. P., C. H. Peterson, J. T. DeAlteris, and J. Catena. 2003. Restoration that targets function as opposed to structure: Replacing lost bivalve production and filtration. *Marine Ecology Progress Series* 264:197–212.

Friedman, J. M., W. R. Osterkamp, and W. M. Lewis Jr. 1996. Channel narrowing and vegetation development following a Great Plains flood. *Ecology* 77:2167–81.

Frissell, C. A., W. J. Liss, C. E. Warren, and M. D. Hurley. 1986. A hierarchical framework for stream habitat classification: Viewing streams in a watershed context. *Environmental Management* 10:199–214.

Frothingham, K. M., B. L. Rhoads, and E. E. Herricks. 2002. A multiscale conceptual framework for integrated ecogeomorphological research to support stream naturalization in the agricultural Midwest. *Environmental Management* 29:16–33.

Gergel, S. E., M. G. Turner, J. R. Miller, J. M. Melack, and E. H. Stanley. 2002. Landscape indicators of human impacts to riverine systems. *Aquatic Sciences—Research Across Boundaries* 64:118–28.

Gessner, M. O., and E. Chauvet. 2002. A case for using litter breakdown to assess functional stream integrity. *Ecological Applications* 12:498–510.

Gibert, J., M.-J. Dole-Olivier, P. Marmonier, and P. Vervier. 1990. Surface water–groundwater ecotones. In *The ecology and management of aquatic-terrestrial ecotones*, ed. R. J. Naiman and H. Décamps, 199–226. Paris, France: UNESCO and Carnforth, UK: Parthenon Publishers.

Gibert, J., J. Mathieu, and F. Fournier, eds. 1997. *Groundwater–surface water ecotones: Biological and hydrological interactions*. Cambridge, UK: Cambridge University Press.

Gresh, T., J. Lichatowich, and P. Schoonmaker. 2000. An estimation of historic and current levels of salmon production in the northeast Pacific ecosystem. *Fisheries* 25:15–21.

Grimm, N. B. 1995. Why link species and ecosystems? A perspective from ecosystem ecology. In *Linking species and ecosystems*, ed. C. G. Jones and J. H. Lawton, 5–15. New York, NY: Chapman and Hall.

Hancock, P. J. 2002. Human impacts on the stream-groundwater exchange zone. *Environmental Management* 29:763–81.

Hancock, P. J., and A. J. Boulton. 2005. The effects of an environmental flow release on water quality in the hyporheic zone of the Hunter River, Australia. *Hydrobiologia* 552:75–85.

Hancock, P. J., A. J. Boulton, and W. F. Humphreys. 2005. The aquifer and its hyporheic zone: Ecological aspects of hydrogeology. *Hydrogeology Journal* 13:98–111.

Harper, D., and M. Everard. 1998. Why should the habitat-level approach underpin holistic river survey and management? *Aquatic Conservation: Marine and Freshwater Ecosystems* 8:395–413.

Harvey, L. W., and C. C. Fuller. 1998. Effect of enhanced manganese oxidation in the hyporheic zone on basin-scale geochemical mass balance. *Water Resources Research* 34:623–36.

Horton, R. E. 1945. Erosional development of streams: Quantitative physiography factors. *Bulletin of the Geological Society of America* 56:275–370.

Hynes, H. B. N. 1970. *The ecology of running waters*. Liverpool, UK: Liverpool University Press.

Hynes, H. B. N. 1975. The stream and its valley. *Verhandlungen Internationale Vereinigung für Theoretische und Angewandte Limnologie* 19:1–15.

Hynes, H. B. N. 1983. Groundwater and stream ecology. *Hydrobiologia* 100:93–99.

Illies, J. 1961. Versuch einer allgemeinen biozonotischen Gleiderung der Fliessgewässer. *Internationale Revue Gesamten Hydrobiologie* 46:205–13.

Illies, J., and L. Botosaneanu. 1963. Problèmes et méthods de la classification et de la zonation écologique des eaux courantes, considérées surtout du point de vue faunistique. *Mitteilungen Internationale Vereinigung für Theoretische und Angewandte Limnologie* 19:1–57.

Jeffrey, K. A., F. W. Beamish, S. G. Ferguson, R. J. Kolton, and P. D. MacMahon. 1986. Effects of the lampricide, 3-trifluoromethyl-4-nitrophenol (TFM) on the macroinvertebrates within the hyporheic region of a small stream. *Hydrobiologia* 134:43–51.

Jones, J. B., and P. J. Mulholland, eds. 2000. *Streams and ground waters*. San Diego, CA: Academic Press.

Kentula, M. E., 2000. Perspectives on setting success criteria for wetland restoration. *Ecological Engineering* 15:199–209.

Kondolf, G. M., A. J. Boulton, S. O'Daniel, G. C. Poole, F. J. Rahel, E. H. Stanley, E. Wohl, Å. Bång, J. Carlstrom, C. Cristoni, H. Huber, S. Koljonen, P. Louhi, and K. Nakamura. 2006. Process-based ecological river restoration: Visualizing three dimensional connectivity and dynamic vectors to recover lost linkages. *Ecology and Society* 11:Article 5. http://www.ecologyandsociety.org/vol11/iss2/art5/ (accessed December 16, 2007).

Kobayashi, S., and T. Kagaya. 2004. Litter patch types determine macroinvertebrate assemblages in pools of a Japanese headwater stream. *Journal of the North American Benthological Society* 23:78–89.

Leopold, L. B., M. G. Wolman, and J. P. Miller. 1964. *Fluvial processes in geomorphology*. San Francisco, CA: W. H. Freeman and Co.

Loreau, M. 2000. Biodiversity and ecosystem functioning: Recent theoretical advances. *Oikos* 91:3–17.

Malcolm, I. A., C. Soulsby, A. F. Youngson, and J. Petry. 2003. Heterogeneity in ground water–surface water interactions in the hyporheic zone of a salmonid spawning stream. *Hydrological Processes* 17:601–17.

Marmonier, P., M.-J. Dole-Olivier, and M. Creuzé des Châtelliers. 1992. Spatial distribution of interstitial assemblages in the floodplain of the Rhône River. *Regulated Rivers: Research and Management* 7:75–82.

Marmonier, P., D. Fontvieille, J. Gibert, and V. Vanek. 1995. Distribution of dissolved organic carbon and bacteria at the interface between the Rhône River and its alluvial aquifer. *Journal of the North American Benthological Society* 14:382–92.

McKee, K. L., and P. L. Faulkner. 2000. Restoration of biogeochemical function in mangrove forests. *Restoration Ecology* 8:247–59.

Merriam, G. 1984. Connectivity: A fundamental ecological characteristic of landscape patterns. *Proceedings of the International Association for Landscape Ecology* 1:5–15.

Naiman, R. J., H. Décamps, and M. M. Pollock. 1993. The role of riparian corridors in maintaining regional biodiversity. *Ecological Applications* 3:209–12.

Newson, M. 1994. *Hydrology and the river environment*. Oxford, UK: Clarendon Press.

Packman, A. I., and K. E. Bencala. 2000. Modeling surface-subsurface hydrological interactions. In *Streams and ground waters*, ed. J. B. Jones and P. J. Mulholland, 45–80. San Diego, CA: Academic Press.

Packman, A. I., and M. Salehin. 2003. Relative roles of stream flow and sedimentary conditions in controlling hyporheic exchange. *Hydrobiologia* 494:291–97.

Parsons, M., M. C. Thoms, and R. H. Norris. 2003. Scales of macroinvertebrate distribution in relation to the hierarchical organization of river systems. *Journal of the North American Benthological Society* 22:105–22.

Pedroli, B., G. de Blust, K. van Looy, and S. van Rooij. 2002. Setting targets in strategies for river restoration. *Landscape Ecology* 17:5–18.

Poole, G. C. 2002. Fluvial landscape ecology: addressing uniqueness within the river discontinuum. *Freshwater Biology* 47:641–60.

Pringle, C. M. 2001. Hydrological connectivity and the management of biological reserves: A global perspective. *Ecological Applications* 11:981–98.

Pringle, C. M. 2003. What is hydrologic connectivity and why is it ecologically important? *Hydrological Processes* 17:2685–89.

Robertson, A. I., S. E. Bunn, P. I. Boon, and K. F. Walker. 1999. Sources, sinks and transformations of organic carbon in Australian floodplain rivers. *Marine and Freshwater Research* 50:813–29.

Ryder, D. S., 2004. Response of epixylic biofilm metabolism to water level variability in a regulated floodplain river. *Journal of the North American Benthological Society* 23:214–23.

Ryder, D. S., and W. Miller. 2005. Setting goals and measuring success: Linking patterns and process in stream restoration. *Hydrobiologia* 552:148–56.

Ryder, D. S., R. J. Watts, E. Nye, and A. Burns. 2006. Can flow velocity regulate epixylic biofilm structure in a floodplain river? *Marine and Freshwater Research* 56:32–40.

Schälchli, U. 1992. The clogging of coarse gravel river beds by fine sediment. *Hydrobiologia* 235/236:189–97.

Schumm, S. A. 1977. *The fluvial system*. New York, NY: Wiley.

SER (Society for Ecological Restoration Science and Policy Working Group). 2004. *The SER primer on ecological restoration*. www.ser.org (accessed March 30, 2006).

Stanford, J. A., and A. R. Gaufin. 1974. Hyporheic communities of two Montana rivers. *Science* 185:700–702.

Stanford, J. A., and J. V. Ward. 1993. An ecosystem perspective of alluvial rivers: Connectivity and the hyporheic corridor. *Journal of the North American Benthological Society* 12:48–60.

Stanford, J. A., J. V. Ward, and B. K. Ellis. 1994. Ecology of the alluvial aquifers of the Flathead River, Montana. In *Groundwater ecology*, ed. J. Gibert, D. Danielopol, and J. A. Stanford, 367–90. San Diego, CA: Academic Press.

Stanley, E. H., and A. J. Boulton. 2000. River size as a factor in river conservation. In *Global perspectives on river conservation: Science, policy and practice*, ed. P. J.Boon, B. R. Davies, and G. E. Petts, 339–409. London, UK: Wiley.

Strayer, D. L., H. M. Malcolm, R. E. Bell, R. M. Carbottel, and F. O. Nitsche. 2006. Using geophysical information to define benthic habitats in a large river. *Freshwater Biology* 51:25–38.

Stream Solute Workshop. 1990. Concepts and methods for assessing solute dynamics in stream ecosystems. *Journal of the North American Benthological Society* 9:95–119.

Thibodeaux, L. J., and J. D. Boyle. 1987. Bed-form generated convective transport in bottom sediment. *Nature* 325:341–43.

Thorp, J. H., M. C. Thoms, and M. D. Delong. 2006. The riverine ecosystem synthesis: Biocomplexity in river networks across space and time. *River Research and Applications* 22:123–47.

Townsend, C. R. 1989. The patch dynamics concept of stream community ecology. *Journal of the North American Benthological Society* 8:36–50.

Townsend, C. R. 1996. Concepts in river ecology: Pattern and process in the catchment hierarchy. *Archiv für Hydrobiologie, Supplementband* 113:3–21.

Valett, H. M., S. G. Fisher, N. B. Grimm, and P. Camill. 1994. Vertical hydrological exchange and ecological stability of a desert stream ecosystem. *Ecology* 75:548–60.

Valett, H. M., S. G. Fisher, and E. H. Stanley. 1990. Physical and chemical characteristics of the hyporheic zone of a Sonoran Desert stream. *Journal of the North American Benthological Society* 9:201–15.

Vannote R. L., G. W. Minshall, K. W. Cummins, J. R. Sedell, and C. E. Cushing. 1980. The river continuum concept. *Canadian Journal of Fisheries and Aquatic Sciences* 37:130–37.

Vaux, W. G. 1962. Interchange of stream and intragravel water in a salmon spawning riffle. *United States Fisheries and Wildlife Services Research Report Fisheries* 405:1–11.

Vinebrooke, R. D., K. L. Cottingham, J. Norberg, M. Scheffer, S. I. Dodson, S. C. Maberly, and U. Sommer. 2004. Impacts of multiple stressors on biodiversity and ecosystem functioning: the role of species co-tolerance. *Oikos* 104:451–57.

Wall, D. H., M. A. Palmer, and P. V. R. Snelgrove. 2001. Biodiversity in critical transition zones between terrestrial, freshwater, and marine soils and sediments: Processes, linkages, and management implications. *Ecosystems* 4:418–20.

Ward, J. V., K. Tockner, D. B. Arscott, and C. Claret. 2002. Riverine landscape diversity. *Freshwater Biology* 47:517–39.

Ward, J. V., K. Tockner, and F. Schiemer. 1999. Biodiversity of floodplain river ecosystems: Ecotones and connectivity. *Regulated Rivers: Research and Management* 15:125–39.

Wheatley, M. J. 2002. *Turning to one another: Simple conversations to restore hope to the future*. San Francisco, CA: Berrett-Koehler. http://www.margaretwheatley.com/articles/listeninghealing.html (accessed December 17, 2007).

White, D. S. 1993. Perspectives on defining and delineating hyporheic zones. *Journal of the North American Benthological Society* 12:61–69.

Wiens, J. A. 2002. Riverine landscapes: Taking landscape ecology into the water. *Freshwater Biology* 47:501–15.

Chapter 7

Principles of River Condition Assessment

Kirstie A. Fryirs, Angela Arthington, and James Grove

> Managing river health should not be confused with measuring it.
>
> Rogers and Biggs 1999, 439

River rehabilitation is ultimately about improving river condition and sustaining the benefits derived from healthy river systems. Frameworks for assessing river condition constitute a core part of the river management process, providing a critical platform for environmental decision-making and associated actions (Brierley and Fryirs 2005). The information gathered through the application of such frameworks provides a critical basis for identifying, and then targeting, abiotic and biotic attributes of the ecosystem that require manipulation through rehabilitation practice with the objective of improving ecosystem condition. Adoption of a truly integrative approach, based on the principles advocated in this chapter, allows a practitioner to assess the causes of deterioration in condition rather than simply treating the visible symptoms. River rehabilitation strategies can then be implemented that are proactive, cost-effective, and more likely to achieve their intended goal—improved river condition.

In this chapter we provide a set of guiding principles for assessing the biophysical (i.e., abiotic and biotic) state of rivers from a scientific perspective. We have adopted the term *river condition* to highlight the biophysical approach explored herein, as opposed to notions of *river health* that often also include societal perceptions of a river's status and its values to society (see chapter 1). This review of biophysical condition assessment sets the context within which to appraise societal perspectives on river health by examination of human perceptions and values, and care for rivers in chapter 8.

A variety of frameworks, methods, and approaches for examining river condition has been developed within a number of environmental disciplines (hydrology, geomorphology, water quality, biology, ecology). However, the integration of these approaches and guidance on which indicators and methods to use, where to use them, and how to interpret the results, tends to be limited. In this chapter we do not aim to develop an integrative framework for assessing river condition or review an exhaustive range of existing methods and protocols for examining river condition. These can be sourced from elsewhere (e.g., Downes et al. 2002; Gordon et al. 2004). Rather, we highlight the key concepts and scientific principles that should underlie assessments of river condition from an integrated physical and ecological perspective. We then present various case studies from Australia that apply such frameworks to assess river condition.

Purposes of River Condition Assessments

Understanding the dynamics of river condition should not be confused with simply recording and measuring it (Rogers and Biggs 1999; Palmer and Bernhardt 2006). A clear distinction should be made between audits and condition assessments. An audit, such as the UK River Habitat Survey (Raven et al. 1998) or the Australian Sustainable Rivers Audit (NLWRA 2001), provides information on the current situation or status of a river. Essentially, these approaches are data collection exercises designed to generate information that can be used to monitor compliance, or track change in status over time. Compliance indicators are useful for ensuring that certain predetermined conditions (e.g., a stream discharge or water quality target) have been achieved, or if not, how much deviation has occurred from the compliance level. However, this chapter advocates that compliance indicators on their own are not sufficient to gauge ecosystem integrity.

Condition assessments measure both "pressure" and "response" variables (indicators) and provide the means to develop a clear understanding of pressure-response (i.e., cause-effect) relationships that regulate the observed changes in system condition. Condition assessments place much greater emphasis on the integration of diverse indicators and interpretation of response variables to examine causes rather than simply recording the symptoms of contemporary river condition.

River degradation is often considered socially unacceptable and the desire to improve river condition leads to management responses that aim to rectify particular problems. River condition assessments should have the potential to change a management choice or to initiate a different management activity. Hence, river condition assessments are often undertaken in order to:

- Identify reference conditions against which to measure condition in impaired reaches.
- Rank reaches against one another, from intact to threatened and degraded.
- Set river rehabilitation priorities and objectives.
- Set conservation priorities for different river types or threatened species.
- Track improvements in condition in space and time against rehabilitation goals or target conditions.
- Determine the effectiveness of management actions over time and adjust strategies.
- Warn of potential pressures or threshold conditions that could be exceeded, resulting in deterioration in condition, and therefore signal the need to carefully monitor these sites or design proactive strategies for management of stressors.

Ancillary outcomes of condition assessment, over and above the goals listed above, should be to develop a clearer understanding of cause-effect (i.e., pressure-response) relationships that regulate changes in system condition. This may be achieved by measurement and comparison of the environmental controls ("drivers" or "fluxes") that regulate ecosystem structure and function at reference and impacted sites (Baron et al. 2002; Clarke et al. 2003; Sullivan et al. 2004) or through tracking the response of condition variables to pressure variables over time. While these may be considered scientific procedures, the knowledge and conceptual or quantitative models generated should inform the design of management strategies and rehabilitation procedures (Bernhardt et al. 2005).

Ecosystem Integrity as a Basis for Assessing Biophysical River Condition

It is well established that physical structure and dynamics provide the habitat template in aquatic ecosystems (Newson and Newson 2000; Southwood 1977; Sullivan et al. 2004; Townsend and Hildrew 1994). However, ecosystem structure and functioning can only be maintained if habitats are viable (see Chapter 1; Hilderbrand et al. 2005) and key natural drivers are maintained. Any assessment of *river condition* should be framed in the context of river structure *and* function, availability *and* viability of habitat, and river character *and* behavior as defined in chapter 1 (Figure 7.1). These concepts are often captured by the notion of *ecosystem integrity*, which implies the maintenance of both physical and ecological integrity, thus ensuring an appropriate level of integration between hydrological, geomorphic, and biological processes (see chapter 1; Karr 1995, 1999; Petts 2000).

The interaction of fundamental environmental drivers (flow regime, water quality, sediment flux, chemical/nutrient flux and thermal/light flux) must be considered when evaluating ecosystem integrity (Baron et al. 2002; Poff et al. 1997). Focusing on one driver at a time will not yield a comprehensive assessment of the condition of a river ecosystem, nor a meaningful understanding of the threats to its ecological integrity. Human activities

Reach 1: Good condition	
Bemboka River, Bega catchment, NSW	***Physical integrity*** A good condition reach is characterized by significant heterogeneity in the geomorphic structure of the channel. Geomorphic units include bedrock pools, runs, riffles, and cascades. An extensive, diverse and functional riparian corridor is evident, and water is of high quality.
	Ecological integrity Aquatic communities are likely to be diverse and composed mainly of native plant, invertebrate, and fish species. The ecosystem is likely to function as a heterotrophic system, deriving energy from upstream processing of organic matter and from the surrounding riparian zone and catchment.
Reach 2: Poor condition	
 Sandy Creek, Bega catchment, NSW	***Physical integrity*** A poor condition reach is characterized by a homogenous sand sheet that infills pools and smothers the channel bed. A scattered or nonexistent riparian corridor is evident, and water quality is likely to be of low quality.
	Ecological integrity Aquatic communities are likely to be of low diversity, confined to a few 'refuge' aquatic habitats (e.g. isolated pools); alien species are likely to be present in the riparian zone and in the remaining aquatic habitats. The ecosystem is likely to function as an autotrophic system, deriving energy from in-stream primary production, with low inputs of organic matter from upstream, the surrounding riparian zone and catchment.

FIGURE 7.1. The physical and ecological integrity of two reaches of a confined valley with occasional floodplain pockets river that are in different condition, Bega catchment, NSW (modified from www.riverstyles.com)

that alter one or more of these environmental drivers can greatly affect biotic assemblage structure, species interactions, system productivity, and ecosystem resilience. Excessive stress (e.g., prolonged drought, eutrophication) and/or simplification of natural complexity (e.g., flow regulation, disconnection of channel-floodplain linkages, channelization, riparian loss) have the potential to push the functionality of intact freshwater ecosystems beyond the bounds of their seasonal cycles and/or interannual fluctuations, thus impairing their resilience or sustainability, and eventually threatening their ability to provide important goods and services in both the short- and long-term (Arthington et al. 2006; Baron et al. 2002).

By adopting such definitions of *physical and ecological integrity*, we promote integration not only of the physical components of a river system and its flora and fauna across a range of scales, but also examine the environmental controls ("drivers" or "fluxes") that regulate ecosystem functioning over time (Baron et al. 2002; Bunn and Arthington 2002; Clarke et al. 2003; Sullivan et al. 2004). This approach requires examination of how factors that limit or stress the physical and/or ecological structure and functions of habitat will impact on the condition of a river system. The severity of the impact and the degree to which physical or ecological function is compromised or impaired will dictate the potential for that system to resist the forces of change and to improve in biophysical condition (Brierley and Fryirs 2005; Hobbs and Harris 2001). To achieve this level of understanding requires that the contemporary state of a system is placed in context of the physical/chemical drivers and ecological interactions that have occurred in the past, and that operate in the system today (Clarke et al. 2003; Gilvear 1999). This approach to the analysis of river condition has been described in geomorphic and ecological terms as interpreting system "memory" such that each system carries with it imprints from its past and the impacts of current disturbance/processes (Petts 2000; Trofimov and Phillips 1992). Therefore, assessments of contemporary river condition must *interpret* and *explain* the causes and resulting manifestations of past processes and present disturbances rather than adopt a static, descriptive audit approach that provides little understanding of cause and effect. This approach provides a powerful basis on which to address the causes of impaired river condition and how best to address them through river management practice, rather than Band-Aiding visible symptoms through piecemeal and poorly informed rehabilitation activities (Hobbs and Harris 2001).

We use the term *river condition* in an integrative sense to describe the integrity of the contemporary system set within the context of its geomorphological, hydrological, and ecological history/evolution. Measures of river condition therefore record deviations from a natural or expected state in any particular reach given this evolutionary and historical perspective. In this context, river condition is defined as "a measure of the capacity of a river to perform ecosystem functions that are expected for that river within the valley-setting that it occupies" (Brierley and Fryirs 2005, 297).

Integrating Abiotic and Biotic Factors in Assessments of River Condition

River reaches are sections of river along which there is little change in the flow or sediment load such that the river maintains a near consistent structure (Brierley and Fryirs 2005). Comparison of reaches of the same river type provides an ideal basis upon which to assess river condition. A framework within which to interpret and explain the range of river forms and adjustments, and the condition of a river, is outlined from a geomorphic perspective in Brierley and Fryirs (2005) through use of a "river evolution diagram" (figure 7.2).

Rivers adjust within a range of *imposed boundary conditions* that are set by valley morphology (width and slope) and geological setting (figure

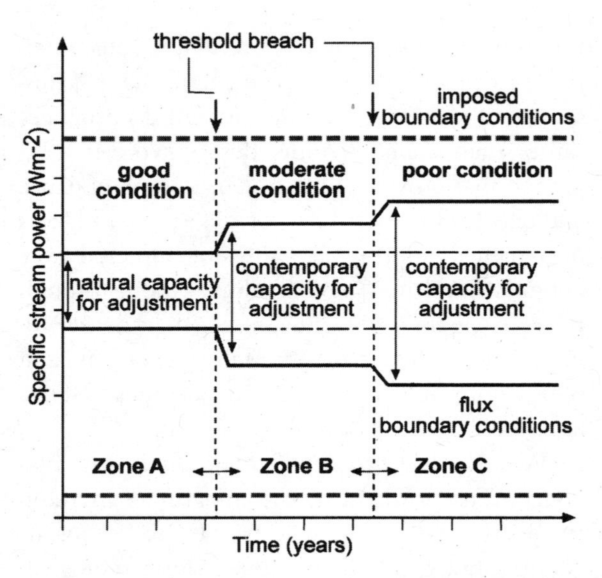

FIGURE 7.2. The river evolution diagram, modified to incorporate assessments of river condition (based on Brierley and Fryirs 2005).

7.2). These imposed boundary conditions can remain unchanged over geomorphic timeframes (i.e., tens to thousands of years). Within this set of imposed boundary conditions, the operation of water, sediment, and vegetation fluxes (i.e., *flux boundary conditions*) enables a river to adopt a range of morphologic variants. Hence, river form adjusts within a particular range, influenced by contemporary flux boundary conditions and historical legacies and imprints. Collectively, these factors and legacies determine the *natural capacity for adjustment* of a river (Brierley and Fryirs 2005).

The natural capacity for adjustment is a measure of the ability of a river to perform biophysical functions that are expected for that river type given the prevailing catchment boundary conditions (i.e., fluxes of water, sediment and vegetation) and historical legacies. Accordingly, measures of river condition record deviations from this expected state in any given reach.

The upper and lower limits of the natural capacity for adjustment define the range of expected ecosystem structure, function, and composition; these limits have been termed *thresholds of potential concern* by du Toit et al. (2003) and Rogers and Bestbier (1997). They represent a multidimensional envelope within which variation of the ecosystem is considered acceptable. Within this zone of adjustment, marked differences in types, patterns, and rates of geoecological adjustment are expected for differing reaches in differing settings. Once the capacity for adjustment is defined, an assessment can then be made of whether ongoing adjustments (and the character of the river) are part of the "expected" behavioral regime for that type of river, or fall beyond the boundaries of expected river behavior.

Within the capacity for adjustment a range of abiotic and biotic processes operate, resulting in a range of ecosystem structures (either physical or biological). By determining the dimensions of adjustment and the resultant structures, a series of attributes can be identified that reflect the expected character and behavior of that river. These measures reflect the *degrees of freedom* of the river (Brierley and Fryirs 2005; Montgomery 1999) or natural range of variation for a given abiotic or biotic measure or process (Arthington et al. 2006; Baron et al. 2002). Within these degrees of freedom, a range of geoindicators or vital ecosystem attributes can be identified and measured to explain river condition (Aronson and Le Floc'h 1996; Arthington et al. 2006; Brierley and Fryirs 2005; Elliott 1996). In geomorphic terms, these degrees of freedom may be bed character, geomorphic units, channel morphology, and channel planform, and geoindicators may be the size and shape of channels, the geomorphic unit assemblage of the reach, riparian vegetation cover, or bed material size and organization (Fryirs 2003).

In many instances, forms of adjustment are mutually interrelated and certain attributes will record a range of underlying adjustments (Meyerson et al. 2005). Identifying indicators that provide a reliable and relevant signal about the behavior of a particular river is key to robust and informative assessments of river condition. Structures and functions that occur within the natural capacity for river adjustment are considered to reflect rivers that are in good condition. They are operating in a way that is expected for that type of river, in that landscape and historical setting, and will sit in Zone A in figure 7.2. In this case, river management strategies should look to conserve, maintain, and enhance these reaches.

If anomalous or accelerated rates of change are detected, or anomalous structures formed, the river is considered to be deviating from its expected character and behavior. In this case the river will be operating outside the limits set by the natural capacity for adjustment (represented in figure 7.2 by a widening of the capacity for adjustment in Zones B and C) and the condition of the river is considered to be deteriorating. Thresholds of potential concern will be breached and the further away from the natural capacity for adjustment a river operates, the poorer its condition (Arthington et al. 2006; Brierley and Fryirs 2005; Rogers and Bestbier 1997; Rogers and Biggs 1999). In these cases, river rehabilitation may be required to arrest this deterioration in condition. However, it should be noted that not all adjustment to changes in drivers and fluxes are likely to take the form of linear responses, and an appreciation of the types of response (e.g., linear, threshold, or complex) is needed to interpret the data derived from a condition assessment and to best make use of the assessment in the design of rehabilitation strategies. Anomalous physical structures and behaviors are often the limiting factors that stall or arrest improvement in condition. These are the factors that should be targeted first in management practice. Depending on the level of intervention required, measures can be implemented to alter fluxes, structures, or processes to instigate an improvement in condition. This is termed a "recovery enhancement" approach to river management (Fryirs and Brierley 2000; Gore 1985).

An approach that can be used to explain the transitions between good, moderate, and poor river condition states (i.e., from Zone A to B to C in figure 7.2) that integrates biotic and abiotic components of systems was developed by Whisenant (1999) and modified by Hobbs and Harris (2001) (figure 7.3). Transitions from good to moderate to poor condition can be framed in terms of structural or process-driven thresholds that reflect the functional attributes of rivers. In many ways this is analogous to depicting transitions in condition across abiotic and biotic thresholds.

Figure 7.3 highlights a pressure-state-response approach for examining how biotic and abiotic components of a river system will interact to dictate its biophysical condition. States 1 and 2 represent a river that is fully functional in biophysical terms and is operating within the natural capacity for adjustment for that river type. The slight deterioration in condition indicated by the two states merely reflects the natural physical adjustments occurring after disturbances such as flood events or dry periods (i.e., the natural range of seasonal or interannual variation or response to natural abiotic extremes). In this model, such disturbances do not severely threaten the natural ecological functioning of the system. Indeed these are the disturbance events to which ecosystems and their biotic components are adapted (Baron et al. 2002; Bunn and Arthington 2002; Poff et al. 1997). Adjustments within the zone represented by States 1 and

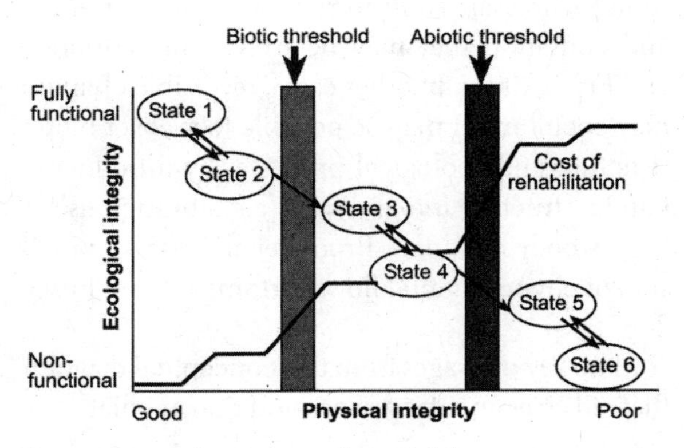

FIGURE 7.3. A conceptual model for system transitions across biotic and abiotic thresholds. This is used to define various degrees of physical and ecological integrity and river condition. Modified from Whisenant (1999) and Hobbs and Harris (2001).

2 reflect internal resilience in geoecological structure and ecological function. However, once a biotic threshold has been breached, the ecological integrity of the river deteriorates as reflected in the transition from Zone A to Zone B in figure 7.2. This may occur in response to introduction of an exotic fish species that changes community dynamics, for example. The potential for recovery (and improvement in river condition) toward States 1 and 2 is constrained until the limiting biotic factor is addressed (e.g., eradication of exotics and reintroduction of native fish species).

In States 3 and 4, the river ecosystem is expected to behave/function within an expanded capacity for adjustment defined by the physical drivers of geomorphic, biogeochemical processes and habitat condition. However, once an abiotic threshold is breached, and a shift to States 5 and 6 occurs, the river is considered to be in poor physical and ecological condition. The physical integrity of the river is compromised and a range of abiotic limiting factors has severely constrained the potential for river recovery and improvement in condition. This threshold shift may be representative of a severe disturbance such as the passage of a sediment slug that smothers habitat, destroys biotic assemblages, and impairs ecological functioning. In severe cases, these threshold breaches may result in wholesale shifts in river structure, function, and condition that may be irreversible (Brierley and Fryirs 2005). In other cases, reversible changes may occur and it may be possible to recover biotic structure and ecological processes by introducing habitat structure into the river (e.g., through use of large woody materials, Brooks et al. 2006; or provision of environmental flows, Arthington and Pusey 2003).

The key messages from this conceptual diagram that aid assessments of river condition are that:

1. abiotic and biotic components are key natural drivers/controls on the condition of rivers,
2. identifying abiotic and biotic thresholds that, once crossed, result in degradation is key to detecting mechanisms that cause changes in condition, and
3. identifying the proximity of system condition to these thresholds of concern provides a basis for preventing further deterioration in condition and developing rehabilitation strategies that address the causes of deterioration.

The potential cost of rehabilitation has also been incorporated into figure 7.3. The potential for rehabilitation of biotic factors (e.g., removal of exotic species) may be more readily achievable (and less costly) if physical structure and function are sufficiently similar to the natural range of variability for the river type/reach. Therefore addressing issues such as water quality or organic matter input and retention are likely to be successful. However, manipulation of physical structure and function (abiotic factors) by either direct intervention (e.g., reconstruction of channel substrate diversity after passage of a sand slug) or natural processes that have been disturbed by human activity (e.g., flows from dams) is much more difficult and usually far more costly.

What Is Natural or Expected? Defining Reference Conditions

Any assessment of river condition must be framed relative to some benchmark or reference reach, thereby providing a determination of the extent to which changes to river character, behavior, and biological characteristics fall outside the long-term pattern, that is, fall outside the *natural capacity for adjustment* (figure 7.2) (Brierley and Fryirs 2005; Arthington et al. 2006). Measures must be tied to appraisals of “naturalness” (whether real or virtual). A natural river is dynamically adjusted so that biophysical structure and function operate within an expected range of variation or capacity for adjustment (Baron et al. 2002; Brierley and Fryirs 2005). Therefore, a reference condition must reflect the expected character and behavior of the river given its position in the catchment, the catchment boundary conditions, and historical legacies. It should represent the permissible range of change in the character of the physical template for that type of river and the likely range of biotic responses that operate within that physical template given the range of variation of key drivers or fluxes in the catchment (Arthington et al. 2006; Brierley and Fryirs 2005; Rogers and Biggs 1999).

In most cases there is no one ideal reference condition (Hilderbrand et al. 2005; Stoddard et al. 2006). Identification of a single reference reach for individual river types runs the risk of homogenizing river character and behavior for that river type. Identifying a range of reference conditions that reflects the range of appropriate character, behavior, and biological characteristics for the river type under investigation is critical. For example, recognizing that bank erosion is within the *natural capacity for adjustment* in some river types and occurs at certain rates gives an indication of the "reference" condition of a river. Lateral migration of river bends (and erosion of concave banks) instigates pool development that provides habitat for aquatic fauna. Similarly, certain vegetation associations occur along these river bends, and meander migration recruits woody debris along many rivers. If however the rate of change is anomalous (e.g., occurring in unexpected places along a river bank) or is accelerated relative to natural rates for that part of the river system and river type, then the condition of that river reach can be considered impaired to some degree.

In many landscape settings, two main options for identifying a reference condition are available:

1. An intact, pristine condition. For example, in Australia, putting aside the impacts of Aboriginal practices (e.g., use of fire), the time of colonization in 1788 provides a suitable reference point. If patterns and rates of river change fall outside the natural capacity for adjustment observed in the pristine state, a reach is considered to have deviated from its "intact" condition. An alternative to the use of pristine condition is the use of some other historical condition (e.g., before all dams were built), or to use "least disturbed" reaches as the benchmark for assessments (Stoddard et al. 2006).

2. An assessment of the best condition that can be attained by a river that has been altered by human disturbance, given the prevailing catchment boundary conditions and historical legacies (Stoddard et al. 2006). This "expected" reference condition must attempt to separate reaches that have been subjected to reversible and irreversible changes in response to direct and indirect human disturbance (Brierley and Fryirs 2005).

Although the first set of options may be preferred from a scientific perspective, the second option often provides a far more practical and realistic perspective from which to appraise river condition (cf. Cairns 1989; Gore and Shields 1995). Given the extent of human disturbance to river systems, whether direct or indirect, comparison with a pristine condition may seem little more than an academic exercise. This is *not* to say that insights from near-intact or least-disturbed reaches do not provide fundamental guidelines about the "natural" structure and functioning of rivers, and associated implications for geomorphological diversity, biodiversity, aquatic ecosystem functioning and conservation. Unfortunately, however, relatively few pristine remnants remain, and these typically form an unattainable goal for river management practices in highly disturbed rivers (Hilderbrand et al. 2005; Stoddard et al. 2006). Nevertheless, aspirations for more demanding river restoration programs should not be abandoned, and there are certainly good examples of major reconstructions of physical and ecological river structure and function (Postel and Richter 2003).

When identifying a physical reference condition, the objective is to determine a morphological configuration that is compatible with the prevailing catchment flux conditions. Viewed in this way, channel attributes, river planform (including the assemblage of geomorphic units), and bed character must be appropriate for the river type under investigation (cf. Hughes et al. 1986; Rhoads and Herrick 1996). A procedure for identifying a physical reference condition is reviewed in Brierley and Fryirs (2005).

Similar and equally critical considerations apply to definition of the reference condition for assessment of biological condition. Most approaches use reference reaches derived from extensive surveys of "undisturbed" or "least-disturbed" systems that incorporate spatial and temporal variation in the physical and biological characteristics of interest. For example, issues involved in accurately defining reference conditions for fish indicators of river health are explored in Kennard et al. (2006a) who compare and contrast four different models for predicting summary fish metrics from abiotic variables. This modeling approach enables a more precise and site-specific comparison of disturbed and reference conditions by focusing on the fauna expected at each test site at any point within a catchment, rather than relying on comparisons of each test site with one or a few reference reaches. Such approaches require the presence of river reaches in relative undisturbed condition. In contrast, accurate description or prediction of the biological characteristics of pristine or least-disturbed reference aquatic systems is often difficult in landscapes that have already been substantially altered by anthropogenic activities (e.g., lowland rivers) and for which little historical information exists. Chessman and Royal (2004) outline the use of environmental filters to predict natural assemblages of river macroinvertebrates when suitable reference sites are not available.

With these considerations and constraints in mind, a number of options are available for selecting or constructing a reference reach against which to undertake the assessment of river condition. These include:

- Using available remnant reaches that occur at a similar position in the catchment and operate under near-equivalent catchment boundary conditions.
- Where remnant reaches do not exist, identifying reference reaches using the evolutionary sequence for the river type and ergodic reasoning.
- Deriving an "expected" range of structure and function that represents natural condition (i.e., integrity) based on expert analyses or a predictive model of river character and behavior for that river type.
- Defining a target condition that is biotically and abiotically attainable but within risk limits set by society. Risk is defined as the consequence of an event resulting in "undesirable" change multiplied by the probability of it occurring.

Identifying Indicators That Provide a Reliable and Relevant Measure of the Biophysical Condition of Rivers

Certain abiotic and biotic indicators provide reliable and relevant insights into system structure and function and the condition of a river (cf. Elliott 1996; Osterkamp and Schumm 1996; Rogers and Biggs 1999; Meyerson et al. 2005). Abiotic indicators have been called "geoindicators" by Brierley and Fryirs (2005). Biotic indicators can be considered analogous to "vital ecosystem attributes" as suggested by Aronson et al. (1995), and functional indicators as the "process indicators" of Bunn et al. (1999).

From an *abiotic* perspective, frameworks for assessing the physical condition of rivers use geomorphic attributes (e.g., channel geometry, planform, and geomorphic unit assemblage) and interpretation of river behavior and evolution to determine the condition of various types of river (e.g., Brierley and Fryirs 2005). From a *biotic* perspective, measures of riparian vegetation and aquatic plant communities, biofilms, algae, and macroinvertebrate composition have all been explored as structural measures of river condition (e.g., Burns and Ryder 2001; Mackay et al. 2003; McCormick and Cairns 1994; Werren and Arthington 2002; Whitton and Kelly 1995; Wright et al. 1993). The biotic attributes of fish assemblages have proven particularly useful as indicators of ecological condition in streams (Karr 1998; Oberdorff et al. 2001). Indicators include fish species richness, community composition, number of alien species and the proportion of individuals in the total fish population that are alien, and for native species, which occupy various functional guilds such as habitat, trophic and reproductive guilds (Kennard et al. 2005; Kennard et al. 2006a, 2006b). *Functional* attributes of streams have also received attention and include production rates, measures of benthic metabolism, and the structure of stream foodwebs (Bunn et al. 1999; Rapport et al. 1997; Smith and Storey 2001).

Bunn (1995) noted that measures of biological structure (e.g., species richness, assemblage composition, alien versus native fish abundance) typically reveal that detrimental ecological change has already occurred, whereas measures of ecological processes (production rates, benthic metabolism, foodweb structure) can provide a much earlier indication of modified function that may ultimately alter biological structure and ecosystem integrity. Hence, a measure of a good indicator is one that provides an early warning signal and direct insight into how a particular river type adjusts to disturbance (i.e., each indicator needs to be diagnostic of a certain adjustment in structure and/or function, and ultimately, must yield a measure of the integrity of the system). The particular indicators used will be specific to each distinctive river type or biotic setting and must be sensitive enough to provide signals of condition along a degradation gradient (Cairns et al. 1993; Fairweather 1999; Arthington et al. 2006). The monitoring framework adopted must be flexible enough to measure and interpret a range of indicators across a range of river types and biotic settings while avoiding the use of redundant variables. For example, from an abiotic perspective, channel geometry would not be measured for rivers in a confined valley setting as this indicator provides no signal about the condition of that river type. However, channel geometry is a useful metric for an alluvial river where channel size and shape are key indicators of how these river types adjust to changes in discharge, for example. Indicators can also be used to identify the system components that require adjustment for improvements in condition to occur. These system components can then be targeted and modified in management practice.

Response gradients provide managers with specific series of spatially and temporally bounded "biophysical indicators" that can measure system response to pressures, whether they are natural or management-induced (Rogers and Biggs 1999). *Pressure indicators* examine *causes* or drivers of change and can be used to explain the contemporary condition of a river. These can operate at a range of spatial scales and must include indicators that assess catchment-scale patterns of rivers and biophysical fluxes. A consideration of the cumulative or offsite impacts that will limit or enhance improvements in condition can be assessed using pressure indicators. Reach-specific pressure indicators should also reflect how the river has evolved/changed over time. *Response indicators* record the *symptoms* of changes in condition from the reference state, and indicate whether the contemporary structure and function of the river is expected or anomalous compared to the reference condition for the river type. Each indicator must be sensitive enough to reflect the response gradient from natural to degraded and vice versa for the river type under investigation (cf. Chessman 2003). In practice, this means that bioindicators must first be tested for consistency of response to known gradients of change in the primary abiotic drivers such as flow regulation, habitat modification, or water quality (Chessman 2003; Arthington et al. 2006; Kennard et al. 2006a, 2006b).

In such a framework, understanding the behavioral and evolutionary dynamics of river systems is just as important as assessing contemporary structural and functional attributes (Clarke et al. 2003; Gilvear 1999). Unfortunately, many river condition assessment frameworks focus solely on measuring response indicators, while failing to consider pressure indicators as drivers (and causes) of contemporary river condition. Including both pressure and response indicators provides a basis to move beyond static, descriptive assessments of contemporary condition and toward a deeper understanding (and explanation) of river condition based on the influence of changes in particular drivers and stressors on the behavioral and evolutionary dynamics of rivers. In this context, prediction of future river condition can occur, making the use of pressure-response indicators for assessing condition a powerful tool for informing river management and rehabilitation practice.

Considerations in the Design and Application of Integrative Frameworks for Assessing Biophysical Condition

When undertaking a river condition assessment, a generic set of principles should be adopted and applied consistently. Ten key scientific principles are presented to guide the design and application of integrative frameworks for assessing river condition.

Principle 1. *Understanding river character and behavior must be specific to the river type under investigation.* This provides the physical template on which to design sampling strategies to appraise physical and ecological integrity. This must include channel and floodplain characteristics and functionality.

Principle 2. *Compare like with like.* Assessment of river character, behavior, and distribution throughout a catchment, or between catchments if required, provides a basis for meaningful comparisons to be made across the range of river types, as well as a means to identify reference conditions against which to assess river condition.

Principle 3. *Each reach must be placed within an evolutionary context* to determine the nature, timing, and extent of change to river structure and function, and to assess whether this change has been irreversible. This is critical for identification of a reference condition against which to assess condition. Underlying causes of change (whether natural or human-induced) can then be isolated. Antecedent knowledge and insight about river character and behavior allows explanation of contemporary condition, identification of lag effects and threshold breaches, identification of biotic or abiotic factors that will limit improvements in condition, and targeting of causes rather than symptoms of deterioration through management.

Principle 4. *Tie assessments to appraisals of "naturalness" or "expected" conditions for the type of river under investigation.* Whether real or virtual, the identification of conditions that are "expected" or "natural" for the type of river under investigation provides a clear sense of what baseline to measure condition against (i.e., the reference condition).

Principle 5. *Identification of reference sites* can be drawn from available examples, historical datasets, or expert insights, but must be specific to river type.

Principle 6. *Measures used to assess condition must be appropriate for the type of river under investigation.* Each indicator must be sensitive enough to reflect the response gradient from the reference condition (i.e., the expected character and behavior) to degraded conditions for the river type under investigation, and must also respond to that gradient in a predictable manner. Therefore, indicators are likely to be river type specific. Specific indicators can then be used to target management actions.

Principle 7. *Consider structural and functional attributes.* This recognizes the premise that physical integrity (abiotic) interactions with ecological integrity (biotic) determine ecosystem integrity.

Principle 8. The framework must be *flexible* enough to encompass *qualitative insights, quantitative measurements, and integrative interpretations*.

Principle 9. *Move away from static, visual assessments of condition and toward analysis of behavior and identification of thresholds of probable concern*. Indicators or measures that limit or enhance improvements in river condition can be identified. In most cases a combination or mix of parameters will need to be treated to improve condition and enhance recovery.

Principle 10. *Application of the framework should be regionally focused* and framed in terms of landscape/catchment conditions and associated geological and climate influences. Clear statements should be made about the transferability of findings from one river type to another, both within and between bioclimatic regions.

Integrating Tools for Assessing River Condition

Much research has focused on generating sets of indicators with which to measure river condition (Rogers and Biggs 1999; Downes et al. 2002). Much less thought has been given to developing frameworks that examine the structure *and* function of systems by integrating indicators and developing a deeper knowledge of system behavior, evolution, and integrity. An integrative approach to the assessment of river condition encompasses the guiding principles set out above and attempts to predict what is expected at the site given the historical and contemporary context. It is against these criteria that monitoring and evaluation of river condition should take place. Such an approach moves away from simply measuring state, and toward an adaptive approach to the appraisal of biophysical status that takes into consideration pressure-response issues, and how to design rehabilitation actions.

A truly interdisciplinary approach that fulfils the guiding principles set out above does not currently exist, although recent effort and research has been focused on developing multidisciplinary approaches that do fulfil most of the proposed criteria. With such systems in place, developing a new integrative framework from scratch may be unnecessary. Rather, the challenge now faced by scientists and managers lies in combining and selecting existing approaches and datasets to achieve a framework that fulfils the ten principles set out above. This will involve testing the available frameworks against these guiding principles and scientifically demonstrating that the "new method" is an improvement on the "old method" (e.g., DSE 2006; Kennard et al. 2006a, 2006b). Where gaps emerge, frameworks may require modification to best suit the conditions under which they are being applied. Where existing methods are to be adapted to make them more relevant and integrative, care should be taken to ensure that previously gathered information can be compared with new monitoring data.

The tools used and the approach adopted will be dictated by the aims for which the assessment is being undertaken and the environmental setting in which it is used. Table 7.1 summarizes some of the tools that are commonly used to undertake assessments of river condition. These are separated into abiotic, biotic, and integrative tools. Of particular importance is the analysis of whether the framework uses pressure and/or response indicators. This provides a sense of whether the approach has the capacity to detect and interpret the causes as well as symptoms of contemporary river condition, or whether the approach simply examines the symptoms of degraded condition.

Although a wide range of methods has been developed to examine river ecosystem condition/integrity, four key issues plague their usefulness. First, few frameworks have actually recognized the

TABLE 7.1

Some Commonly Used Biophysical Approaches For Assessing River Condition

Tool	Characteristics (use of pressure and/or response indicators)	Source
Abiotic		
River Styles geomorphic condition approach	Examines geomorphic condition based on analysis of the character and behavior of different river types. Is embedded within a hierarchical and evolutionary framework so causes of condition are identified. ***Pressure and response***	Brierley and Fryirs (2005) Fryirs (2003)
River Habitat Survey (RHS)	Uses a large database of reference sites to assess the physical condition of a river using visual descriptors. ***Response***	Raven et al. (1998) Walker et al. (2002)
South African River Health Programme (Index of Stream Geomorphology)	Classifies rivers based on their geomorphological characteristics and assesses habitat availability at sites based on flow hydraulics and substrate conditions. ***Response***	Rowntree and Wadeson (1998)
Biotic		
Pressure-Biota-Habitat (PBH)	Uses six criteria: physical diversity, biological diversity, vigor, resilience, rarity, and risk across a range of ecosystem components such as fish, diatoms, and macroinvertebrates. ***Pressure and response***	Chessman (2002)
Index of Biotic Integrity (IBI)	System-specific framework used to compare presence/absence or relative abundance of taxa (fish, macroinvertebrates, vascular plants) with expected species richness. ***Response***	Karr (1998)
Index of Stream Condition	Examines a range of parameters within the five subindices of hydrology, streamside zone, physical form, water quality, and aquatic life. ***Response***	Ladson et al. (1999)
Rapid bioassessment protocol (US EPA)	Periphyton, macroinvertebrates, and fish in low- and high-gradient streams ***Response***	Barbour et al. (1999)
River Invertebrate Prediction and Classification Scheme (RIVPACS)	Detects patterns of variability within groups of taxa by classifying unpolluted sites and predicting community types from environmental data. An observed versus predicted index gives a measure of condition. Based on macroinvertebrate surveys. ***Response***	Wright et al. (1993)
Australian River Assessment Scheme (AusRivAS)	Based on the RIVPACS method, it is used to compare observed versus expected macroinvertebrate community structure as an index of biological condition. ***Response***	Smith et al. (1999)
Oregon Water Quality Index	Integrates eight water quality parameters (temperature, dissolved oxygen, biochemical oxygen demand, pH, Ammonia + Nitrate nitrogen, total phosphates, total suspended solids, faecal coliforms). ***Response***	Barmuta et al. (1998)
ANZECC Water quality guidelines	Defines expected range of metabolism, macroinvertebrate, macrophyte, fish, algae, bacteria, and fungi community structures for a given water quality guideline. ***Response***	ANZECC (2000)

TABLE 7.1

(Continued)

Tool	Characteristics (use of pressure and/or response indicators)	Source
Integrative (or have been used in combination)		
River Styles + PBH	River Styles + PBH ***Pressure and response***	Chessman et al. (2006)
EU Water Framework Directive	Eco-hydro-morphic approach that includes measures of physical structure, a range of flora and fauna (e.g., phytoplankton, macro-invertebrates, fish), thermal conditions and water chemistry (e.g., oxygenation, nutrients, salinity, pollutants), and continuity. Uses reference reaches to examine stream condition. ***Response***	European Commission (2002)
Ecosystem Health Monitoring Program (EHMP)	The EHMP applies five sets of physical/chemical, biological, and processes indicators known to respond in consistent ways to measured gradients of pressure variables (landuse type and extent, riparian condition, bed and bank structure). Response variables include water temperature, turbidity, nutrients, algal biomass, invertebrate, and fish community structure (including alien species indices), and ecological processes (gross PP, respiration, foodweb structure, and so on). ***Pressure and response***	Smith and Storey (2001) Kennard et al. (2005; 2006a, 2006b)
System for Evaluating Rivers for CONservation value (SERCON)	River habitat survey + RIVPACS ***Response***	Boon et al. (2002)

need for *separation of classification procedures* (which merely group like with like) from *condition assessments* (which compare disturbed with reference condition in similar river reaches and river types, and interpret the drivers of change over time and space) (Brierley and Fryirs 2005). While the measures and indicators used to classify river types and assess river condition may overlap to some degree, the indicators used to measure river condition will often be quite different to those used in classification (Brierley and Fryirs 2005). Second, as a result, some river condition assessment frameworks fail to *use a rigorous physical template* for comparing like with like and selecting reference reaches. One example that addresses both these issues is the combined River Styles and Pressure-Biota-Habitat approach presented in case study 1 (box 7.1).

Third, many frameworks only focus on spot measures of *physical or ecological structure* that are used to detect deterioration in condition (response), and do not interpret causes of condition (pressures) within a historical context. These methods develop a model for predicting biological community structure based on certain abiotic characteristics of undisturbed sites and reaches, and then compare the observed with the expected structure. They provide a capacity to determine differences from the expected biological characteristics of the assemblage such that the difference represents a measure of ecological condition. Examples of these frameworks include RIVPACS (Wright et al. 1993) and its Australian equivalent, AusRivAs (Smith et al. 1999), the U.S. Index of Biotic Integrity (Karr 1998), and the Index of Stream Condition in case study 2 (box 7.2).

Fourth, it is only recently that frameworks have been developed wherein potential bioindicators are first tested for sensitivity and consistency of response to known gradients of change in primary

Box 7.1. An Integrative Approach for Examining the Geoecological Condition of Rivers in Bega Catchment, NSW, Australia: River Styles and Pressure-Biota Habitat (see Chessman et al. 2006; Fryirs 2003; Fryirs and Brierley 2005; www.riverstyles.com for further details).

Aims

A study was conducted in the Bega catchment (south coast of NSW, Australia) that combined the River Styles framework with the Pressure-Biota-Habitat approach, in order to:

- Determine the role that geomorphic condition (i.e., physical structure and function) plays in dictating the ecological condition (i.e., the composition and abundance of riverine biological assemblages) of rivers in this system.
- Test an integrated system that would allow the identification of any factors that are limiting to biological recovery in a river system, so that management can target removal of these constraints.

Methods

Geomorphic Condition

Using Stage 2 of the River Styles framework, the condition of each reach was rated as either "good" (close to the natural reference condition), "moderate" (somewhat modified by human impact), or "poor" (severely modified) on the basis of variables that are considered relevant for each River Style and its geomorphic evolution.

Ecological Assessment

Using the Pressure-Biota-Habitat approach, forty-one sites were surveyed for four biological assemblages: diatoms, aquatic and semi-aquatic macrophytes, aquatic macroinvertebrates, and fish. The sampling strategy covered the full range of River Styles and geomorphic conditions of each River Style.

Integration

Ecological results were compared to River Style and geomorphic condition categories to detect statistically significant associations between biota and geomorphic structure and condition.

Box 7.1. (Continued)

Results

Ecological Relationships to River Style (Habitat Type)

- Nine River Styles were identified in Bega catchment, ranging from steep headwater, to partly confined with bedrock-controlled discontinuous floodplain, to low sinuosity sand bed river types.
- Each biological assemblage differed significantly among River Styles, suggesting that geomorphic structure (habitat availability) may be used as a physical template for correlatively assessing and predicting biological assemblages.
- For macrophytes and macroinvertebrates, geomorphic river type exerted a direct influence via variation in physical habitat characteristics.
- For diatoms and fish, this association could also be accounted for by geographic clustering of sites in the same River Style, and a tendency for River Styles to occupy particular altitudinal zones.

Ecological Relationship to Geomorphic Condition (Habitat Availability and Viability)

- Twice as many taxa favored sites in good geomorphic condition compared to poor condition reaches.
- Many of the taxa associated with sites in poor condition were alien taxa introduced to Australia since European settlement.
- Geomorphic condition was significantly associated with differences in macrophyte and macroinvertebrate assemblages. Significant difference occurred between sites in good and moderate geomorphic condition rather than between sites in moderate and poor condition.
- Fish and diatom assemblages were not significantly associated with reaches of variable geomorphic condition. This suggests that the overriding influences of water quality on diatoms and of altitude-related variation in water temperature and distance from the ocean on fish are key drivers.

Key Messages

- The taxa that differed significantly in abundance among condition categories were most abundant in sites of good condition. Alien taxa were strongly associated with sites in poor condition, reflecting a preference of invasive species for disturbed environments. Thus, reaches in good geomorphic condition are not merely biologically different from reaches in poor condition, but contain native biodiversity.

Box 7.1. (Continued)

- Macrophytes and macroinvertebrates are sensitive to geomorphic deterioration from good to moderate condition, but further deterioration from moderate to poor condition may have little additional impact on biodiversity. Once physical/geomorphic integrity is impaired (and abiotic thresholds breached) ecological integrity is compromised.
- Protection of reaches that are in good geomorphic condition is critical to the maintenance of indigenous biodiversity and ecological function. Once the process of deterioration in geomorphic condition begins, it is often difficult and costly to arrest. For reaches that are already in moderate or poor geomorphic condition, any rehabilitation works should aim to achieve good geomorphic condition in order to engender a high level of biological recovery.

Box 7.2. The Victorian Index of Stream Condition (ISC) (see DSE 2006; Ladson et al. 1999, 2006).

Aims

The ISC was set up to report on the biotic condition of waterways in the State of Victoria, in order to assist broad-scale management of waterways. The key to the index is the selection and combination of variables, both abiotic and biotic, that are useful for the definition of stream condition at both a river catchment and State scales. They must also able to detect changes over a five-year interval. Applications have been focused on rural lowland streams that are highly degraded, have the highest pressures, and could be effectively managed. Only biotic and abiotic components that had some relevance to management practices were chosen. This clearly defined constraint meant that the indicators used are rarely contextual but make the best use of existing data.

Methods

The State of Victoria was divided up into reaches, expert opinion was used to delineate abiotically homogenous reaches that, after rigorous field testing, contained three sites for optimizing field measurements (Ladson et al. 2006). The index is comprised of four pressure subindices and one response subindex.

Hydrology (Pressure)

Five ecologically significant indicators are used: (1) low flows; (2) high flows; (3) zero flows; (4) seasonality; and (5) variability. Using more than than fifteen years of monthly streamflow data, each of these metrics are referenced against an unimpacted flow series. The reference flow series is created

Box 7.2. (Continued)

by extracting urban, farm dam, and private diverters from the current stream flow record. The REALM model is also used to gain current and unimpacted flow series. The outputs are uncorrelated scores that define an ecologically significant flow stress ranking that can be regionalized.

Water Quality (Pressure)

Four indicators best represented pressures on Victorian streams: (1) total phosphorous; (2) turbidity; (3) salinity; and (4) pH. The State Environment Protection Policy (Waters of Victoria) is used to set the reference conditions for each of these indicators. The State is regionalized into five bioregions based on macroinvertebrates. For each bioregion, unimpacted reference sites are used to set reference conditions, and expert judgement is used to define the scoring levels from 0 to 4 (extreme to no modification). The values are based on percentile deviations of the monitoring sites data from the reference values, under low flow conditions.

Physical Form (Pressure)

Like the Streamside Zone, this subindex is collected every five years under low flow conditions. Bank condition, large wood, and fish barriers are included in the score. Reference conditions for these abiotic components are not regionalized, and are based on expert panel opinion. To avoid the concept of a static reference condition, banks are only assessed in areas where erosion was not expected.

Aquatic Life (Response)

Macroinvertebrates are used to detect a response to pressures at measurement sites. The scores for the deviations of AUSRIVAS and SIGNAL indicators from reference, for each of the five bioregions used in water quality, are based on ecological goals. The reference conditions are designed to protect the existing aquatic environment of healthy streams, and to provide goals for impacted streams. It is recognized in the scores that a return to near pristine conditions is neither practical nor achievable.

Streamside Zone (Pressure)

Only those vegetation communities that occur within a limited riparian width are assessed. The reference condition is the Ecological Vegetation Class (EVC) for Victoria that defines the pre-European vegetation condition across the state. Nine different indicators are scored against the EVC but the indicators used vary depending on the EVC region. Deviation from reference provides the basis for the score.

BOX 7.2. (CONTINUED)

Results

Each subindex is scored out of ten, and the overall index out of fifty. However, the overall score is not a simple aggregate of the five subindices; the ISC is weighted to recognize that a particularly low score in one subindex may have a limiting effect on river health even if the other subindices score highly. The results are available online (http://www.vicwaterdata.net/isc/index.html) for both the 1999 and 2004 indices. Scores are available for site, reach, catchment, and the State. This allows the user to take the measured data and aggregate it in different ways. However, inappropriate use of the index for reach scale consultancy work has often occurred due to the discrepancy between the ISC endpoints and the consultancy aims.

Key Messages

- The ISC has been successfully adopted by the government and the public.
- By setting clear objectives, the ISC has been able to develop better ways of monitoring and/or evaluating stream conditions. This has been shown, for example, by the much more sensitive indicators of hydrology, tailored to ecological condition.
- While the reporting of the ISC achieved most of it goals, it was weak at identifying changes induced by stream management projects-or isolating interannual variation. This may be resolved by having sentinel sites that are measured every year, and by making some of the metrics more sensitive to specific stream management practices. This is currently being developed.

abiotic drivers such as flow regulation, habitat modification, and water quality (Chessman 2003; Arthington et al. 2006; Kennard et al. 2006a, 2006b). This approach allows rigorous *identification of those indicators sensitive enough to reflect the response gradient* from degraded to natural for the river type under investigation (Smith and Storey 2001). Indicators selected in this way provide the most rigorous scientific basis for examining thresholds of ecological change and their relationship to the causes of degraded ecological condition. The Ecosystem Health Monitoring Program (EHMP) (case study 3, box 7.3) outlines an application of the disturbance gradient approach for testing, selecting, and using indicators in a large scale river condition assessment in southeast Queensland.

Conclusion

This chapter has explored the concept of river condition, defining condition in biophysical terms, as opposed to notions of river "health" that often also include societal perceptions of a river's status and its value to society. Notions of physical integrity, ecological integrity, and ecosystem integrity have been presented as the preferred basis for assessing river condition. If integrative biophysical analyses of river condition are to be successfully developed

Box 7.3. Ecosystem Health Monitoring Program (EHMP) (see Ecosystem Health Monitoring Program 2002–2003 Annual Technical Report, Moreton Bay Waterways and Catchments Partnership, Brisbane, www.healthywaterways.org).

Aims

The Ecosystem Health Monitoring Program (EHMP) was developed by a team of researchers engaged in the project "Design and Implementation of a Baseline Monitoring" (DIBM3), designed to develop a suite of biophysical indicators of stream ecosystem health in agricultural and urban catchments of southeast Queensland.

Methods

The DIBM3 study involved a phase of indicator development followed by testing of promising indicators against known disturbance gradients using the approach originally developed by the U.S. GEEP (Group of Experts on the Effects of Pollutants). Potentially useful indicators (initially fifty-two, reduced to twenty-one) were selected for field trials based on their perceived sensitivity to land-use pressures. Relatively novel indicators were tried in pilot studies and a further six were eliminated, leaving fifteen in five groups that were tested against gradients of disturbance in a large-scale field trial involving six catchments. The chosen basins were classified into four groups (upland, lowland, coastal, and wallum heathland) within each of which it was valid to compare the performance of indicators by comparing each index with the reference condition (based on least-disturbed sites/reaches for each type of stream). Fish assemblage reference condition was derived by linear regression modeling of the expected assemblage for each stream type.

Physical condition. Three geomorphic indices were derived using the Rosgen (1996) classification metrics (sensitivity score, erosion score, geomorphic condition). Habitat permanence and the role of riparian vegetation in channel stability were also assessed.

Water quality (surface and groundwater). Indicators selected to describe overall water quality were: conductivity, pH, diel change in temperature and DO, TSS, nutrient concentrations, discharge (for loads), and depth of aquifer.

Ecosystem processes. These indices included benthic metabolism (measured as gross primary productivity (GPP) or diel DO change as a surrogate, and twenty-four-hour respiration), algal biomass (measured on artificial substrates), and delta-13 carbon isotope levels in growing tips of aquatic plants (representing a surrogate for GPP).

Biological indicators tested were: Invertebrate assemblage structure (SIGNAL score—Stream Invertebrate Grade Number—Average Level; PET index—[Plectoptera, Ephemeroptera, and Trichoptera]) and total taxa. Fish assemblage metrics were composition and O/E scores; presence of alien species and percentage of alien individuals in total population (Kennard et al. 2005; 2006a, 2006b).

Box 7.3. (Continued)

Using the fifteen indicators, fifty-three sites across the four stream types were assessed in the winter/spring (dry period) of 2000. Descriptors of the disturbance gradient included percentage of catchment cleared and percentage cropped, channel condition, riparian condition, water chemistry, instream habitat, and flow-related variables, especially downstream barriers.

Results and Integration

All five groups of indicators demonstrated responses to the disturbance gradient and all fifteen were included in the final conceptual model and ecosystem health monitoring protocols for southeast Queensland. Details of the EHMP and the methods of presentation of results can be found at www.healthywaterways.

Key Messages

- Many of the streams assessed were in poor ecological condition due to the effects of catchment land use, loss of vegetation cover, and compositional change in the riparian corridor along streams, diffuse chemical and sediment inputs, water quality deterioration, and modification of habitat structure and stability.
- Macroinvertebrates were sensitive to habitat and water quality deterioration reflected in low family richness, a low number of PET taxa and low SIGNAL scores.
- Fish assemblages declined in species richness and their composition changed with increasing land-use stress (Kennard et al. 2005; 2006a, 2006b). Alien taxa were typically (but not always) strongly associated with sites in poor condition, reflecting the capacity of invasive alien species to live in disturbed environments that in general are unsuitable for native species. Thus, reaches in good ecological condition were not only biologically different and more diverse than reaches in poor condition, but sustained more native biodiversity.
- Ecosystem process indicators were also sensitive to the disturbance gradients. GPP and algal biomass provided direct measures of the total primary productivity (ecological vigor) of the stream. Degraded sites had high GPP levels usually reflecting land use and riparian condition (more light), high nutrient levels, and unacceptably high algal biomass for the stream type. High delta-13 carbon isotope levels in aquatic plants provided a rapidly measured surrogate for GPP. When high levels are found, direct measure of GPP should be taken and remedial measures implemented to reduce nutrient levels and to restore riparian vegetation.
- The DIBM3 field study demonstrated that many indices distilled down to five groups can give a robust and internally consistent assessment of stream condition with strong linkages to stressors and fluxes such as diffuse nutrients coming off rural lands. By including both stressor and response variables ranging from geomorphological to biological to ecosystem-level responses of a stream to disturbance in the overall assessment, the EHMP program provides one of the most integrated pressure-response frameworks available for stream condition assessment.

Box 7.3. (Continued)

- The EHMP developed on the basis of convincing outcomes from the DIBM3 study is now employed routinely to assess stream health throughout southeast Queensland and has strong institution and public support The results of each annual assessment are publicly reported in the form of grades from A to E. These are taken very seriously by local governments and the assessment, grading, and reporting process has helped to identify problem areas for remediation. The EHMP is being adapted for application in the Wet Tropics region of north Queensland and has attracted significant attention internationally.

and implemented, analyses of river behavior and evolution in relation to historical legacies and contemporary fluxes and disturbances must be integrated with the measurement of pressure and response variables (indicators) for different river types and reaches within river types. Indicators may be river type specific but are couched within a framework that is generic. Ten guiding principles for analyzing river condition are proposed and supported with a brief review of existing condition assessment frameworks. Three Australian case studies highlight the potential for development of an integrative, crossdisciplinary biophysical approach to analyse and interpret river condition.

Our final recommendation is that river condition assessment should always be nested within an adaptive management framework such that results can be linked to the design and implementation of management activities that aim to rehabilitate critical drivers and fluxes and thereby improve river ecosystem condition. Although full-scale river restoration to a former, pristine state will seldom be possible in much of the developed world, our understanding of the functionality of near pristine and remnant river ecosystems should form the scientific basis for the planning, design, and implementation of river management strategies. Such approaches go some way toward recovery of valuable ecosystem components (e.g., fish and riparian trees) and the drivers and processes that sustain them.

Acknowledgments

The authors wish to thank Richard Hobbs and Darren Ryder for constructive review comments on this chapter.

References

ANZECC. 2000. *Core environmental indicators for reporting on state of the environment*. Australian and New Zealand Environment and Conservation Council, State of the Environment Reporting Task Force 2000, Environment Australia, Canberra, ACT.

Aronson, J., S. Dhillion, and E. Le Floc'h. 1995. On the need to select an ecosystem of reference, however imperfect: A reply to Pickett and Parker. *Restoration Ecology* 3:1–3.

Aronson, J., and E. Le Floc'h. 1996. Vital landscape attributes: Missing tools for restoration ecology. *Restoration Ecology* 4:327–33.

Arthington, A. H., S. E. Bunn, N. L. Poff, and R. J. Naiman. 2006. The challenge of providing environmental flow rules to sustain river ecosystems. *Ecological Applications* 16:1311–18.

Arthington, A. H., and B. J. Pusey. 2003. Flow restoration and protection in Australian rivers. *River Research and Applications* 19:377–95.

Barbour, M. T., W. F. Swietlik, S. K. Jackson, D. L. Courtemanch, S. P. Davies, and C. O. Yoder. 1999. Measuring the attainment of biological integrity in the USA: A critical element of ecological integrity. *Hydrobiologia* 422/423:453–64.

Barmuta, L. A., B. Chessman, and B. T. Hart. 1998. *Interpreting the outputs from AUSRIVAS*. Occasional Paper 02/98. Canberra, Australia: Land and Water Resources Research and Development Corporation.

Baron, J. S., N. L. Poff, P. L. Angermeier, C. N. Dahm, P. H. Gleick, N. G. Hairston Jr., R. B. Jackson, C. A.

Johnston, B. D. Richter, and A. D. Steinman. 2002. Meeting ecological and societal needs for freshwater. *Ecological Applications* 12:1247–60.

Bernhardt, E. S., M. A. Palmer, J. D. Allan, G. Alexander, K. Barnas, S. Brooks, J. Carr, S. Clayton, C. Dahm, J. Follstad-Sgah, D. Galat, S. G. Loss, P. Goodwin, D. Hart, B. Hassett, R. Jenkinson, S. Katz, G. M. Kondolf, P. S. Lake, R. Lave, J. L. Meyer, T. K. O'Donnell, L. Pagano, B. Powell, and E. Sudduth. 2005. Synthesizing U.S. river restoration efforts. *Science* 308:636–37.

Boon, P. J., N. T. H. Holmes, P. S. Maitland, and I. R. Fozzard. 2002. Developing a new version of SERCON (System for Evaluating Rivers for Conservation). *Aquatic Conservation: Marine and Freshwater Ecosystems* 12:439–55.

Brierley, G. J., and K. A. Fryirs. 2005. *Geomorphology and river management: Applications of the River Styles framework*. Oxford, UK: Blackwell.

Brooks, A. P., T. Howell, T. B. Abbe, and A. H. Arthington. 2006. Confronting hysteresis: Wood based river restoration in highly altered riverine landscapes of south-eastern Australia. *Geomorphology* 79:395–442.

Bunn, S. E. 1995. Biological monitoring of water quality in Australia. Workshop summary and future directions. *Australian Journal of Ecology* 20:220–27.

Bunn, S. E., and A. H. Arthington. 2002. Basic principles and ecological consequences of altered flow regimes for aquatic biodiversity. *Environmental Management* 30:492–507.

Bunn, S. E., P. M. Davies, and D. T. Mosisch. 1999. Ecosystem measures of river health and their response to riparian and catchment degradation. *Freshwater Biology* 41:333–45.

Burns, A., and D. S. Ryder. 2001. The emergence of biofilms as a monitoring tool in Australian riverine systems. *Ecological Restoration and Management* 2:53–63.

Cairns, J. J. 1989. Restoring damaged ecosystems: Is predisturbance condition a viable option? *The Environmental Professional* 11:152–59.

Cairns, J. J., P. V. McCormick, and B. R. Niederlehner. 1993. A proposed framework for developing indicators of ecosystem health. *Hydrobiologia* 263:1–44.

Chessman, B. C. 2002. *Assessing the conservation value and health of New South Wales rivers: The PBH (Pressure-Biota-Habitat) project*. Parramatta, Australia: Centre for Natural Resources, New South Wales Department of Land and Water Conservation, NSW Government.

Chessman, B. C. 2003. New sensitivity grades for Australian river macroinvertebrates. *Marine and Freshwater Research* 54:95–103.

Chessman, B. C., K. A. Fryirs, and G. J. Brierley. 2006. Linking geomorphic character, behaviour and condition to fluvial biodiversity: Implications for river rehabilitation. *Aquatic Conservation: Marine and Freshwater Research* 16:267–88.

Chessman, B. C., and M. J. Royal. 2004. Bioassessment without reference sites: Use of environmental filters to predict natural assemblages of river macroinvertebrates. *Journal of the North American Benthological Society* 23:599–615.

Clarke, S. J., L. Bruce-Burgess, and G. Wharton. 2003. Linking form and function: Towards an ecohydromorphic approach to sustainable river restoration. *Aquatic Conservation: Marine and Freshwater Ecosystems* 13:439–50.

Downes, B. J., L. A. Barmuta, P. G. Fairweather, D. P. Faith, M. J. Keough, P. S. Lake, B. D. Mapstone, and G. P. Quinn. 2002. *Monitoring ecological impacts: Concepts and practice in flowing waters*. Cambridge, UK: Cambridge University Press.

DSE 2006. *Index of stream condition: Users manual (2nd edition)*. Victoria, Australia: Department of Sustainability and Environment.

du Toit, J. T., K. H. Rogers, and H. C. Biggs, eds. 2003. *The Kruger experience: Ecology and management of savanna heterogeneity*. Washington, DC: Island Press.

Elliott, D. C. 1996. A conceptual framework for geoenvironmental indicators. In *Geoindicators: Assessing rapid environmental changes in earth systems*, ed. A. R. Berger and W. J. Iams, 337–49. Rotterdam, Netherlands: A.A. Balkema.

European Commission. 2002. *European Water Framework Directive*. www.europa.eu.int/comm/environment/water/ (accessed September 20, 2006).

Fairweather, P.G. 1999. State of environment indicators of river health: Exploring the metaphor. *Freshwater Biology* 41:211–20.

Fryirs, K. 2003. Guiding principles for assessing geomorphic river condition: Application of a framework in the Bega catchment, South Coast, New South Wales, Australia. *Catena* 53:17–52.

Fryirs, K., and G. J. Brierley. 2000. A geomorphic approach for the identification of river recovery potential. *Physical Geography* 21:244–77.

Fryirs, K. A., and G. J. Brierley. 2005. *Practical applications of the River Styles framework as a tool for catchment-wide river management: A case study from Bega catchment, NSW, Australia*. E-book published on www.riverstyles.com (accessed November 17, 2006).

Gilvear, D. J. 1999. Fluvial geomorphology and river engineering: Future roles utilising a fluvial hydrosystems framework. *Geomorphology* 31:229–45.

Gordon, N. D., T. A. McMahon, B. L. Finlayson, C. J. Gippel, and R. J. Nathan. 2004. *Stream hydrology: An introduction for ecologists (second edition)*. Chichester, UK: Wiley.

Gore, J. A., ed. 1985. *The restoration of rivers and streams: Theories and experience*. Boston, MA: Butterworth.

Gore, J. A., and F. D. Shields Jr. 1995. Can large rivers be restored? *BioScience* 45:142–52.

Hilderbrand, R. H., A. C. Watts, and A. M. Randle. 2005. The myths of restoration ecology. *Ecology and Society* 10(1): Article 19. http://www.ecologyandsociety.org/vol10/iss1/art19/ (accessed December 16, 2007).

Hobbs, R. J., and J. A. Harris. 2001. Restoration ecology: Repairing the Earth's ecosystems in the new millennium. *Restoration Ecology* 9:239–46.

Hughes, R. M., D. P. Larson, and J. M. Omernik. 1986. Regional reference sits: A method for assessing stream potential. *Environmental Management* 10:629–35.

Karr, J. R. 1995. Ecological integrity and ecological health are not the same. In *Engineering within ecological constraints*, ed. P. Schulze, 97–109. Washington, DC: National Academies Press.

Karr, J. R. 1998. Rivers as sentinels: Using the biology of rivers to guide landscape management. In *River ecology and management: Lessons from the Pacific coastal ecoregion*, ed. R. J. Naiman and R. E. Bilby, 502–23. New York, NY: Springer-Verlag.

Karr, J. R. 1999. Defining and measuring river health. *Freshwater Biology* 41:221–34.

Kennard, M. J., A. H. Arthington, B. J. Pusey, and B. D. Harch. 2005. Are alien fish a reliable indicator of river health? *Freshwater Biology* 50:174–93.

Kennard, M. J., B. D. Harch, B. J. Pusey, and A. H. Arthington. 2006a. Accurately defining the reference condition for summary biotic metrics: A comparison of four approaches. *Hydrobiologia* 572:151–70.

Kennard, M. J., B. J. Pusey, A. H. Arthington, B. D. Harch, and S. L. Mackay. 2006b. Development and application of a predictive model of freshwater fish assemblage composition to evaluate river health in southeastern Australia. *Hydrobiologia* 572:33–57.

Ladson, A. R., R. B. Grayson, B. Jawecki, and L. J. White. 2006. Effect of sampling density on the measurement of stream condition indicators in two lowland Australian streams. *River Research and Applications* 22:853–69.

Ladson, A. R., L. J. White, J. A. Doolan, B. L. Finlayson, B. T. Hart, P. S. Lake, and J. W. Tilleard. 1999. Development and testing of an Index of Stream condition for waterway management in Australia. *Freshwater Biology* 41:453–68.

Mackay, S .J., A. H. Arthington, M. J. Kennard, and B. J. Pusey. 2003. Spatial variation in the distribution and abundance of submersed aquatic macrophytes in an Australian subtropical river. *Aquatic Botany* 77:169–86.

McCormick, P.V., and J. Cairns. 1994. Algae as indicators of environmental change. *Journal of Applied Phycology* 6:509–26.

Meyerson, L. A., J. Baron, J. Melillo, R. J. Naiman, R. I. O'Malley, G. Orians, M. A. Palmer, A. S. P. Pfaff, and O. E. Sala. 2005. Aggregate measures of ecosystem services: Can we take the pulse of nature? *Frontiers of the Ecological Environment* 3:56–59.

Montgomery, D. R. 1999. Process domains and the river continuum. *Journal of the American Water Resources Association* 35:397–410.

Newson, M. D., and C. L. Newson. 2000. Geomorphology, ecology and river channel habitat: Mesoscale approaches to basin-scale challenges. *Progress in Physical Geography* 24:195–217.

NLWRA. 2001. *National land and water resources audit: An initiative of the Natural Heritage Trust*. http://www.nlwra.gov.au/default.asp (accessed November 17, 2006).

Oberdorff T., D. Pont, B. Hugueny, D. Chessel. 2001. A probabilistic model characterizing fish assemblages of French rivers: A framework for environmental assessment. *Freshwater Biology* 46:399–415.

Osterkamp, W. R., and S. A. Schumm. 1996. Geoindicators for river-valley monitoring. In *Geoindicators: Tools for assessing rapid environmental changes*, ed. A. R. Berger and W. J. Iams, 97–114. Rotterdam, Netherlands: A.A. Balkema.

Palmer, M. A., and E. S. Bernhardt. 2006. Hydroecology and river restoration: Ripe for research and synthesis. *Water Resources Research* 42:W03S07, doi:10.1029/2005WR004354.

Petts, G. E. 2000. A perspective on the abiotic processes sustaining the ecological integrity of running waters. *Hydrobiologia* 422/423:15–27.

Poff, N. L., J. D. Allan, M. B. Bain, J. R. Karr, K. L. Prestegaard, B. D. Richter, R. E. Sparks, and J. C. Stromberg. 1997. The natural flow regime: A paradigm for river conservation and restoration. *BioScience* 47:769–84.

Postel, S., and B. D. Richter. 2003. *Rivers for life: Managing water for people and nature*. Washington, DC: Island Press.

Rapport, D. J., R. Costanza, and A. J. McMichael. 1997. Assessing ecosystem health. *Trends in Ecology and Evolution* 13:397–402.

Raven, P. J., P. J. Boon, F. H. Dawson, and A. J. D. Ferguson. 1998. Towards an integrated approach to classifying and evaluating rivers in the UK. *Aquatic Conservation: Marine and Freshwater Ecosystems* 8:383–93.

Rhoads, B. L., and E. E. Herrick. 1996. Naturalization of headwater streams in Illinois: Challenges and possibilities. In *River channel restoration: Guiding principles for sustainable projects*, ed. A. Brookes and F. D. Shields Jr., 331–67. Chichester, UK: Wiley.

Rogers, K. H., and R. Bestbier. 1997. *Development of a protocol for the definition of the desired state of riverine systems in South Africa*. Pretoria, South Africa: Department of Environmental Affairs and Tourism.

Rogers, K. H., and H. Biggs. 1999. Integrating indicators, endpoints and value systems in strategic management of rivers of the Kruger National Park. *Freshwater Biology* 41:439–51.

Rosgen, D. L. 1996. *Applied river morphology*. Pagosa Springs, CO: Wildland Hydrology.

Rowntree, K. M., and R. A. Wadeson. 1998. A geomorphological framework for assessment of instream flow requirements. *Aquatic Ecosystem Health and Management* 1:125–41.

Smith, M. J., W. R. Kay, D. H. D. Edward, P. J. Papas, K. S. J. Richardson, J. C. Simpson, A. M. Pinder, D. J. Cale, P. H. J. Horwitz, J. A. Davis, F. H. Yung, R. H. Norris, and S. A. Hales. 1999. AusRivAS: Using macroinvertebrates to assess ecological condition of rivers in Western Australia. *Freshwater Biology* 41:269–82.

Smith, M. J., and A. M. Storey. 2001. *Design and implementation of a baseline monitoring (DIBM3)* Phase 1 Final report. Brisbane, Australia: Report to the SEQR-WQMS, November 2000.

Southwood, T. R. E. 1977. Habitat, the templet for ecological strategies. *Journal of Animal Ecology* 46:337–65.

Stoddard, J. L., D. P. Larsen, C. P. Hawkins, R. K. Johnson, and R. N. Norris. 2006. Setting expectations for the ecological condition of streams: The concept of reference condition. *Ecological Applications* 16:1267–76.

Sullivan, S. M. P., M. C. Watzin, and W. C. Hession. 2004. Understanding stream geomorphic state in relation to ecological integrity: Evidence using habitat assessments and macroinvertebrates. *Environmental Management* 34:669–83.

Townsend, C. R., and A. G. Hildrew. 1994. Species traits in relation to a habitat templet for river systems. *Freshwater Biology* 31:265–75.

Trofimov, A. M., and J. D. Phillips. 1992. Theoretical and methodological premises of geomorphological forecasting. *Geomorphology* 5:203–11.

Walker, J., M. Diamond, and M. Naura. 2002. The development of physical quality objectives for rivers in England and Wales. *Aquatic Conservation: Marine and Freshwater Ecosystems* 12:381–90.

Werren G., and A. H. Arthington 2002. The assessment of riparian vegetation as an indicator of stream condition, with particular emphasis on the rapid assessment of flow-related impacts. In *Landscape health of Queensland*, ed. J. Playford, A. Shapcott, and A. Franks, 194–222. Brisbane, Australia: Royal Society of Queensland.

Whisenant, S. G. 1999. *Repairing damaged wildlands: A process-oriented, landscape-scale approach*. Cambridge, UK: Cambridge University Press.

Whitton, B. A., and M. G. Kelly. 1995. Use of algae and other plants for monitoring rivers. *Australian Journal of Ecology* 20:45–56.

Wright, J. F., M. T. Furse, and P. D. Armitage. 1993. RIVPACS: A technique for evaluating the biological quality of rivers in the UK. *European Water Pollution Control* 3:15–25.

Chapter 8

Social and Biophysical Connectivity of River Systems

Mick Hillman, Gary J. Brierley, and Kirstie A. Fryirs

> Place is the location . . . where the social and the natural meet.
>
> Dirlik 2001, 18

Successful integrative river management requires an understanding of the links between natural and cultural landscapes, ensuring that institutional and community values are meaningfully incorporated in the process of environmental repair (Harris 2006). Coherent approaches to the assessment of river health integrate biophysical and social dimensions of environmental condition, building on the relationship between healthy rivers as products of, and in turn promoting, healthy societies. Understanding and working with the concept of connectivity across both the biophysical and social dimensions is a core component of this relationship. A connected approach to integrative river management aims for a dialogue between scientific understanding and community values. Applying this principle therefore means understanding both catchment-scale biophysical linkages and community perceptions of what constitutes a "healthy river." Such understandings are specific to time and place, militating against the application of generic models and assumptions.

Contemporary perceptions of river health are contingent upon present and past connections between people and their river. They encompass a range of potential uses and values. However, in the large body of literature influenced by Eurocentric ideas of landscape, healthy rivers are often romanticized as single, continuous, constantly flowing channels (Kondolf 2006). Postcolonial societies have afforded rivers a limited range of uses, and systems are expressed as healthy only if conditions suitable for these uses are maintained. Disconnection in a river system, whether it is the presence of isolated pools in river channels or ephemeral tributaries, has been portrayed as an undesirable and unsustainable state (Kondolf et al. 2006). This contrasts with the recognition of variable and changing forms of connection and disconnection in many indigenous cultures (Smolyak 2001; James 2006). Recognition of the variable and changing patterns of connectivity across time and space, and between social and biophysical dimensions, is a core component of integrative river management.

Whether appraised in biophysical or social terms, landscapes, ecosystems, and communities can be relatively connected or disconnected. For river management to be successful and relevant it is important to recognize that biophysical disconnection may be natural and healthy at a given time and place. For this reason we use the term *(dis)connectivity* to refer to dynamic patterns of connection and disconnection. For example, disconnected and isolated parts of river systems may shelter distinct genetic populations of species and unique floristic and faunal attributes (Sheldon et al. 2002; Bunn et al. 2006). Likewise, human (dis)connectivity with rivers has often been mediated by cultural factors, particularly in indigenous societies through taboos, totems, and sacred sites that are the result of co-evolution with landscapes over many generations (Rose 1999; Townsend et al. 2004). However, in more recent history, biophysical disconnection has often been imposed through the construction of barriers, while social disconnection has resulted from appropriation and enclosure of riparian land or in response to rapidly developing perceptions by local communities of the river as polluted, unhealthy, or as bringer of damaging floods. Present day disconnection is often the legacy of an earlier focus on narrow and exclusive uses of the river for irrigation, discharge of effluent, or navigation. This type of disconnection is referred to by Ward (2001) as "geo-environmental disconnection," the product of technocentric efforts to forge landscapes for agriculture, industry, and recreation. Conversely, biophysical and social connections may have been imposed through the development of irrigation systems in semi-arid landscapes.

In this chapter, we argue that imposed, arbitrary (dis)connectivity based on a narrowly defined or exclusive use of water is unhealthy in river management and is ultimately unsustainable and unjust. Such (dis)connectivity reduces community understanding of, and concern for, our rivers, while allowing dominant and environmentally damaging uses and practices to continue. It also promotes feelings of inequity, distributing costs and benefits through top-down decisions or decrees. On the other hand, broad and holistic (dis)connectivity in its many forms strongly implies the convergence of social and biophysical perspectives and acknowledgment of a wider range of values as a vital step in the process of river repair. Transdisciplinary work on links between ecological and community health and well-being indicates that the healthy appreciation of the inherent diversity and variability of river systems is an integral part of healing our relationship to the natural world (Costanza and Mageau 1999; Connor et al. 2004).

Based on these guiding principles, this chapter uses a transdisciplinary and place-based analysis of biophysical and social (dis)connectivity to:

1. Examine the broad links between connectivity and river health.
2. Describe the biophysical and social forms and changing patterns of connection and disconnection within a river system and between people and that system.
3. Analyze key themes in the interrelationship between the biophysical and social dimensions of (dis)connectivity through case examples.
4. Highlight implications of these themes in the development of just and sustainable approaches to the management of healthy rivers.

Connectivity and River Health

The view of river health outlined in chapter 7 focuses on external, biophysical, and verifiable indicators of river condition. Often such indicators are specific to particular disciplines and scales. However, given that rivers epitomize the links between landscapes and ecosystems (Jungwirth et al. 2002), a practical understanding of biophysical linkages is crucial in producing the "mature knowledge" that is increasingly required for effective integrated ecosystem management (Lake 2001; Dunn 2004). This is in itself a major challenge, since complex indicators of river condition, such as connectivity itself, have proved difficult to quantify both for conceptual reasons and because of scientific concern over valid descriptors and rigor of resultant data (Dunn 2004).

A complementary but distinct view of river health is one based on the idea of social connection, in which health is about people developing, maintaining, or losing interaction with the river. Integral to thinking about the way people connect and disconnect with rivers is the broad geographical notion of place as socially constructed (Massey 2005), and of a sense of place as individually interpreted rather than having one particular "essence." The notion of "place-identity" has been used to describe this social dimension of connection: "it can be said to represent the physical settings' importance for a person's identity. Research in place-identity suggests that an individual has more complex relations to the environment than simply living in it" (Wester-Herber 2004, 111).

Connection through place-identity is fundamental to community engagement in river management programmes, fostering a sense of commitment, building social capital, and allowing local knowledge a role in planning (Hillman et al. 2003; Thompson and Pepperdine 2003). Place-identity also highlights the need to think of health in terms of both biophysical indicators and human-nature relationships (Brierley et al. 2006b). However, the establishment of place-identity is a *necessary* prerequisite rather than an inherently *sufficient* condition for river health—there is no reason to argue that some good use-values create place-identity and others do not. Our point here is that without such connection, the relationships that sustain integrative river management cannot be forged and that narrow and exclusive values will prevail.

The nexus between biophysical and community health in major river rehabilitation strategies has been expressed in practice in several forms. For instance, the Loire Vivante (Living Loire) program aims to "lead people back to the river," while the Mersey Basin Campaign includes in its vision the goal of increasing the valuing of the river by its community. The Dutch policy of "Space for the River" aims to maintain flood protection in the face of increased design discharges, while at the same time conserving landscape, ecological, and historical features (Cals and van Drimmelen 2000). The Victorian River Health Strategy includes the objective of "maintaining the rivers' place in our collective history" with the overall aim that "our communities will be confident and capable, appreciating the values of their rivers, understanding their dependency on healthy rivers and actively participating in decision-making" (Victorian Government 2002, 3).

The next sections develop a conceptual framework for exploring forms of biophysical and social (dis)connectivity in river systems, providing a lens on this integrative approach and on a condition-connection notion of health.

Forms, Patterns, and Changes to Physical (Dis)Connectivity

Biophysical elements of disconnectivity or decoupling in river systems refer to factors that constrain the operation of biophysical fluxes, such that the linkages of processes and conveyance of materials is affected. Disconnectivity may be "natural," such that some areas of landscapes are effectively isolated from other compartments. Alternatively, natural variability in precipitation regimes may induce a significant range in flow conditions at any given location within a river system, such that hydrological connectivity varies markedly over differing timescales. Impediments to functional linkages may be permanent, temporary or intermittent. Linkages may adjust over time in response to disturbance events. Human disturbance has significantly disrupted the natural patterns of (dis)connectivity of river systems, through measures such as drainage of wetlands, irrigation and flow regulation schemes, removal of riparian vegetation and woody debris, and introduction of exotic vegetation. As noted by Sheldon and Thoms (2006, 219): "Rivers . . . differ from terrestrial systems in that they are naturally fragmented in both space and time. . . . Therefore, it is not fragmentation itself that may impact on diversity in large rivers *but changes to the degree and nature of spatial and/or temporal fragmentation*."

Changes to flow regime are often considered to be the most profound form of human disturbance to river systems, whether induced by indirect impacts (e.g., land use change) or direct impacts (e.g., dams) (Postel and Richter 2003; Wohl 2004). Changes to the hydrological connectivity of rivers alters not only the quantity, seasonality, and quality of water, but also changes ecosystem functioning through its effect on habitat availability/viability, fish passage, thermal regulation, algal growth, and sediment/nutrient flux (Kondolf et al. 2006). As an example, the biophysical role of wetlands and swamps highlights the importance of landscape disconnectivity. For most of the year wetlands that make up discontinuous watercourses may be functionally disconnected from trunk stream processes in terms of surface transfer of water, sediment, nutrients, biota, seeds, and so on (though subsurface transfers may occur). However, sustained rainfall or storm events may modify the nature and extent of connectivity, as short-term increases in runoff and overland flow link biophysical fluxes to the trunk stream. As a second example, the distinctive ecology of dryland rivers is intimately linked to their variability in flow. During floods, dryland rivers are highly connected, with little fragmentation. However, during droughts, fragmentation is substantial as systems are highly disconnected, and aquatic community composition is strongly influenced by the interval since waterbodies were last connected (Sheldon and Thoms 2006).

Human disturbance has extensively modified these "natural" linkages of biophysical processes in landscapes. The quest to make wetlands drier and drylands wetter has permanently connected many discontinuous watercourses, increasing the conveyance of flow, sediments, nutrients, and seeds from tributaries to the trunk stream. The filtering effects of former wetland zones have been negated, as peak flows are enhanced while maintenance of base flows is diminished. Alternatively, flow regulation has augmented the connectivity of dryland rivers, increasing the homogenization of habitat diversity in both space and time (Sheldon and Thoms 2006).

Over longer timeframes, geomorphic changes to river structure alter the capacity of the channel to store and transport sediment, modifying the longitudinal (upstream-downstream), lateral (channel-floodplain), and vertical (surface-subsurface) connectivity of the river system (cf. Ward 1989; chapters 4 and 6). Impediments to sediment conveyance may disconnect areas of a catchment from the primary sediment conveyor belt, as landforms store sediment for various periods of time. These impediments have been termed buffers, barriers, and blankets (Fryirs et al. 2007a, 2007b) (table 8.1). *Buffers* are landforms such as alluvial fans or floodplains that prevent sediment from entering the channel network. *Barriers* are features that impede sediment delivery along a channel, whether natural (e.g., sediment slugs) or human-induced (e.g., dams or weirs). *Blankets* are sediment accumulations that smother other landforms, impeding vertical linkages (see chapter 6). Only those areas downstream of impediments actively contribute to sediment delivery at the river mouth. These forms and patterns of physical (dis)connectivity vary in different parts of catchments and in differing landscape settings (Lane and Richards 1997; Brierley and Fryirs 2005; Fryirs et al. 2007a).

Landscape change, whether natural or human-induced, may alter the extent and pattern of physical (dis)connectivity. For example, changes to the magnitude, frequency, duration, or sequencing of flow events affect the operation of biophysical fluxes. Formation or breaching of a buffer, barrier, or blanket alters the pattern and rate of biophysical linkages in a catchment (Fryirs et al. 2007a). Landscape (dis)connectivity impacts significantly upon the propagation of disturbance responses throughout catchments. Highly connected catchments manifest change, such that offsite disturbances (e.g., in the upper parts of a catchment) produce a signal elsewhere (e.g., in the downstream parts of a catchment). In disconnected catchments, changes

TABLE 8.1

Forms and mechanisms of connection and (dis)connection across the biophysical and social dimensions of river systems.

Biophysical forms	Social forms of connection (use-values)	Mechanisms of social (dis)connection
Buffer: landforms that prevent sediment *from entering the channel network*.	Resource: river as generic placeless commodity (water) or trade route.	Barrier: *social and institutional obstacle* to an authentic connection between people and rivers.
Barrier: disrupt sediment moving *along the channel*.	Recreation: river as place with physical characteristics allowing water-based activities.	Boundary: *externally imposed* scale, category, or distinction delimiting the scale at which participation, planning, and decision-making occur.
Blanket: features that *smother other landforms*, protecting those forms from reworking, temporarily removing those stores from the sediment cascade.	Aesthetic: more explicit and place-based recognition of the intrinsic value of a river and its effect on human well-being.	Edges: spaces *away from centers of decision-making* that provide potential for discourse and engagement. For this very reason these may be critical points of reconnection.
Implications: *identifying critical points for reconnection, timeframe of (dis)connectivity (effective timescales)*. Affects interpretation of sensitivity, recovery, and prediction of likely future trajectory (prediction).	Implications: *Identifying critical points for disconnection and reconnection over adequate timeframes*. Appraisal of link to place affected by diversity of values.	Implications: *Institutional change, adaptivity, and resilience required to address disconnecting characteristics and recognize potential for reconnection in hidden places.*

are absorbed or suppressed within a certain part of the system (Brierley et al. 2006a). In some instances, reinstating disconnectivity (e.g., through the formation of swamps) may be a desirable management goal.

Effective description and explanation of landscape (dis)connectivity throughout a catchment provides a basis to predict the future trajectory of geomorphic change (Brierley and Fryirs 2005). Hence, management applications must emphasize explicitly whether they strive to connect, disconnect, or re-disconnect biophysical relationships within any given river system. Understanding the spatial and temporal variability in physical (dis)connectivity is critical to integrative place-based management. This biophysical variability is both a prerequisite to, and an outcome of, social connection, as discussed in the following section.

Social (Dis)Connectivity

In this chapter, we use the term "social connection" to describe a potential or actual link between people (as individuals, groups, or communities) and the river environment. One way of framing this connection is through the idea of a *use-value*, an expression of the various ways in which humans appreciate or utilize the river (Walmsley 2002; Johnson et al. 2004). In many instances this separation of use-values is a product of the imposition of colonial land and water management practices. Jackson (2006, 19) argues that the compartmentalization of indigenous values toward water as "cultural" has effectively marginalized indigenous connection as "cultural heritage," as an object of interest for anthropology or archaeology, at the expense of valuing ongoing "relationships, processes, and connections between social groups, people, and place, and people and non-human entities." The use-values discussed in this section must therefore be viewed in this light as culturally specific and exclusive but nevertheless as forming an important context for contemporary river management (Palmer 2006).

For Higgs (2003) and Wester-Herber (2004), "social disconnection" implies a loss of place and place-identity for individuals and communities. In many cases, a sense of loss and alienation has been associated with the physical deterioration of rivers (Parkes and Panelli 2001; Michaelidou et al. 2002). Albrecht (2005, 45) uses the term *solastalgia* to describe "the distress caused by the lived experience of the destruction of one's home and sense of belonging." Such distress creates feelings of psychological or even physical hurt induced by pervasive change. Often associated with this is a deep sense of injustice based upon a belief that the costs of such change are being unevenly distributed, exacerbated by a feeling of political powerlessness (Connor et al. 2004). Colonial dislocation of indigenous connections with land and water is a particularly intense exemplar, creating a long-lasting legacy of environmental degradation and institutional mistrust (Rose 1999; Coombes 2003). However, experiencing such a sense of loss is contingent upon having formed a strong sense of place and connection in the first place. For many, a connection with the river may not have been established. Such alienation (Hillman 2005) or "riparian anomie" (Gandy 2006) may be increasingly common, generated by lack of physical access, cultural change, or by the dominance of noncompatible consumptive uses.

Competing use-values for fresh water throw up major challenges for integrative management in negotiating trade-offs and for assessing the impact of biophysical changes and rehabilitation works across a range of consumptive, recreational, and broader community values (Bennett and Whitten 2002).Therefore, the realization of use-values in specific contexts is a matter of procedural justice, of the extent of fair and inclusive decision-making processes that in turn influence how and where rehabilitation takes place and to what purpose, a question of distributive justice (Tisdell 2003). Use-values also form the basis of authentic catchment-framed visions (chapter 2). These assessments take place in a framework of longer-term changes in the use of land and water and in the increased recognition of nonconsumptive values. In this chapter, we identify three broad use-values (table 8.1).

Use-Values as Forms of Connection

In the industrial age the dominant, and in many cases the only, use-value of water in river systems was as a *resource* or commodity, whether for irrigation, power generation, mining, or waste disposal, or as a route for navigation and trade (King and Woolmington 1960; Evenden 2006). The belief in inexhaustible water resources and associated land management practices has been a principal contributing factor to land and water degradation (Monteith 1952; chapter 1). Metaphors of conquest and progress were frequently invoked in celebrating and promoting these usages and beliefs (Powell 2000; Webster and Mullins 2003).

A second key use-value is that of *recreation*. Rivers have always been focal points for a range of social activities. Recreational uses of rivers for fishing or swimming now compete for space with more technologically driven activities such as powerboats and jet-skis, challenging and reconstituting a community's traditional connections. Many people are excluded from one or more of these activities by their perception of the degraded condition of the river, insufficient flows and depth, or lack of physical or legal access.

Thirdly, and more difficult to articulate, is an *aesthetic* or *spiritual* value of the river, and its role in promoting a sense of place-identity, appreciation of nature, and psychological health and well-being. The term "cultural" is sometimes used in this context but is often inappropriately restrictive, since although culture can include resource and recreational use, the term has been used more narrowly with indigenous societies to imply that active uses are no longer significant and only the spiritual memory remains as heritage (Jackson 2006; Palmer 2006). Despite these semantic issues, the pleasure of just being "down by the river" is very real and palpable for many people of diverse cultural backgrounds.

Rivers also serve as connectors of differing spatial and temporal communities. That is, they function as both a link and a memory. It is common to hear rivers spoken of as the "lifeblood" of a region or catchment, referring not only to their role as conveyors of nutrients and waste but also as a common point of reference and linkage. One comment from the Hunter Valley, Australia, captures this sense of connection: "I've been associated with the river ever since I was a kid, this is the life blood of the area, if the river dies, everything else dies with it" (Ashley-Brown 2003).

It is immediately apparent that these use-values may either overlap or be mutually exclusive. These differing relations between use-values constitute a further source of cultural variability, mirroring the patterns of variability in biophysical (dis)connectivity. For example, use of the river for waste treatment, as cooling for coal-fired power generation, or extensive pumping for irrigation, generally makes recreational use less likely and aesthetic appreciation near impossible. However, caution must be exercised in attributing single or exclusive use-value connections to people. For instance, there is a danger of stereotyping farmers of riparian land exclusively as resource users, or indigenous people solely as cultural users. This reductionism detracts from everyday real-life experience and oversimplifies the stake of particular groups based on some commonsense but simplistic notion of how people connect to their river.

Despite these individual overlaps, fundamental differences between use-values remain important. The biophysical connectivity of a river means that these differences may be more than site-specific. Connectivity for one section of the community may be disconnectivity for others, as riparian land is appropriated or rivers are made unusable for recreation by pollution and the presence of infrastructure. The associated sense of injustice is due in no small part to the perception of winners and losers, and of the distribution of costs and benefits. It may of course be possible for one particular group (e.g., riparian landholders) to enjoy all the use-values. In consequence, other groups may suffer virtually complete alienation and lack of connection. What is critical is the way that use-value connections frame community perspectives on river health, on ideas of fairness, and on the goals of river management.

Mechanisms of Social Disconnection

While social disconnection from rivers may be viewed as a matter of individual choice, broader political and institutional mechanisms constrain or facilitate these choices. It is important for river managers to be aware of these sometimes hidden factors. Literature from human geography and environmental management can help in understanding how these mechanisms operate. Three key concepts—barriers, boundaries, and edges (Table 8.1)—are considered in the remainder of this section. While not in themselves forms of (dis)connectivity, they shed light on the processes through which connection and disconnection occur and act as tools to analyze patterns of (dis)connectivity in the case studies reported later in this chapter.

A *barrier* is an institutional characteristic that acts as an obstacle to an authentic connection between people and rivers (Rhodes et al. 2002; Lachapelle et al. 2003). However, Dovers (2001) and Jennings and Moore (2000) emphasize that institutions are not only formal structures, but also include the practices, cultures, and values that drive them. Institutions are therefore social constructs loaded with their own histories and agendas. If these agendas fail to acknowledge the diversity of community connections, the adaptivity and flexibility of such institutions is limited. A number of institutional barriers can hinder such connection. Examples include:

- Concern for a narrowly defined "policy community" (Hillier 2003), where only one or two interest groups and use-values are represented;
- Use of top-down, noninclusive decision-making processes (Holling and Meffe 1996);
- Focusing management programs at scales that do not reflect the experience of many sections of the community (Brunkhorst and Reeve 2006); and
- An emphasis on particular ecosystem components (Dunn 2004).

A *boundary* represents the imposition of categorical, fragmented ways of seeing and thinking upon what is in essence a highly relational and interconnected system (Howitt 2001b). At its worst, boundary thinking is tantamount to drawing a line that doesn't exist around an area that doesn't matter. Scale fundamentalism is a critical component of categorical boundary thinking—where one "suitable scale" (Kauffman 2002) or set of criteria for a suitable scale of management (Brunkhorst and Reeve 2006) is put forward—whether that be a site, reach, catchment, or region. Such choices, while at times a necessary construct (e.g., a "riparian zone"), may prescribe management at one particular scale. As a result, the knowledge of unscaled groups is effectively excluded. One key impact of rigid boundary thinking is the removal, through social or institutional prescription, of a connected "sense of place" from management considerations (Roe 2006). Comparable epistemic boundaries (Mejia 2004) are created by definitions of what constitutes "scientific" (and hence useful) knowledge, determination of what is included or excluded, and how it is used in environmental assessment (Townsend et al. 2004; Peterson 2006). In this sense a boundary may include intergenerational loss or nontransmission of knowledge and memory.

In contrast to a boundary, an *edge* is a "space" (*sensu* Massey 2005) located away from power centers or hubs of decision-making. Unlike boundaries, edges are not clearly delineated or prescribed. Rather, they are "zones of transformation, transgression, and possibility," characterized in ecological terms by enormous diversity and complexity (Howitt 2001a, 240). As such, they form areas of open, and probably contested, views of river health and management. While appearing as disconnected from a centralist top-down perspective—the word "feral" is sometimes used and resources are frequently withheld from such areas—edges can provide a space in which new perspectives and knowledge may emerge or be sustained. These applications may be uninhibited and uninhabited by commonsense views of the world, views which in reality usually reflect dominant but unacknowledged paradigms and practices. Edges may therefore have high potential for *reconnection*—establishing new forms of connection or reestablishing historical connections.

Dynamics of Social (Dis)Connectivity

As with biophysical factors, patterns of social (dis)connectivity vary across and between catchments. Population pressures, land use, and water use-values create particular patterns in any given system. Understanding these relationships requires in-depth quantitative and qualitative knowledge that is both expert and experiential. Pressures for change may come from within the catchment or may be driven (increasingly) by global factors, such as export markets or appeals to the national interest. In this context, changes to connectivity in turn create new opportunities and threats, new ways of working riverine landscapes, and changes to use-values. This in turn creates new tensions and trade-offs, and removes or inserts new "stakes" into the equation that may or may not be recognized at the formal institutional level. This new social and institutional mix in turn creates community and political pressure to undertake forms of river works that intentionally or inadvertently generate further biophysical barriers and social boundaries.

This dynamic and iterative nature of (dis)connectivity is now examined through case study material. First, work conducted by the authors on biophysical (dis)connectivity in selected sub-catchments of the Hunter Valley, New South Wales, is set alongside sociohistorical themes. Second, an Australian icon, the Snowy Mountains Hydro Scheme, is considered in terms of imposed interbasin connectivity and competing uses between formerly disconnected biophysical and social catchments.

Contrasting Sub-Catchments from the Hunter Valley, New South Wales

In this case study we compare two sub-catchments of the Hunter Valley to illustrate differing types of (dis)connectivity and changing relationships between biophysical and social dimensions.

Wollombi Brook

Wollombi Brook is the main southern tributary of the Hunter River. It drains a catchment area of 2,150 square kilometers, or approximately 10 percent of the total Hunter Catchment. Valleys are relatively confined within sandstone. Prior to 1948, upper reaches of Wollombi Brook comprised a near-discontinuous watercourse in which a low-capacity, tortuous channel occasionally broke down into swampy ponds (Erskine, 1994; figure 8.1). Downstream reaches comprised a low-sinuosity compound channel with extensive benches (Erskine 1994). At this time, the system was longitudinally disconnected. The meandering chain of ponds acted as an effective buffer to downstream sediment transfer. In 1948 the largest flood on record triggered a headcut that transformed the upper reaches into a continuous channel with a trench-like morphology. This increased longitudinal connectivity through the system. Over time, a sediment slug generated by the liberation of large volumes of sediment from the upper reaches has infilled the channel in mid-catchment. This sediment slug is trapping sediment within the system, acting as a barrier to sediment conveyance. Hence, the Wollombi channel network could be considered to be re-disconnecting. The infilling of the channel and the reconstruction of benches means that channel-floodplain connectivity and inundation frequency of the floodplain has increased since 1948. These changes, and their effect on the (dis)connectivity dynamics of this system, are similar to those described for other sand bed systems in the region (e.g., Fryirs and Brierley 2001).

Socioeconomic change has also created a cycle of connection and disconnection and alteration in the mix and intensity of use-values (figure 8.1). The valley floor has been extensively cleared since European settlement, initially for wheat growing and subsequently for dairy farming (Parkes et al. 1979). Dairying has now given way to beef cattle grazing but more importantly to an influx of retirees and weekenders moving up from Sydney, one to two hours to the south, taking up rural subdivisions from long-time farming families (Mahony 1994). Wollombi is thus becoming a "bush change" or lifestyle sub-catchment. This has meant disconnection and connection by differing social groups, with consequent changes in dominant use-values and attitudes to conservation and management, particularly over the appropriate rehabilitation of the river following the impacts of the 1948 flood. While recovery from this event has been a slow process mainly due to limited sediment supply, channel narrowing is occurring and is aided by less intensive land use and increased replanting as full-time farming disappears from this part of the Hunter.

There is contention within and between local landcare groups over the desired target condition

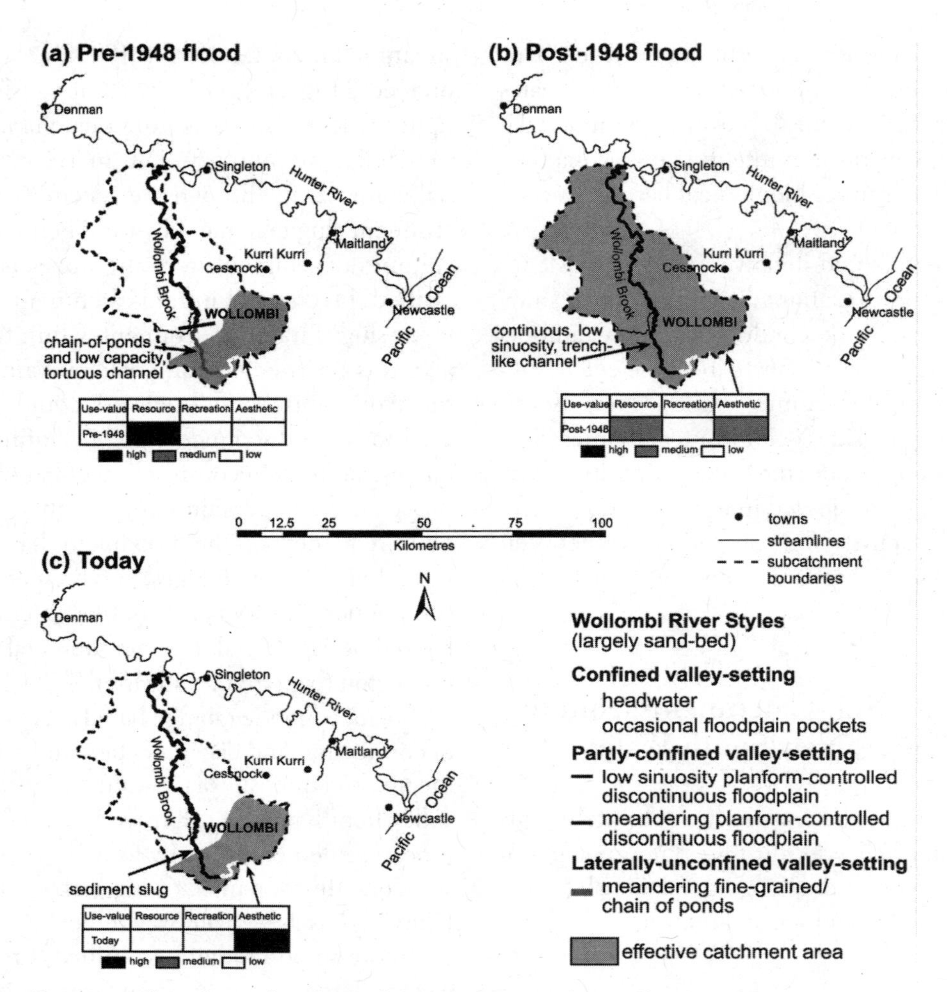

FIGURE 8.1. Patterns of biophysical and social connectivity in the Wollombi sub-catchment of the Hunter valley.

for river recovery. For example, concerns have been expressed over the extent to which in-channel vegetation should be allowed to grow, and over related loss of channel connectivity and potential for bank erosion. This "connectivity battle" is enhanced by overlapping institutional responsibility with different boundaries and scales (e.g., local council, state government agencies, Catchment Management Authority). In this instance, none of these agencies is situated within the Wollombi sub-catchment itself, making Wollombi something of an edge in management terms and providing a space in which contested visions for the river, and what is seen as natural and desirable, can be played out.

Upper Hunter Trunk Stream

The Upper Hunter sub-catchment illustrates the dynamic nature of biophysical and social (dis)connectivity driven by major dam construction and land use change (figures 8.2 and 8.3). Biophysically, the bedload sediment transfer regime (i.e., longitudinal connectivity) has become more disconnected since the largest flood on record in 1955 and the construction of Glenbawn Dam in 1967. Prior to 1955, most sediment supplied to the primary channel networks on the eastern side of the catchment (Hunter River, Pages River, Rouchel Brook, and Isis River) was conveyed to the Hunter River at Muswellbrook. At that time, the only sub-catchment that was entirely disconnected was Dart Brook, where a discontinuous channel and wide alluvial plain acted as a significant buffer. The 1955 flood caused significant transformation of river courses along the alluvial reaches of the Hunter River and lower Pages River. In the latter system, a large gravel sediment slug was formed. This effectively clogs the lower Pages River, acting as a barrier to sediment conveyance. In addition, the construction of Glenbawn Dam effectively disconnected a third of the upper Hunter catchment (Erskine et al. 1999), resulting in reduced sediment conveyance to the Hunter River at Muswellbrook (Fryirs et al. 2007b). The wide alluvial plain and competence-limited channel along lower Dart Brook still act as significant buffers and barriers to sediment conveyance from that sub-catchment. As a result of these physical changes to the river system, the linkage of biophysical processes in the upper Hunter catchment has become more disconnected over time.

Social disconnection, loss, and alienation in the Upper Hunter have also been substantial, driven by dramatic change in the regional economy and land and water use. In the last twenty-five years, the Upper Hunter has become one of the world's leading producers of black coal. The area generates 70 percent of New South Wales' electricity, using substantial volumes of water in the process (Hunter River Management Committee

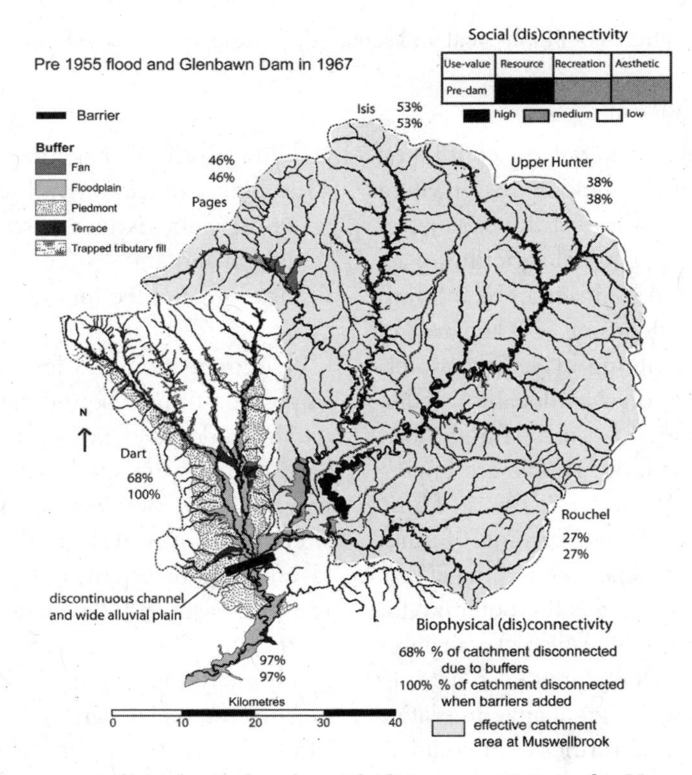

FIGURE 8.2. Pre-1955 patterns of biophysical and social (dis)connectivity in the Upper Hunter (based on Fryirs et al. 2007b)

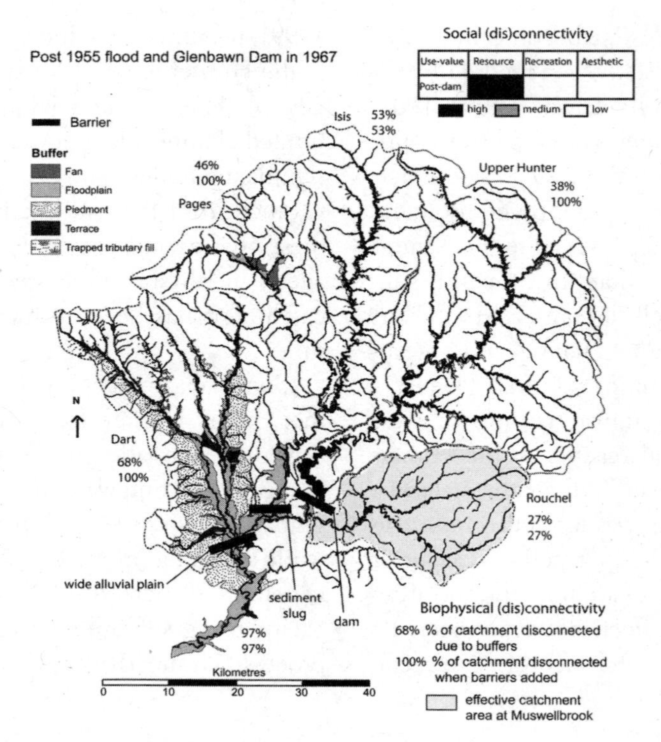

FIGURE 8.3. Post-1955 patterns of biophysical and social (dis)connectivity in the Upper Hunter (based on Fryirs et al. 2007b)

2002). Other important and potentially conflicting trends in land and water use include the growth of horse stud farms and viticulture at the expense of dairy farming and other traditional agricultural activities (Healthy Rivers Commission of New South Wales 2003). Irrigation licensing is a key area of conflict: "People are gambling by selling [water] allocations and relying on off-allocation [additional extractions permitted at times of high flow] but losing value on their farm. People are buying properties because of the water licenses. Those without licenses will have to stop farming" (former dairy farmer, Pages River, pers. comm. June 2003).

Given these changes, there is the potential for "sleeper" and "dozer" irrigation licenses (licenses granted but wholly or partially unused to date) to be traded and activated with significant cumulative impacts on river health and further commodification of water distinct from established use-values (Hunter River Management Committee 2002). This commodification and separation of land and water use-values is a major social boundary, institutional barrier, and source of social disconnection, dislocation, and feelings of injustice in the Upper Hunter.

Disconnection and loss of access through the imposition of new boundaries also occurred as riparian land was locked up for agriculture, as dam catchments were enclosed, and as mining and power generation companies acquired large stretches of river. As a result, many long-used and taken-for-granted parts of the river are now out of reach. The sheer dollar value of the new industries marginalizes those parts of the ecosystem (natural and human) that are not part of the core economy, creating financial barriers for river management institutions in terms of triple bottom line decisions. Those parts of the river that are accessible are often ignored, particularly in urban environments. This is exemplified by, but not restricted to, the town of Maitland along the Lower Hunter River, which literally turned its back on the river following the 1955 flood through the building of flood protection structures and the physical orientation of buildings (Bargwanna 1976). The result of this alienation has been a barrier in the form of decreased awareness of the multiple dimensions of river health and a reinforcement of the limited recognition of the range of interests in the health of the river. Many new boundaries and edges have being created in this way, as recreational use-values have been excluded, and as townspeople's knowledge of the river has decreased.

Interbasin Transfers: The Snowy Hydro Scheme

The Snowy River has become an Australian icon through Banjo Paterson's famous poem (and subsequent movie) "The Man from Snowy River." However, in terms of (dis)connectivity, its recent history is less illustrious. The construction of a massive hydroelectric power and irrigation scheme in the 1950s and 1960s reduced flows along the Snowy by 99 percent. Diversion of flows into the Murray-Darling system connected previously separate river basins (figure 8.4). Biophysically, disconnection along the Snowy itself resulted in a transformation from a river with reliable flow to a river characterized by a series of pools and sandbars, an altered flow regime, infestation of weeds such as willows, and increased salinity and algal blooms (Erskine et al. 1999; Pigram 2000). Hydrological changes to the recipient river, the Murrumbidgee, have modified flow regimes and thermal regimes, impacting upon floodplain flora and fauna (Jansen and Healey 2003).

The iconic significance of the Snowy Mountain Scheme extends well beyond the catchments directly affected, being tied closely to migration programs associated with economic and social revival following World War II. Such nation-building enterprises represent a major form of social connection in their own right. Irrigation communities such as Griffith and Renmark developed and thrived, although socially disconnected from the Snowy source water supply areas (and their disaffected communities) that fostered this economic development. In contrast, a distinct sense of social disconnection and loss of place-identity emerged from communities along the Snowy River with loss of recreational facilities, in particular swimming and fishing, and a decrease in tourism potential (Tennant-Wood 2006).

Belated recognition of the demise of a literary and environmental icon, and ongoing community pressure fueled by a clear vision of what the Snowy had been and could be again, led in 2002 to the adoption by the Victorian and New South Wales Governments of a staged agreement to return a mean of 28 percent of flows to the Snowy. However, transboundary tensions have returned to haunt the heirs of the Snowy project. Economically, the project connected farmers and other stakeholders in the Snowy catchment to the huge irrigation industry in the Murray and Murrumbidgee valleys. Within each catchment, patterns of (dis)connectivity are therefore contingent upon interbasin links. Policies to return water to the Snowy as environmental flows have thus needed to consider offsets, compensation, and provision of infrastructure for irrigators (Pigram 2000), while inquiries into such proposals typically receive hundreds of submissions from a wide range of groups in both basins. At times social boundaries break down and these interests coalesce. For example, recent attempts to privatize the Snowy Hydro Authority met with widespread public opposition and ultimately collapsed following withdrawal of the Commonwealth Government from the process. Recognition of the national historic significance of the Scheme, along with concerns for ecological values in both basins and economic well-being of irrigators, were crucial factors in these decisions. Such proposals have to consider a wide range of related costs and benefits across government jurisdictions and boundaries, including electricity supply, impacts on agriculture, recreation, and tourism, together with environmental and social issues.

(Dis)Connectivity: Themes for Integrative River Management

This section uses three themes to bring out comparisons and contrasts from the case studies with particular implications for river management. These themes are scale, naturalness, and change.

Scale: Measuring (Dis)Connectivity

Recognizing and articulating the various forms of biophysical and social (dis)connectivity is contingent upon a choice of scale for research, planning, and management. Nothing is obvious or given in this choice. While the catchment is now generally seen as *the* unit of river management, the case studies in this chapter demonstrate that patterns of (dis)connectivity occur in diverse and complex forms both at sub-catchment and supra-catchment scales. For instance, from a whole catchment scale, the Upper Hunter sub-catchment acts as a focal point of both biophysical and social (dis)connection that determines the connectivity and aquatic health of the Hunter Valley in its entirety. In

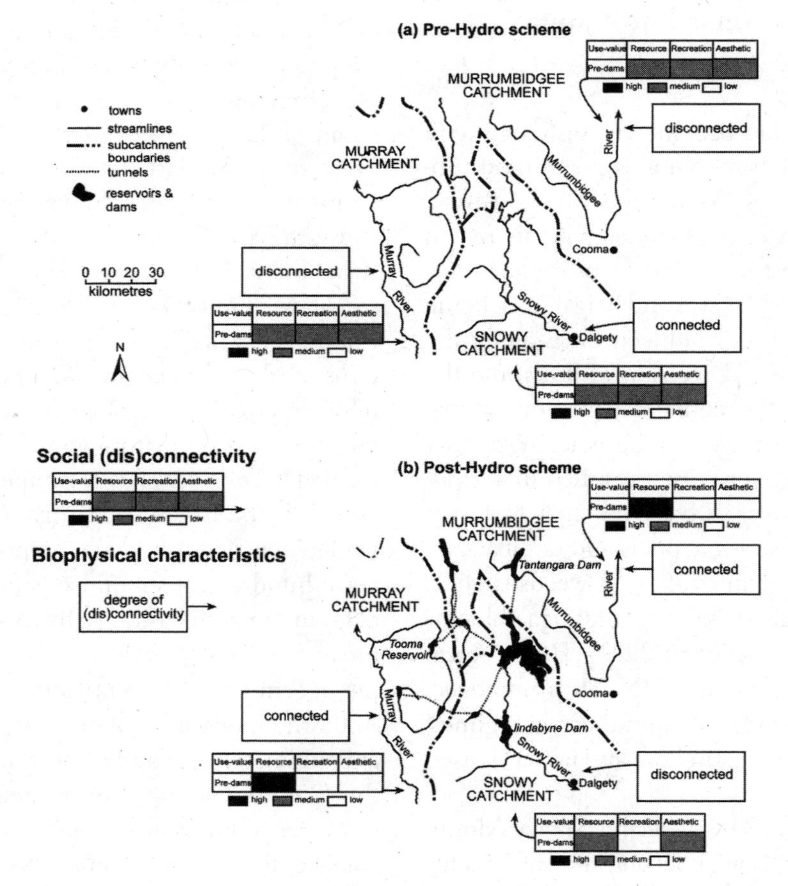

FIGURE 8.4. Pre- and post-Hydro scheme patterns of (dis)connectivity in the Snowy and Murrumbidgee catchments

contrast along Wollombi, another sub-catchment of the Hunter Valley, a slow physical recovery of the river through downstream disconnection and lateral reconnection is creating management tensions and perceptions of injustice between groups with long-standing or more recent social connections. Thus what happens at the sub-catchment scale in the formation of boundaries may not be reflected at larger spatial scales.

Temporal scale is also a critical factor in determining patterns of (dis)connectivity. For example, the scales at which physical reconnection, or in the case of Wollombi Brook, re-disconnection, may occur will generally not match those at which people think, act, plan, or relate to the river, and most certainly not in terms of political and electoral cycles. This is a key issue in terms of sustainability and integrative management. Physical changes or loss of access to a river over time may lead to social disconnection and alienation and to the creation of edge zones where marginalized groups either ignore the river or find new ways to reconnect. One important avenue for reconnection is through engagement in rehabilitation work itself. However, each case is different. For example, the Snowy demonstrates ongoing place-identity and attachment due to its strong cultural connections, despite massive biophysical disconnection, institutional barriers, and a decline in environmental health.

In light of these considerations, rehabilitation efforts may appear to be fruitless or go unrewarded, fostering feelings of frustration, injustice, and powerlessness. Institutional barriers also abound where local government and state agency boundaries cannot respond in a holistic manner or at the appropriate scale to this complexity. This is a major issue in each case study although for different reasons, reflecting multiple local interests, geographical marginalization, and cross-boundary issues. Scale is therefore a necessary lens through which to view (dis)connectivity. An authentic and applicable understanding of (dis)connectivity must be accompanied by an appraisal of multiple scales as nonlinear, socially constructed, and related if river management actions are to achieve just and sustainable outcomes (Howitt 1998; Brierley et al. 2006b).

Naturalness: Is Connectivity Good?

Earlier in this chapter, we challenged the assumption that connectivity per se is inherently good or healthy, pointing to examples of naturally disconnected systems and widely accepted cultural restrictions on access to rivers in indigenous societies. Equally, we have argued that the presence of artificial barriers and boundaries such as dams and weirs and social disconnection arising from enclosure of land and poor water quality may be an impediment to effective rehabilitation of damaged systems. This serves to emphasize the need for place-based management and understanding that the natural variability of rivers is a defining component and indicator of (dis)connectivity. Gibbs (2006, 79) puts forward the notion of "valuing variability as a new framework for valuing water" as a way of understanding both the diversity of values in water management and that the variability of water itself is a community value. However, it has proved difficult for most colonial societies to adjust to, or even to recognize, new patterns of variability (see chapter 14). In many cases, acclimatization societies established a form of "social connectivity" to the landscape by establishing alien patterns of connection from the homeland. This has had disastrous impacts on ecosystems and made establishing new social connections and place-identity in a new homeland very difficult.

Addressing this distorted sense of connection based upon imported knowledge and values requires an appropriate information base on how changes to (dis)connectivity influence the variability of systems. Within a given catchment such as the Hunter, differing natural forms of (dis)connectivity have evolved across sub-catchments. This means that the biophysical and social impact of floods, artificial barriers, and sediment movement have been very different in terms of disconnecting and reconnecting biophysical and social interactions. Other systems described in the Upper Hunter are radically different. This wide variety of biophysical forms of (dis)connectivity provides the spatial mosaic that is critical to catchment-scale biodiversity but also makes uniform management a misguided and damaging approach. This heterogeneity may well contrast significantly with the widely held community view of a natural and connected Snowy River. Reconnecting naturalness therefore means addressing the natural variability and connectivity of a system, since this is a key driver of river health. Attempts to derive a one-size-fits-all approach are ultimately doomed to failure. This is so whether the intent or effect is to disconnect rivers through downstream impoundment, lateral flood barriers, or channel-bed protection, or to reconnect through soft engineering, removal of engineering structures, or removal of sediment by dredging or flushing flows.

Connectivity and Change

An essential message from the case studies is that the forms and patterns of (dis)connectivity are inherently dynamic and nonlinear and that biophysical and social patterns are *related* but not necessarily positively *correlated*. On Wollombi Brook, changing social connections are altering the perception of physical (dis)connectivity along the river, while physical disconnection has itself created a new raft of stakeholder tensions. Social and biophysical changes are interacting to produce a new cycle of (dis)connectivity that poses major planning challenges and institutional barriers. The Upper Hunter example illustrates a strong pattern of social disconnection imposed upon a biophysical system that exhibits markedly varied patterns. Underpinning social understanding in all the Hunter systems was a historically long flood-free period in which old assumptions about who is connected to what have become fuzzy and tenuous, particularly in the light of demographic change, a new "river community" and loss of "river memory." Generational boundaries have been placed around what counts as useful knowledge given a failure to acknowledge the environmental history of natural variability, patterns of biophysical (dis)connectivity, and social change. Acknowledging this generational and epistemic boundary as a contemporary barrier to integrative river management is a crucial step.

An integrative approach recognizes that changing patterns of (dis)connectivity are key drivers of physical and cultural landscapes (Burel and Baudry 2005) and actively seeks to understand these patterns using a diversity of sources. This suggests strongly that an evolutionary and sociohistorical perspective is an important part of thinking about river health in these terms. "Thinking like an ecosystem" requires an understanding not only of the physical structure and function of the system itself, but also of how past practices have influenced both landscape evolution and changed human-nature interactions as river management and use-values change over time. Biophysical drivers provide a broad platform upon which human agents use land and water and within which institutional forms of planning and decision-making are established and reestablished. Baseline knowledge of biophysical limiting factors provides a basis for discussion and debate over notions of recovery. For instance, issues of sedimentary justice must be addressed in determining whether sediment release from upstream should be encouraged through manipulation of natural or human barriers to flow/sediment conveyance. The answer as to who benefits or loses from such forms of connectivity depends upon context, reflecting prevailing patterns of river condition and land use.

Synthesis: Sustainability, Health, Justice, and Policy in Addressing (Dis)Connectivity

River scientists, managers, and the broader society will continue to debate exactly what constitutes a healthy river (see chapters 1 and 7). In this chapter, we have explored the idea of (dis)connectivity as both an indicator of biophysical health and as part of a community's understanding of what it means to live in a healthy riverine ecosystem. One aim of this focus has been to promote a holistic notion of health in the sense of stewardship, or in Australian Aboriginal terms, of caring for country. From this position, the interdependence of human communities and environmental health is central rather than viewing the nonhuman world merely as a resource. This is crucial when, as is increasingly the case, an assessment of river health is used as a key tool for setting priorities. As Rapport (1998, 4) has argued, "The very concept of 'health' is a value-driven, mission-oriented notion." However, the examples and themes raised in this chapter also emphasize the need to look back to understand ecological baselines, changing understandings, and institutional lessons for current practice. Exploring changing patterns of (dis)connectivity is a strong catalyst for pursuing this goal, so that environmental history in its broadest sense should be a key policy tool for river managers.

A second reason for identifying holistic and connected notions of river health is to emphasize the importance of democratic and fair procedures for making judgments as to what is a healthy river and therefore what an inclusive community is prepared to accept. We have pointed to examples of how biophysical and social disconnection and reconnection can bring about the reality, or at the very least the perception, of injustice. Changing patterns of connectivity are selective in their impacts on ecosystems and communities, either by accident or design. Physical (dis)connectivity produces winners and losers along the channel as biophysical fluxes transport sediments and weeds that choke channels, reducing the potential for activities such as irrigation pumping or recreation. Resulting perceptions (and realities) of inequity can only reinforce the institutional problems that bedevil management. While river managers might formerly have seen this as not their problem, integrative approaches have to explicitly address issues of distributive and procedural justice. Even more broadly, an ecologically informed notion of environmental justice (Gleeson and Low 2003; Hillman 2006) implies the connection of social and environmental justice to sustainability and health. From our holistic perspective, none of these elements can exist independently. Addressing justice toward the environment, along with the distribution of costs and benefits and fair decision-making processes, entails a concern both for damage to the biophysical ecosystem and loss of place-identity.

Attempts to manipulate (dis)connectivity and downstream or floodplain sediment supply may be seen in some quarters as little more than an extension of the physical and social engineering paradigm. However, such manipulation is ongoing. Rather than viewing this purely as a technical challenge, the multiple interests and use-values involved must be heard. Otherwise, any claim to the establishment of a new era in river management will ultimately be seen as hollow, superficial, and nonsustainable. A connected view provides, and indeed requires, an opportunity for these interests to be recognized and restored. Negotiating the science-management-community link is crucial to practical success in these areas. Connected science must frame its research questions and methods with community values and management needs in mind (see chapter 3). This does not mean "soft science" in terms of its rigor, but rather recognition of its social context and responsibilities, and at times its incapacity to provide certain solutions. Managers likewise will recognize that working across temporal and spatial scales is central to the task of enhancing and maintaining participatory bottom-up approaches to rehabilitation. A robust approach to sustainability in river management will incorporate these concepts into a holistic approach to river health along with social and environmental justice.

Finally, it is important to consider how the rehabilitation and management process itself can act as a means of reconnecting communities with their landscapes. To date, rehabilitation has generally focused solely on biophysical condition, where restoration to a previous condition misses the essential importance of social connection and the understanding that restoration will not produce the same river. This suggests that the process of rehabilitation must incorporate a link between condition and connection, drawing both on scientific knowledge of river condition and on local knowledge of, and feelings toward, *particular* rivers or sections of rivers. Rehabilitation and management activities provide a tool for such a dialogue and engagement to occur, promoting both ecological and cultural regeneration (Higgs 2003). That is not to hold up place-based values and local knowledge as the Holy Grail since impartial and objective rigor is still required to generate good science. However, connection to place ensures a meaningful basis with which to promote on-the-ground engagement with and ownership of such science. This provides a sense of rehabilitation as conversation, prospectively reconnecting people to place, thereby invigorating the process of river repair.

Prospectively, this is a key component in the process of healing our relationship to the natural world—an integral step in the era of river repair—and in re-engendering a sense of identity and distinctiveness that is part of our relationship to place. This entails reciprocity between restorationists and ecosystems, and among researchers, managers, and community. What is needed is a form of "situated engagement" (Suchet 2002), a call to create spaces, or to identify edges, within which singular and distorted views of the natural environment and the boundaries and barriers between nature and culture can be challenged. Ecosystems cannot speak in a conventional sense, but engagement with the specific needs of ecosystems allows restorationists to represent those interests. Ideally, this input reflects understanding of system dynamics, the natural range of variability in both structural and functional terms, and appreciation of the trajectory of system evolution.

There are many places to look for such forms of engagement and dialogue, both in indigenous societies and in emerging forms of river management around the globe. Some of these are canvased in the chapters that follow. These developments involve the promotion and sustaining of stewardship and caring for country, and recognition of the spiritual significance of water, in whatever way that connection is framed across cultures. Inclusion and diversity must form a core component of river management if authentic and productive forms of (dis)connectivity are to be developed and sustained.

Conclusion

This chapter has identified biophysical and social forms of connection and disconnection and pointed to environmental and institutional mechanisms that promote or restrict connectivity. Case studies have emphasized the complexity of relationships between social and biophysical connections, in particular their scaled, variable, and dynamic characteristics. It is only through the integration of these biophysical and social dimensions of (dis)connectivity that managers can develop policy beyond narrow technocentric attempts to restore biophysical components of river systems. For example, while removing physical barriers *may* be a crucial step forward in rehabilitation, in the absence of a catchment-scale understanding of patterns of biophysical and social (dis)connectivity, such actions may actually represent a backward step. Recognizing this difference is of necessity a value-based task, but one framed within an understanding of biophysical linkages in any given landscape. A focus on questions of sustainability, in particular elements of health and justice is, we argue, essential for this task.

Acknowledgments

We would like to thank Brad Coombes and Andrew Boulton for constructive review comments on this manuscript. Gary J. Brierley and Kirstie A. Fryirs acknowledge financial support from the Australian Research Council.

References

Albrecht, G. 2005. Solastalgia: A new concept in human health and identity. *PAN: Philosophy Activism Nature* 3:41–55.

Ashley-Brown, P. 2003. *River stories: Exploring the Hunter River and its people*. Newcastle, Australia: Australian Broadcasting Corporation.

Bargwanna, S. 1976. "The Hunter: Use or abuse." *Maitland Mercury*, February 5.

Bennett, J. W., and S. M. Whitten. 2002. *The private and social values of wetlands: An overview*. Canberra, Australia: Land and Water Australia.

Brierley, G. J., and K. A. Fryirs. 2005. *Geomorphology and river management: Applications of the River Styles framework*. Oxford, UK: Blackwell.

Brierley, G. J., K. Fryirs, and V. Jain. 2006a. Landscape connectivity: The geographic basis of geomorphic applications. *Area* 38:165–74.

Brierley, G. J., M. Hillman, and K. Fryirs. 2006b. Knowing your place: An Australasian perspective on catchment-framed approaches to river repair. *Australian Geographer* 37:131–45.

Brunkhorst, D., and I. Reeve. 2006. A geography of place: Principles and application for defining "eco-civic" resource governance regions. *Australian Geographer* 37:147–66.

Bunn, S. E., M. C. Thoms, S. K. Hamilton, and S. J. Capon. 2006. Flow variability in dryland rivers: Boom, bust and the bits in between. *River Research and Applications* 22:179–86.

Burel, F., and J. Baudry 2005. Habitat quality and connectivity in agricultural landscapes: The role of land use systems at various scales in time. *Ecological Indicators* 5:305–13.

Cals, M. J. R., and C. van Drimmelen. 2000. *Space for the river in coherence with landscape planning in the Rhine-Meuse Delta: River restoration in Europe: Practical approaches*. Wageningen, The Netherlands: European Centre for River Restoration.

Connor, L., G. Albrecht, N. Higginbotham, S. Freeman, and W. Smith. 2004. Environmental change and human health in Upper Hunter communities of New South Wales, Australia. *EcoHealth* 1(Supplement 2). New York, NY: Springer. DOI 10.1007/s10393-004-0053-2.

Coombes, B. 2003. The historicity of institutional trust and the alienation of Maori land for catchment control at Mangatu, New Zealand. *Environment and History* 9:333–59.

Costanza, R., and M. Mageau. 1999. What is a healthy ecosystem? *Aquatic Ecology* 33:105–15.

Dovers, S. 2001. Institutions for sustainability. *Tela: Environment, Economy and Society* 7. Carlton, Australia: Australian Conservation Foundation.

Dirlik, A. 2001. Place-based imagination: globalism and the politics of place. In *Places and politics in an age of globalization*, ed. R. Prazniak and A. Dirlik, 15–51. Lanham, MD: Rowman & Littlefield.

Dunn, H. 2004. Defining the ecological values of rivers: The views of Australian river scientists and managers. *Aquatic Conservation: Marine and Freshwater Ecosystems* 14:413–33.

Erskine, W. D. 1994. Flood-driven channel changes on Wollombi Brook, NSW since European settlement. In *The way of the river: Environmental perspectives on the Wollombi*, ed. D. Mahoney and J. Whitehead, 41–69. Wollombi, Australia: Wollombi Valley Landcare Group.

Erskine, W. D., L. M. Turner, N. Terrazzolo, and R. F. Warner. 1999. Recovery of the Snowy River: Politics and rehabilitation. *Australian Geographical Studies* 37:330–36.

Evenden, M. 2006. Precarious foundations: Irrigation, environment, and social change in the Canadian Pacific Railway's Eastern Section, 1900–1930. *Journal of Historical Geography* 32:74–95.

Fryirs, K., and G. J. Brierley. 2001. Variability in sediment delivery and storage along river courses in Bega catchment, NSW, Australia: Implications for geomorphic river recovery. *Geomorphology* 38:237–65.

Fryirs, K., G. J. Brierley, N. J. Preston, and M. Kasai. 2007a. Buffers, barriers and blankets: The (dis)connectivity of catchment-scale sediment cascades. *Catena* 70:49–67.

Fryirs, K., G. J. Brierley, N. J. Preston, and J. Spencer. 2007b. Landscape (dis)connectivity in the upper Hunter catchment, New South Wales, Australia. *Geomorphology* 84:297–316.

Gandy, M. 2006. Riparian anomie: Reflections on the Los Angeles River. *Landscape Research* 31:135–45.

Gibbs, L. M. 2006. Valuing water: Variability and the Lake Eyre Basin, central Australia. *Australian Geographer* 37:73–85.

Gleeson, B., and N. Low. 2003. Environmental justice. In *A companion to political geography*, ed. J. A. Agnew, K. Mitchell, and G. Toal, 455–69. Malden, MA: Blackwell.

Harris, E. 2006. Development and damage: Water and landscape evolution in Victoria, Australia. *Landscape Research* 31:169–81.

Healthy Rivers Commission of New South Wales. 2003. *A decision support process for rapid assessment of catchment goals and priorities*. Sydney, Australia: Health Rivers Commission.

Higgs, E. 2003. *Nature by design: People, natural process and ecological restoration*. Cambridge, MA: MIT Press.

Hillier, J. 2003. Fighting over the forests: Environmental conflict and decision-making capacity in forest planning processes. *Australian Geographical Studies* 41: 251–69.

Hillman, M. 2005. Justice in river management: Community perceptions from the Hunter Valley, New South Wales, Australia. *Geographical Research* 43:152–61.

Hillman, M. 2006. Situated justice in environmental decision-making: Lessons from river management in Southeastern Australia. *Geoforum* 37:695–707.

Hillman, M., G. Aplin, and G. J. Brierley. 2003. The importance of process in ecosystem management: lessons from the Lachlan Catchment, New South Wales, Australia. *Journal of Environmental Planning and Management* 46:219–37.

Holling, C. S., and G. K. Meffe. 1996. Command and control and the pathology of natural resource management. *Conservation Biology* 10:328–37.

Howitt, R. 1998. Scale as relation: Musical metaphors of geographical scale. *Area* 30:49–58.

Howitt, R. 2001a. Frontiers, borders, edges: Liminal challenges to the hegemony of exclusion. *Australian Geographical Studies* 39:233–45.

Howitt, R. 2001b. *Rethinking resource management: Justice, sustainability and indigenous peoples*. London, UK: Routledge.

Hunter River Management Committee. 2002. *Draft water sharing plan for the regulated Hunter River and Glennies Creek Water Source and Paterson River Water Source*. Sydney, Australia: Department of Land and Water Conservation.

Jackson, S. 2006. Compartmentalising culture: The articulation and consideration of indigenous values in water resource management. *Australian Geographer* 37:19–31.

James, D. 2006. Re-sourcing the sacredness of water. In *Fluid bonds: Views on gender and water*, ed. K. Lahiri-Dutt, 85–104. Kolkata, India: STREE, National Institute for Environment and Australian National University.

Jansen, A., and M. Healey. 2003. Frog communities and wetland condition: Relationships with grazing by domestic livestock along an Australian floodplain river. *Biological Conservation* 109:207–19.

Jennings, S. F., and S. A. Moore. 2000. The rhetoric behind regionalization in Australian natural resource management: Myth, reality and moving forward. *Journal of Environmental Policy & Planning* 2:177–91.

Johnson, C., J. M. Bowker, and H. K Cordell. 2004. Wilderness values in America: Does immigrant status or ethnicity matter? *Society and Natural Resources* 17:611–28.

Jungwirth, M., S. Muhar, and S. Schmutz. 2002. Re-establishing and assessing ecological integrity in riverine landscapes. *Freshwater Biology* 47:867–87.

Kauffman, G. J. 2002. What if . . . the United States of America were based on watersheds? *Water Policy* 4:57–68.

King, H. W., and E. R. Woolmington.1960. The role of the river in the development of settlement in the Lower Hunter Valley. *Australian Geographer* 8:3–16.

Kondolf, G. M. 2006. River restoration and meanders. *Ecology and Society* 11(2): Article 42. http://www.ecologyandsociety.org/vol11/iss2/art42 (accessed December 2, 2007).

Kondolf, G. M., A. J. Boulton, S. J. O'Daniel, G. C Poole, F. J. Rahel, E. H. Stanley, E. Wohl, Å. Bång, J. Carlstrom, C. Cristoni, H. Hube, S. Koljonen, P. Louhi, and K. Nakamura. 2006. Process-based ecological river restoration: Visualizing three-dimensional connectivity and dynamic vectors to recover lost linkages. *Ecology and Society* 11(2): Article 5. http://www.ecologyandsociety.org/vol11/iss2/art5/ (accessed December 2, 2007).

Lachapelle, P. R., S. F. McCool, and M. A. Patterson. 2003. Barriers to effective natural resource planning in a "messy" world. *Society and Natural Resources* 16:473–90.

Lake, P. S. 2001. On the maturing of restoration: Linking ecological research and restoration. *Ecological Management and Restoration* 2:110–15.

Lane, S. N., and K. S. Richards. 1997. Linking river channel form and process: Time, space and causality revisited. *Earth Surface Processes and Landforms* 22:249–60.

Mahony, D. 1994. Changing patterns of landuse and decision-making in the Wollombi Valley. In *The way of the river: Environmental perspectives on the Wollombi*, ed. D. Mahony and J. Whitehead, 121–32. Wollombi, New South Wales: Wollombi Valley Landcare Group and University of Newcastle Department of Community Programmes.

Massey, D. 2005. *For space*. London, UK: Sage.

Mejia, A. 2004. The problem of knowledge imposition: Paulo Freire and critical systems thinking. *Systems Research and Behavioral Science* 21:63–82.

Michaelidou, M., D. J. Decker, and J. P. Lassoie. 2002. The interdependence of ecosystem and community viability: A theoretical framework to guide research and application. *Society and Natural Resources* 15:599–616.

Monteith, N. H. 1952. The contribution of historical factors to the present erosion condition of the Upper Hunter River. *Journal of the Soil Conservation Service of New South Wales* 9:85–91.

Palmer, L. 2006. "Nature," place and the recognition of indigenous polities. *Australian Geographer* 37:33–43.

Parkes, M., and R. Panelli. 2001. Integrating catchment ecosystems and community health: The value of participatory action research. *Ecosystem Health* 7:85–106.

Parkes, W. S., J. Comerford, and M. Lake. 1979. *Mines, wines and people: A history of Greater Cessnock*. Cessnock, Australia: Cessnock City Council.

Peterson, R. B. 2006. Why Mami Wata matter: Local considerations for sustainable waterpower development policy in Central Africa. *Local Environment* 11:109–25.

Pigram, J. J. 2000. Options for rehabilitation of Australia's Snowy River: An economic perspective. *Regulated Rivers: Research and Management* 16:363–73.

Postel, S., and B. Richter. 2003. *Rivers for life*. Washington, DC: Island Press.

Powell, J. M. 2000. Snakes and cannons: Water management and the geographical imagination in Australia. In *Environmental history and policy: Still settling Australia*, ed. S. Dovers, 47–73. South Melbourne, Australia: Oxford University Press.

Rapport, D. 1998. Need for a new paradigm. In *Ecosystem health*, ed. D. Rapport, R. Costanza, P.R. Epstein, C. Gaudet, and R. Levins, 3–17. Malden, MA: Blackwell Science.

Rhodes, H. M., L. S. Leland, and B. E. Niven. 2002. Farmers, streams, information, and money: Does informing farmers about riparian management have any effect? *Environmental Management* 30:665–77.

Roe, M. 2006. "Making a wish": Children and the local landscape. *Local Environment* 11:163–81.

Rose, D. B. 1999. Indigenous ecologies and an ethic of connection. In *Global ethics and the environment*, ed. N. Low, 175–87. London, UK: Routledge.

Sheldon, F., A. J. Boulton, and J. T. Puckridge. 2002. Conservation value of variable connectivity: Aquatic invertebrate assemblages of channel and floodplain habitats of a Central Australian Arid-Zone River, Cooper Creek. *Biological Conservation* 103:13–31.

Sheldon, F., and M. C. Thoms. 2006. Relationships between flow variability and macroinvertebrate assemblage composition: Data from four Australian dryland rivers. *River Research and Applications* 22:219–38.

Smolyak, A. V. 2001. Traditional principles of natural resource use among indigenous peoples of the Lower Amur River. *Journal of Legal Pluralism* NR 46:173–83.

Suchet, S. 2002. Totally wild? Colonising discourses, indigenous knowledges and managing wildlife. *Australian Geographer* 33:141–57.

Tennant-Wood, R. 2006. Silent partners: The fluid relationship between women and dammed rives in the Snowy Region of Australia. In *Fluid bonds: Views on gender and water*, ed. K. Lahiri-Dutt, 317–34. Kolkata, India: STREE, National Institute for Environment and Australian National University.

Thompson, D., and S. Pepperdine. 2003. *Assessing community capacity for riparian restoration*. Canberra, Australia: Land and Water Australia.

Tisdell, J. G. 2003. Equity and social justice in water doctrines. *Social Justice Research* 16:401–16.

Townsend, C., G. Tipa, L. Teirney, and D. Niyogi. 2004. Development of a tool to facilitate participation of Maori in the management of stream and river health. *EcoHealth* 1:184–95.

Victorian Government. 2002. *Healthy rivers, healthy communities and regional growth—Draft Victorian river health strategy: Detailed summary*. Melbourne, Australia: Victorian Government.

Walmsley, J. J. 2002. Framework for measuring sustainable development in catchment systems. *Environmental Management* 29:195–206.

Ward, E. R. 2001. Geo-environmental disconnection and the Colorado River Delta: Technology, culture, and the political ecology of paradise. *Environment and History* 7:219–46.

Ward, J. V. 1989. The four-dimensional nature of lotic ecosystems. *Journal of the North American Benthological Society* 8:2–8.

Webster, B., and S. Mullins. 2003. Nature, progress and the "disorderly" Fitzroy: The vain quest for Queensland's "noblest navigable river," 1865–1965. *Environment and History* 9:275–99.

Wester-Herber, M. 2004. Underlying concerns in land-use conflicts: The role of place-identity in risk perception. *Environmental Science and Policy* 7:109–16.

Wohl, E. 2004. *Disconnected rivers: Linking rivers to landscapes*. New Haven, CT: Yale University Press.

PART III

International Perspectives on the Process of River Repair

> The responsibility of scientists now lies in practicing integrated science, for our society needs integrated knowledge. From this integration should flow a sense of responsibility, including a responsibility to communicate to nonscientists the extent of destruction of the natural world and the means to mitigate or reverse this destruction.
>
> Wohl 2004, 264

The emerging approach to river repair takes different forms in different parts of the world. However, common trends include an increasing awareness and inclusion of ecosystem values in river management and increasing use of crossdisciplinary river science in guiding this process. Differences reflect the socioeconomic and cultural values of any given society, governance structure, the historical imprint of human disturbance, and past management practices in different biophysical settings.

The international case study chapters presented here reflect perspectives and approaches to river management from Australia, the United States, Europe, Japan, and South Africa. Each chapter frames the condition of rivers in their historical context (in terms of biophysical setting and human disturbance), river management frameworks, and the priorities that drive the process of river repair. Examples of the use of integrative river science are presented. This sets the context for discussion of challenges faced and ways forward in the use of integrative river science in management.

Chapter 9 presents the river management perspective adopted in Australia. This is set within the context of a large continent with low population density and a recent history (around 230 years) of colonial settlement. The chapter outlines the emergence of a participatory approach to river management and reviews crossdisciplinary and transdisciplinary applications. Future challenges in the integration of science, integration within and between institutions, and the integration of biophysical and social sciences in the process of river repair are highlighted.

River management experiences in the United States are documented in chapter 10, highlighting steps taken to address the question posed by Palmer and Allen (2006, 42): "How did the United States reach a point where the majority of our rivers are degraded and ecologically dysfunctional?" The chapter describes a top-down regulatory approach to management. This approach is set within the context of contemporary pressures and social constraints on water resources, and issues that will likely continue to influence river management in the future. Various case studies demonstrate the successful use of river science in management and the value of undertaking large-scale, well-resourced rehabilitation projects.

Chapter 11 highlights the use of integrative river science in management of rivers in Europe, set within the context of the European Water Framework Directive. Challenges faced in delivering consistent, coherent, and useful science within the directive are presented. Case studies from Austria, France, and the United Kingdom demonstrate how ecosystem-based approaches to river rehabilitation now drive the process of river repair across Europe.

Chapter 12 reviews the river management experience in Japan. Ecosystem values are increasingly incorporated into strategies that must address concerns for sediment disasters in a highly regulated and engineering-based approach to river management. The example of Sabo dam construction and modification demonstrates how a focus on infrastructure protection, through ameliorating sediment transfer through catchments in this highly active landscape, can be supplemented by inclusion of ecosystem-based river rehabilitation designs.

Chapter 13 reviews river management experiences in South Africa. Recent political transformation in this country has been accompanied by the development of new policies, legislation, and water laws that incorporate an ecosystem-based approach to management.

Perspectives highlighted in Part III emphasize the diverse array of challenges, pressures, and uncertainties faced in managing the process of river repair, prompting moves toward the adoption of adaptive management principles as outlined in Part IV.

Reference

Palmer, M. A., and J. D. Allan. 2006. Restoring rivers. *Issues in Science and Technology* Winter:40–48.

Wohl, E. 2004. *Disconnected rivers: Linking rivers to landscapes*. New Haven, CT: Yale University Press.

Chapter 9

The Australian River Management Experience

Kirstie A. Fryirs, Bruce Chessman, Mick Hillman, David Outhet, and Alexandra Spink

> The need to establish comprehensive and representative freshwater protected areas is urgent, given increasing concerns about limited water availability for Australia's cities, industries, and agriculture—and the ongoing degradation of aquatic ecosystems. This should be accompanied by effective land and water management that pays more than lip service to the environmental requirements of aquatic ecosystems.
>
> Kingsford and Nevill 2005, 162

Intensive river management in Australia is a relatively recent phenomenon. While Australia has been occupied by humans for at least 60,000 years, agricultural and urban development involving significant intervention in river flows and morphology did not begin until after the first permanent settlement of Europeans in 1788. It is within the subsequent 220-year period, and in particular the twentieth and twenty-first centuries, that this chapter focuses.

Four themes are examined here. First, the biophysical characteristics and state of Australian rivers are outlined. This state determines what is biophysically achievable in river conservation and management over a fifty- to one-hundred-year timeframe. Second, the governmental and institutional context within which river management occurs is summarized. This represents the institutional capacity to enact river repair. Third, the social context is considered in terms of community will to participate in and maintain and monitor the process of river repair. The final part of the chapter outlines issues associated with institutional and social integration, scientific integration, and the use of science in river management practice in Australia.

Setting the Scene: The Australian Landscape and Historical Setting

The physical and ecological characteristics of Australian rivers, and how they have changed since European settlement, provide an important context for discussing biophysical, organizational, and social aspects of river management in Australia. It is to these former themes that we now turn our attention.

Physical and Ecological Characteristics of Australia and Its Rivers

Australia is the world's driest inhabited continent. Its population is only 20 million, of whom more than 70 percent live in urban centers along the coast. The mean altitude of Australia is 330 meters above sea level and 40 percent of the continent lies below 200 meters (Jennings and Mabbutt 1967; Rutherfurd and Gippel 2001). Because of its size and latitudinal extent, Australia has a wide variety of hydro-climatic regions. The tropical north has a monsoon climate with dry winters, wet summers, and annual floods. Central Australia is arid with rivers that flow only after episodic heavy rainfall. Southwestern and southeastern areas have a Mediterranean climate and are wettest in winter and spring and driest in summer and autumn. Eastern Australia has a humid temperate climate (semi-arid inland) and a more even seasonal distribution of rainfall. The non-monsoon regions of Australia are most affected by extended periods of drought caused by irregular El Niño events. Much of the eastern seaboard is dominated by bedrock-controlled rivers with very short coastal plains, the arid interior and tropical settings are characterized by anastomosing river networks, and a range of discontinuous watercourses occurs in low-slope and/or low-energy settings (figure 9.1).

Australian rivers are noted for their high interannual variability in flow, extreme flood behavior, low average runoff per unit area of catchment, and low proportion of base flow (Finlayson and McMahon 1988; McMahon et al. 1991; Poff et al. 2006). They typically have low sediment delivery ratios because of the highly denuded landscape and low rates of rejuvenation (only a small portion of the continent was glaciated in the Pleistocene) (Olive and Rieger 1986). Offshore sediment contributions are trivial at the global scale as sediment is stored in-catchment for extended periods of time (i.e., tens of thousands of years) (Wasson 1994; Fryirs et al. 2007a, 2007b). Sediment transport is often intermittent and geomorphic river change is episodic and unpredictable, with long periods of relative inactivity punctuated by short phases of localized, event-driven activity. Sediment exhaustion is common in highly disturbed catchments (e.g., Fryirs and Brierley 2001; Brooks and Brierley 2004). As a result, sediment availability and supply often limit prospects for geomorphic recovery following anthropogenic disturbance. These antecedent conditions and limiting factors are key influences on what is achievable in river rehabilitation and management practice from a geomorphic or physical perspective.

The biological character of Australian rivers is influenced by the long geographical isolation of the continent and evolutionary adaptation to increasing aridity since the Miocene. Isolation has resulted in a high degree of endemism of the aquatic fauna and to a lesser extent the flora (Crisp et al. 2001; Unmack 2001) (figure 9.2). The aquatic macroinvertebrate fauna is rich by world standards, but the fish fauna is impoverished (Lake 1995). Much of the Australian aquatic biota appears to be well adapted to cope with climatic variability and uncertainty. For example, as many macroinvertebrate species are able to live in both perennial and nonperennial streams, nonperennial streams have a rich invertebrate fauna (Lake 1995). Many Australian aquatic species are also tolerant of elevated salinity (Nielsen et al. 2003). However, freshwater fish fauna do not seem especially adapted to aridity (Lake 1995), with very few species capable of aestivation.

River Metamorphosis and River Management since European Settlement

After European settlement in 1788, intensive agriculture spread rapidly, with many regional areas in southern Australia opened up by the 1820s. The clearing of woody vegetation from riparian zones and floodplains was a priority of pioneer settlers, as these were the most fertile and well-watered lands. Initial desnagging efforts reduced the volume of woody debris in rivers to improve navigation; later programs were applied for flood mitigation or even to remove perceived barriers to fish migration (Gippel et al. 1994; Brooks et al. 2003). These combined activities induced rapid river metamor-

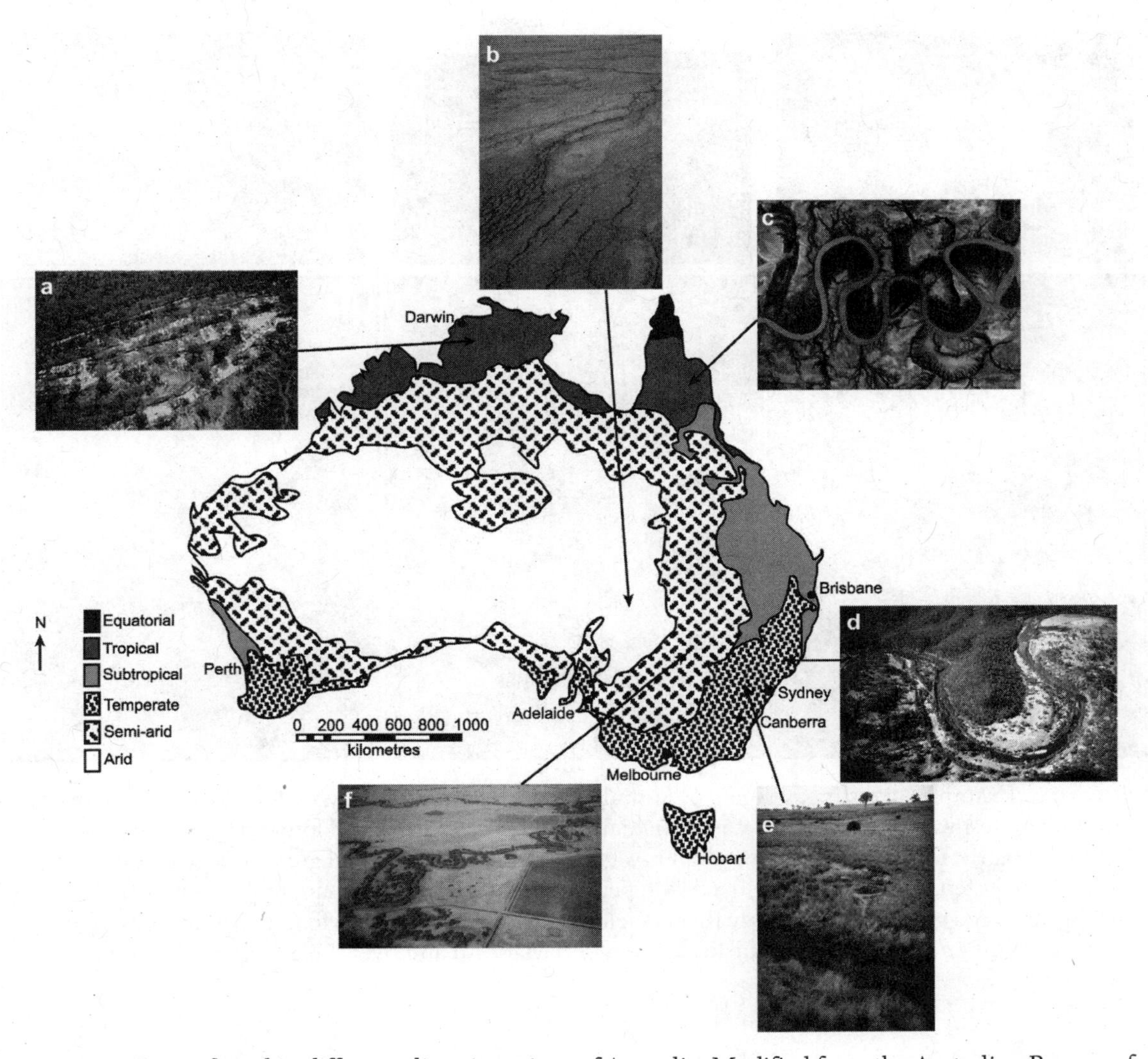

FIGURE 9.1. Rivers found in different climatic regions of Australia. Modified from the Australian Bureau of Meteorology, Köppen classification. Types of rivers include: (a) Anabranching river in a tropical setting (source: Gerald Nanson), (b) Anastomosing river in an arid setting (source: Gerald Nanson), (c) Meandering river in a tropical setting (source: http://www.wilderness.org.au), (d) Partly confined valley with bedrock-controlled discontinuous floodplain river in a coastal, temperate setting (source: Gary J. Brierley), (e) Chain-of-ponds river in a highland, temperate setting (source: Gary J. Brierley), (f) Meandering river in a semi-arid setting (source: Kirstie A. Fryirs).

phosis within the first generation after colonization (Brierley et al. 2005; Rutherfurd 2000). In addition, post-European land use and associated water resources development have resulted in widespread nutrient enrichment of Australian rivers and alteration of flow, sediment, and solute regimes, and consequent changes in biological communities (Young et al. 1996; Harris 2001; Prosser et al. 2001; Barmuta 2003; Brodie and Mitchell 2005).

FIGURE 9.2. Examples of distinctive biota of Australian river systems. (a) Platypus found extensively in eastern Australian freshwaters (source: www.australianfauna.com). (b) Murray cod found through most of the Murray-Darling river system (source: www.dpi.nsw.gov.au). (c) An endangered green and golden bell frog found in southeastern Australia (reproduced with permission from Pavel German, http://www.australiannature.com). (d) River red gums on the Murray River, Victoria (source: Kirstie A. Fryirs). (e) Morton Bay fig tree, Orara River, northern NSW (source: Kirstie A. Fryirs). (f) Mat-rush and river she oak, Orara River, northern NSW (source: Kirstie A. Fryirs).

River management activities in Australia began with modifications by aboriginal peoples, such as the arrangement of boulders to create fish traps and weirs (Bandler 1995). In the early post-European era, the focus was mainly on navigation via dredging, the construction of weirs, and the removal of rock bars and wood. The need for reliable water supplies for urban centers, and subsequently for irrigation industries, led to an extensive program of dam construction and other structural works, which culminated in large-scale engineering programs for hydropower, irrigation, and flood mitigation (e.g., the Snowy Mountains Scheme; Outhet et al. 1999). Most of the large rivers in southern Australia now have some form of regulated flow. In the 1950s and 1960s, the command-and-control paradigm dominated, resulting in extensive channelization and stabilization of channels through installation of a range of engineered bank-control structures. Flood mitigation resulted in extensive removal of wood from channels and the construction of levees. There has been extensive removal of vegetation from riparian zones. Subsequently, willows and other exotic vegetation have been planted along many river courses. Gravel extraction also occurred on a relatively large scale, and continues in some regions.

Demands for the conservation and rehabilitation of riverine ecosystems from the 1970s onward reflect an increased scientific understanding of river ecosystem dynamics, greater community awareness of environmental issues, and corresponding developments in public policy (Allan and Lovett 1997). This paradigm shift was associated with changing institutional and social arrangements that, at least in theory, are better placed to deal with more complex and potentially conflicting objectives. These new arrangements include increasing effort to involve groups of landholders and other volunteers in collective management of natural resources, multi-institutional partnerships, and the emergence of regional bodies (e.g., catchment management authorities) that are progressively assuming responsibilities formerly held by state government departments or single-purpose local bodies (e.g., river improvement trusts). The federal government has also assumed a greater role through direct funding to regional bodies and by fostering intergovernmental agreements and national guidelines and principles for the management of natural resources.

Biophysical Themes in Australian River Management Practice: What Is Achievable?

With expanding recognition that river management decisions must consider biophysical, economic, and social processes at a range of spatial scales and timeframes, river managers have become particularly challenged by information needs for decision-making (McCool and Guthrie 2001). As management is now moving toward an ecosystem-based approach, reliance on science to provide information about biophysical and societal interactions has increased (chapter 1). Unfortunately, in many places, baseline information on the character, behavior, evolution, condition, biophysical linkages, and recovery potential of rivers is lacking. Elsewhere, baseline information is poorly organized and the timeframes for planning and on-the-ground implementation are unrealistically short. Understanding of problems, prioritizing of activities, and determination of good management practices often require detailed analyses and ongoing research and development.

National research and development programs have been established in response to these needs. Examples include the river health and riparian lands programs of Land and Water Australia (www.lwa.gov.au) and frameworks for assessing the character and condition of rivers (e.g., River Styles [www.riverstyles.com] and the Index of Stream Condition [DSE 2006]). A National Land and Water Resources Audit (NLWRA 2001) has been completed (www.nlwra.gov.au) along with various State of the Environment reports (www.environment.gov.au/soe/index.html). Although coarse-grained in its coverage and constrained by the scarcity of nationally consistent data, NLWRA (2001) provided a first-level snapshot of river condition and a preliminary basis upon which to target regions of concern, make conservation/rehabilitation decisions and monitor improvement or deterioration over time.

Nationwide assessments have confirmed that the southern, nontropical river systems have been extensively degraded since European settlement (figure 9.3). For example, 96 percent of stream length in Victoria has been modified or disturbed in some way (Department of the Environment and Heritage 2006). Major river systems such as the Murray-Darling are in a state of ecological crisis and their biological communities are in decline (e.g., floodplain eucalypts, colonially nesting waterbirds, and some fish species) (Kingsford and Nevill 2005). The most common river problems resulting from extensive human-induced degradation and modification are outlined in table 9.1. Many of these problems have major implications for infrastructure and human health, in addition to their environmental impact.

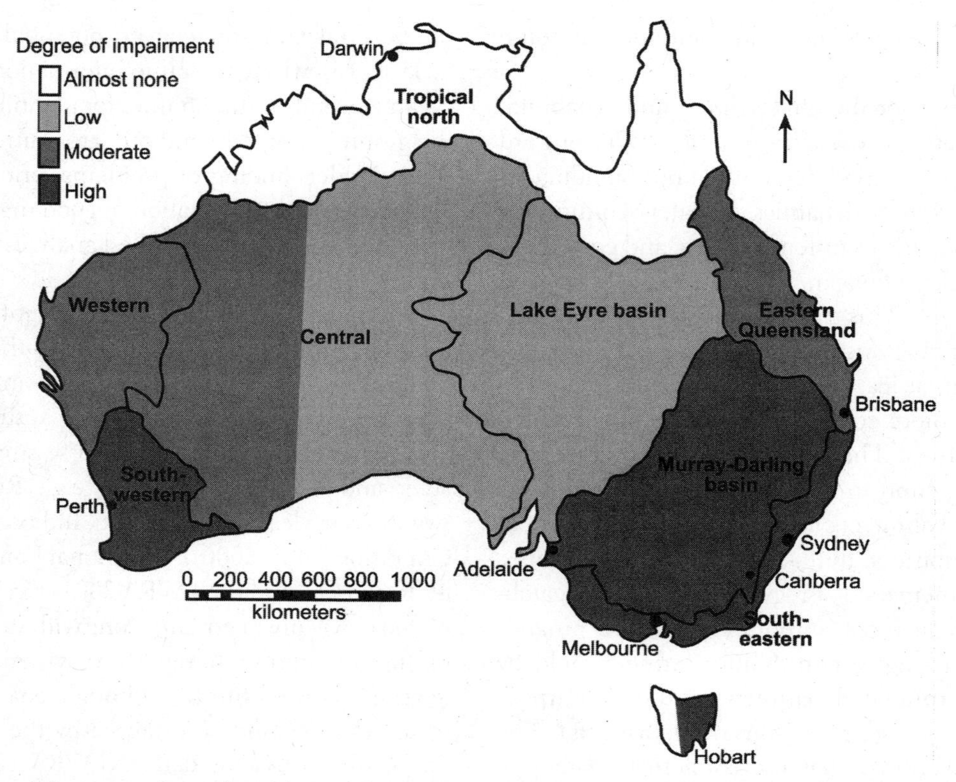

FIGURE 9.3. The geoecological condition of Australian river systems, generalised from a range of sources including the National Land and Water Resources Audit (2001) (accessed November 17, 2006 at http://audit.ea.gov.au). An "impaired" river is one with a highly modified catchment that generates unnaturally high nutrient and suspended sediment loads, that has lost much of its natural riparian vegetation, or that has dams and levees that substantially alter the flow regime and disrupt longitudinal and lateral connectivity.

While this paints a bleak picture of the state of the rivers, particularly in southern Australia, much can still be achieved through conservation and rehabilitation. Australia still has some 463,000 kilometers of undisturbed stream length or "Identified Natural Rivers" (formerly called "Wild and Scenic Rivers") mapped by the Australian Government Department of the Environment and Heritage (Australian Government 2006). "Wild Rivers" or "Heritage River Areas" have been formally designated in legislation for protection in New South Wales, Queensland, and Victoria. Outside of these areas, significant potential remains for preserving unique and rare river types as well as rivers that remain in good condition or support important natural assets (Brierley and Fryirs 2005; Rutherfurd and Gippel 2001). As relatively few Australian rivers have been channelized and engineered, there is significant opportunity to conserve unmodified remnants and to work with the natural character and behavior of rivers that have been disturbed. Kingsford (2007) and Kingsford and Nevill (2005) call for the establishment of a national Heritage Rivers system as a means to protect high-conservation rivers and wetlands. The primary limitations to implementation of such initiatives in river management are often the capacity and will to do something about it, as outlined in the next two sections.

TABLE 9.1

Examples of Human-Induced Changes in Components of Australian Fluvial Systems

Physical	Chemical/thermal/hydrological	Biological
• Gullying and river bed incision resulting in channel expansion • Accelerated rates of bank erosion • Sedimentation of estuaries and the formation of sediment slugs • Wetland drainage and flushing of valley-fill sediments • Loss of rare and valuable river types such as chains of ponds • Channelization and homogenization of urban streams	• Extensive chemical pollution from point and nonpoint sources (acidity, metals, pesticides, excess nutrients) • Highly altered river water temperature regimes • Highly altered rainfall/runoff relationships and regimes of river flow and wetland inundation	• Loss of indigenous riparian/channel vegetation and woody debris • Restriction of migration and dispersal of aquatic organisms by artificial structures • Local and regional extinctions of aquatic species • Invasions of alien plants and animals • Large-scale and localized cyanobacterial blooms, some of which are toxic to humans, livestock, fish, and other aquatic fauna

The Organizational Context of Australian River Management Practice: The Capacity to Do Something

Australia spends a trifling amount of money on river management initiatives compared to other OECD countries. Rutherfurd and Gippel (2001) estimated that around 25 million U.S. dollars are spent each year on river management initiatives in Australia. This is less than the budget for the Kissimmee River project in the United States (Rutherfurd and Gippel 2001). As a proportion of GDP, Australia invests very little per kilometer of stream length (Rutherfurd and Gippel 2001). It has only been in the face of increasing evidence of long-term climate change, the worst drought on record in much of the country, and water crises in major cities, that national initiatives and investment in water resources management have dramatically increased.

Australia has a multilevel system of river management, reflecting its three levels of elected government (federal, state/territory, and local), and private ownership of the bed and banks of many streams. Recent years have seen the creation and consolidation of regional bodies charged with responsibilities for managing rivers, and increasing participation by volunteer groups of riparian landowners and other citizens. As a consequence, complex management arrangements encompass a multitude of agencies and individuals. Many rivers run across jurisdictions. Reaches include both private and public land. River managers must navigate through a maze of legislation, policies, development pressures, conflicting land uses, social issues, economic impacts, and cultural sensitivities, as highlighted below.

Legislative and Institutional Frameworks for River Management Practice

Tan (2002) reviews the development of legislative frameworks for water management in Australia. Early colonial reliance on the English common law of riparian rights proved unworkable because of significant differences between the two countries, both in the quantity of water available for extraction and in river morphology (Pigram 2006). At the time of federation in 1901, the Commonwealth government was given no specific constitutional powers over inland waters, and consequently Commonwealth intervention has relied on external affairs and commercial powers. Except where intergovernmental arrangements take precedence, agencies operating on behalf of individual states and territories have statutory responsibility for the management of rivers, riparian zones, and floodplains. These agencies generally endeavor to work closely with local landowners and community groups in on-the-ground activities. Funding comes from state and territory government budgets and, in recent times, from Australian government initiatives. In most cases, catchment management bodies report directly to a state government minister. The structures of these bodies, their positions relative to other state agencies, and their mandated roles vary considerably from state to state. Some have adopted a model with high levels of community participation whereas in others the state government agencies retain full responsibility for all statutory functions (Pannell et al. 2007). Table 9.2 provides a summary of the catchment management bodies in each state that are responsible for the administration of relevant parliamentary acts and statutory plans associated with natural resources and river management (as in October 2006). Figure 9.4 shows representative examples of the structure of the catchment management bodies across all States of Australia as in April 2007 (Pannell et al. 2007).

In the face of differing legislative instruments and institutional arrangements for river management, the development of integrated approaches that could remedy severe cross-boundary environmental problems, particularly those of the Murray-Darling Basin, has been a lengthy and tortuous process. However, recent years have seen the growth of a coordinated regional approach and of a number of national initiatives. These two trends are now discussed in turn.

Regionalism and Regionalization

The growth of "decentralized regionalism" (Lane et al. 2004) as a platform for natural resource management in Australia is neither a straightforward nor a one-way process. Jennings and Moore (2000) make what is essentially a top-down/bottom-up distinction between *regionalization*—government-initiated establishment of administrative regions to improve management efficiency—and *regionalism*—based on initiatives of local communities, often in partnership with industry and government. Both have been evident in Australia, with regional bodies often being a hybrid of the two processes. The growth of regional bodies is a recognition of both international trends in the 1990s (e.g., Agrawal and Gibson 1999; Curtis et al. 2002), and the Wentworth Group's call for the establishment of regional authorities with statutory powers to implement nationally accredited plans for the management of natural resources (Wentworth Group of Concerned Scientists 2003).

Catchment Management Authorities (CMA) or their equivalents have been formed in every state and territory of Australia in the last decade (with Victoria leading the way with its Catchment and Land Protection Act 1994). This regionalization has been accompanied by frequent and continuing changes in the statutory powers, responsibilities, institutional structures, and reporting arrangements within which river management occurs in each jurisdiction (Pannell et al. 2007). The following summary provides a snapshot of the situation in 2006.

The aim of regionalization of river management is the delivery of long-term, strategic, integrated, coordinated, and collaborative planning and management close to the multitude of stakeholders involved (Farrelly 2005). At the same time there has been a shift away from both central bureaucratic control and individual project-based approaches. This integrated model subscribes to a comprehensive yet goal-focused approach with interconnectivity, interactivity, and inclusiveness of scientists, managers, and stakeholders as essential components (Farrelly 2005; Rogers 2006). The CMAs aim to provide a link between government planning and operational activities undertaken by volunteer groups. The overall aim is to achieve a nexus between bottom-up and top-down processes to strengthen the middle ground in environmental management (Jennings and Moore 2000). Although it has been argued that the adoption of adaptive management will improve management efficiency and on-the-ground implementation (Jennings and Moore 2000), it has been difficult to translate these principles into practice (Allan and Curtis 2005).

In most jurisdictions, regional bodies are still forming or evolving, but strategic planning, community empowerment, and more democratic decision-making has yet to be realized (Jennings and Moore 2000). In particular, critics have challenged the single-scaled and idealistic nature of current regional models (e.g., Lane et al. 2004). Many practical difficulties have arisen through the development phase and problems are also likely in the future (Pannell 2006). First, a central agency often still holds the statutory responsibility for implementation of catchment management and rehabilitation initiatives. Hence, in many cases, the CMAs lack the authority to influence local action and issues of accountability and participation become problematic (Lane et al. 2004; Lane and McDonald 2005). This can create tension between the CMAs with the mandated role to implement management strategies and the central agencies with the authority to grant consent for management activities. Second, there is an ongoing tension, variously dealt with in the different state models, between the perceived need for regional bodies to include representatives of multiple stakeholders, and the desire for scientific and technical expertise in natural resource management, governance, and conflict resolution (Jennings and Moore 2000). A third difficulty arises if state/territory and federal governments do not understand the challenges in design and implementation of cost-effective, outcome-oriented regional plans (Pannell 2006) that are based on good science. An inadequate technical foundation for planning often places reliance on the preferences of stakeholders without adequate assessment of scientific or economic realism (Pannell 2006). It remains to be seen how these regional bodies will perform in the coming years.

Federal and Intergovernmental Initiatives

Regional delivery has been part of the federal government's recent efforts to address serious, widespread problems in natural resource management through various funding programs and intergovernmental agreements. Since 1994, the Council of Australian Governments (COAG) has promoted more efficient use of water, including the full pricing of water, water trading to disconnect property and water rights, the formal allocation of water to the environment, and purchaser-provider models to create a separation of regulators and service providers (Hillman et al. 2003). However, these initiatives have been geographically, politically, and institutionally fragmented, and there is still no strategic national-level framework for the identification, conservation, and protection of riverine ecosystems (Dunn 2003, 2004). This section briefly reviews some of the federal and intergovernmental initiatives that currently shape river management practice in Australia.

The trans-boundary Murray-Darling Basin is a continuing focus of federal government plans for water allocation and river rehabilitation. A Murray-Darling Basin Agreement was signed in 1992 "to promote and coordinate effective planning and management for the equitable, efficient, and sustainable use of the water, land, and other environmental resources of the Murray-Darling Basin," and established a Murray-Darling Basin Commission. A subsequent Murray-Darling Basin Integrated Catchment Management (ICM) Policy for 2001 through 2010 (www.mdbc.gov.au/nrm/icm) established goals, values, and principles to guide community, industry, and government partnerships to improve the health of the basin, but its effectiveness has been limited both by institutional barriers and by widespread and enduring drought

TABLE 9.2

Legislative Responsibilities of Catchment Management Bodies Charged with River Management in Each State and Territory of Australia

State	New South Wales (from Hillman and Howitt, forthcoming)	Victoria (from Victorian Catchment Management Council, www.vcmc.vic.gov.au/Web/vcmc-home.html)	South Australia (Department of Water, Land and Biodiversity Conservation, State Natural Resource Management Plan 2006, www.dwlbc.sa.gov.au/nrm/index.html)	Tasmania (from. Department of Primary Industries, Water and Environment, Natural Resource Management Framework 2002, www.dpiw.tas.gov.au/inter.nsf/WebPages/LBUN-56U6CY?open)	Western Australia (from www.nrm.gov.au/nrm/wa.html)	Queensland (Department of Natural Resources and Water, State policy for Vegetation Management 2006, www.nrm.qld.gov.au/vegetation/legislation.html)	Australian Capital Territory (from ACT Natural Resource Management Board, Plan 2004-2014, www.environment.act.gov.au)	Northern Territory (from Natural Resource Management Board 2006, www.nrmbnt.org.au)
Legislation/Plan	*Catchment Management Authority Act 2003*	*Catchment and Land Protection Act 1994*	*The Natural Resources Management Act 2004*	*Natural Resource Management Act 2002*	*Policy decision*	*Vegetation Management Act 1999*	*Natural Resource Management Plan 2004–2014*	*Integrated Natural Resource Management Plan 2005*
Bodies established and date	Thirteen Catchment Management Authorities across New South Wales (2004)	Ten Catchment Management Authorities (1997) and the Victorian Catchment Management Council	One state-wide Natural Resources Management Council and eight regional NRM boards (2004)	Tasmanian Natural Resource Management Council and three Regional Natural Resource Management Committees (2002)	Natural Resource Management Council (2002)	Department of Natural Resources, Mines and Water	Natural Resource Management Board (2003)	Natural Resource Management Board (2005)
Key functions	• develop catchment action plan • provide funding for NRM activities • provide education and training	• combines the roles of the former River Management Broads, Catchment and Land Protection Boards, and community based ad-	• review NRM • promote integrated management of natural resources • advise the minister on how to achieve ecologically sustainable natural re-	• provide advice to government on NRM issues • to provide a systematic way of integrating NRM • provides a NRM administrative system	• adopt a community leadership role for NRM • provide policy and strategic advice on NRM to the chair of the Cabinet Standing Com-	• develop Regional Vegetation Management Plans • regulate the broad-scale clearing of vegetation on freehold land • identify protec-	• oversight of the implementation and review of the NRM plan; including projects derived from the plan • expenditure of NHT funds; • management	• identify, review, and prioritize NRM issues • implement a strategic approach to NRM

		visory groups • develop regional catchment management strategies • develop regional vegetation management strategies • each CMA reports to the Council and Minister	source management	• integrates the elements of the Resource Management Planning System	mittee on Environmental Policy • foster a consultative approach that ensures broad community involvement in NRM policy development	tion required for regional ecosystems • develop codes of practice for clearing activities	actions and territory targets identified in the plan	
Objects/Guiding principles	• devolution of NRM to catchment level • planning based on state-wide standards • involve communities in catchment planning • balance economic, social, and environmental interests	• to create a whole-of-catchment approach to NRM • protect and restore land and water resources • sustainable development of natural resources-based industries • conservation of natural and cultural heritage	• regional variation • integration and consolidation of existing NRM legislation	• ecosystem approach • balanced decisions • integrated management • priority-based • prevention better than cure • partnerships • shared responsibility	• none stated	• preserve remnant endangered regional ecosystems • preserve vegetation in areas of high nature conservation value and areas vulnerable to land degradation • ensure that clearing does not cause land degradation, maintains or increases biodiversity, maintains ecological processes, and allows for ecologically sustainable land use	• provide a strategic focus for the protection and management of the environment • promote community and government partnerships • encourage integrated, coordinated actions • access investment from multiple sources, including the NHT and NAP • engage and energize the community • provide a link to national and regional agendas in NRM	• promote and support ecologically sustainable development • apply the precautionary principle in the absence of sound data on which to base planning decisions • promote and support adaptive management

TABLE 9.2

(Continued)

State	New South Wales (from Hillman and Howitt, forthcoming)	Victoria (from Victorian Catchment Management Council, www.vcmc.vic.gov.au/Web/vcmc-home.html)	South Australia (Department of Water, Land and Biodiversity Conservation, State Natural Resource Management Plan 2006, www.dwlbc.sa.gov.au/nrm/index.html)	Tasmania (from Department of Primary Industries, Water and Environment, Natural Resource Management Framework 2002, www.dpiw.tas.gov.au/inter.nsf/WebPages/LBUN-56U6CY?open)	Western Australia (from www.nrm.gov.au/nrm/wa.html)	Queensland (Department of Natural Resources and Water, State policy for Vegetation Management 2006, www.nrm.qld.gov.au/vegetation/legislation.html)	Australian Capital Territory (from ACT Natural Resource Management Board, Plan 2004-2014, www.environment.act.gov.au)	Northern Territory (from Natural Resource Management Board 2006, www.nrmbnt.org.au)
Governance	• five to seven board members for three-year term with collective skills and knowledge in the following areas: • primary production • environmental, social, and economic analysis	• up to ten council members • members appointed through advertisement and approval by the minister • more than half of the board's members must have primary production as	• ninemembers • membership based on knowledge, skills, and practical experience identified in the act. • advisory body to minister	• the Council is appointed by Government. • up to sixteen council members • the Council broadly represents regions and stakeholders, and is gender balanced	• fifteen members who have skills and knowledge in NRM areas • majority are not government members • Council meets on a bi-monthly basis, or more regularly if required	n/a	• thirteen board members who are appointed in an individual capacity by the Minister for the Environment • members are appointed for terms of up to three years. • members are selected for their capacity	• six members including a chairman and a NT government representative • members are selected from a broad geographic base across the NT • members must have skills and knowledge in one or more of

• government administration • negotiation and consultation • business administration leadership • community • biodiversity conservation • cultural heritage	their principle occupation, the composition of each authority must reflect the land and water uses of the region • the board must have experience and knowledge of land protection, water resource management, primary industry, environmental conservation, and local government		• regional committees contain twelve members. Their composition varies between regions			to contribute on natural resources management issues, to achieve balance in views, and to achieve equity in representation • board meets at least four times a year	the following areas: business admin and governance; finance and accounting; legal and contractual issues; capacity building, social, and economic development; NRM, systems, research; and NT government policies and strategic direction

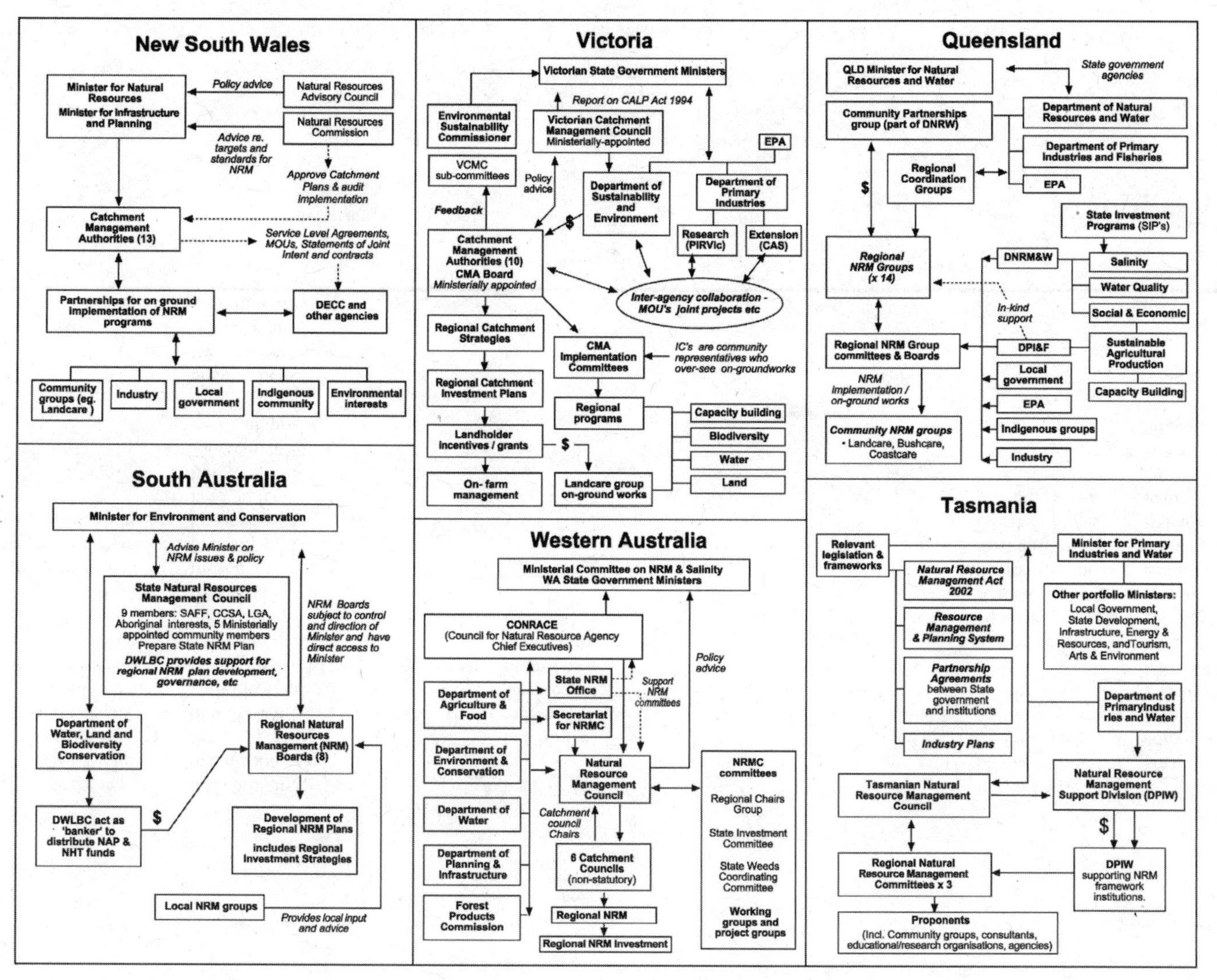

FIGURE 9.4. Structure of regional natural resource management bodies in six States of Australia (from Pannell et al. 2007 and reproduced with permission from David Pannell, University of Western Australia, April 2007).

in southeastern Australia. The intergovernmental Living Murray (http://thelivingmurray.mdbc.gov.au/home), a program established in 2002 to recover environmental water for six icon sites, has paid attention to social impacts, including the needs of indigenous communities along the river (Morgan et al. 2004), but has also been hampered by the impact of drought on water availability.

The Natural Heritage Trust (NHT) (www.nht.gov.au), established by the Australian Government in 1997, is a primary funding source for river management outside of the Murray-Darling Basin. Its investment of 3 billion Australian dollars to 2013 is focused on three overarching objectives: biodiversity conservation, sustainable use of natural resources, and community capacity building and institutional change. These objectives are embedded within four programs: Landcare, Bushcare, Rivercare, and Coastcare. NHT funds environmental activities at a community level, regional level, and national/state level. River rehabilitation projects focus largely on riparian revegetation, fencing off stock access, installation of bed and bank control structures, and instream habitat and native fish restocking (Hassall and Associates 2005). A criterion of NHT funding is that it is matched by contributions from local communities, landowners, and state agencies that support rehabilitation and monitoring activities. Bilateral and regional partnership agreements, investment against accredited regional plans, and the provision of foundation and priority funding are key components of NHT allocations. A report on NHT achievements from 1997 to 2002 reveals that there was a significant impact in terms of raised awareness, changed attitudes, and increased understanding and knowledge of environment protection, NRM, and sustainable agriculture. It also improved the levels of communication and cooperation between stakeholders, and built capacity within both the community and government (Hassall and Associates 2005). However, others have commented that on-the-ground change and improvement in river condition have been limited by the site-specific and piecemeal nature of the projects undertaken (Australian National Audit Office 2001; Bradby 2004).

In 2000, COAG endorsed the National Action Plan for Salinity and Water Quality (NAP) (www.napswq.gov.au), which prioritizes actions to address salinity and deteriorating water quality in selected catchments and regions across Australia. The goal is to motivate and enable regional communities to use coordinated and targeted action to prevent, stabilize, and reverse trends in dry-land salinity, conserve biological diversity, improve water quality, and secure reliable allocations of water for human uses, industry, and the environment. As with NHT, there has been criticism of the failure to take a strategic approach and provide for adequate monitoring in this program, and of delays due to intergovernmental tensions and the extent of institutional changes required (Australian National Audit Office 2004).

In 2003, WWF Australia convened the Wentworth Group of Concerned Scientists (www.wentworthgroup.org) to develop practical scientific and economic ideas for the management of Australia's landscapes and natural resources. The resulting Blueprint for a National Water Plan (Wentworth Group of Concerned Scientists 2003) proposed a consistent national approach to water entitlements, water accounting, and water allocation to ecosystems. A subsequent intergovernmental National Water Initiative (NWI) (www.nwc.gov.au/nwi/index.cfm) provided 500 million Australian dollars for improved freshwater management throughout Australia. Objectives of the NWI focus on water trading, water recovery for environmental outcomes, better monitoring, reporting and accounting of water use, and more sophisticated water planning (including groundwater systems). Implementation plans at the federal and state levels are being developed with a strong focus on regulatory and planning frameworks, water trading, and pricing mechanisms (e.g., Australian Government 2006). In January 2007, the federal government announced a $10 billion National Plan for Water Security (http://pandora.nla.gov.au/pan/10052/20070615-0000/www.pm.gov.au/government/water.html). This focused largely on the Murray-Darling river system, proposing that its governance be passed from the states to the federal government. Criticized by some commentators as precipitate and politically motivated, the future of this plan is uncertain given the change of government in late 2007.

Social Themes in Australian River Management Practice: Community Will to Do Something

Given a low population density and limited resources to undertake river rehabilitation across the vast Australian continent, community participation is a vital component of the process of river repair.

Participatory Frameworks for River Management

While governments have put in place a suite of policies and programs to address the long-term environmental problems of rural Australia, voluntary change has been emphasized. Australia has invested heavily in participatory approaches to river management (Curtis and Van Nouhuys 1999), and almost all river rehabilitation schemes in Australia are implemented with volunteer participation. The NHT and regional authorities have been the main mechanisms for the delivery of government support to volunteer groups and private land managers (Curtis and Van Nouhuys 1999; Curtis and Lockwood 2000). The aim is participatory democracy through regional governance, decentralization, accountability, equity, and empowerment/ownership, tailoring management strategies to regional conditions and circumstances, thus improving the prospects of achievement (Jennings and Moore 2000; Gooch 2004). However, in many cases, the realities of implementation have been far from inclusive. In some cases, the process has been captured by vested interests, important stakeholders have been excluded or sidelined, and stalemates have arisen through difficulties of achieving consensus on tough decisions (Lane et al. 2004).

Nevertheless, the National Landcare Program (www.daffa.gov.au/natural-resources/landcare/national-landcare-programme and www.landcareonline.com) is an example of how a participatory approach can be used to achieve considerable cultural change (Black and Reeve 1993; Baker 1997) for more sustainable management practices (Byron and Curtis 2002). Landcare originated in Victoria in 1986 and has since spread throughout Australia (Curtis and De Lacy 1995). After lobbying by farmer and conservation groups, the federal government committed $360 million to the Decade of Landcare (Hawke 1989), which saw Landcare become a national program under the NHT (Curtis and De Lacy 1996). Landcare aims to combine technical expertise, access to funding sources, commitment of agency staff with local knowledge, and on-ground work by volunteers (Curtis and De Lacy 1995). Landcare groups have no legislative backing and are only informally linked to local government and regional planning bodies such as CMAs (Curtis and Lockwood 2000).

Spin-off programs including Rivercare and Coastcare were also established under the NHT (www.nht.gov.au/index.html). The Rivercare program focuses on activities that contribute to improved water quality and environmental conditions in river systems and wetlands. It built on the growing popularity of the Landcare movement and in practice there was substantial overlap (Baker 1997). Rivercare originated in the Hunter and Manning Valleys in New South Wales, born of recognition of the conceptual limitations of previous approaches and frustration at the lack of effectiveness of reach-scale works on individual properties. Predicated on a collaborative approach across property boundaries, Rivercare takes a strategic view of priorities along a section of river. Plans for rehabilitation are set by the participants, with no prescriptive policy on whether priority should be given to the maintenance of healthy sections of river or to intensive rehabilitation efforts on more degraded reaches. Planning tends to occur at catchment or sub-catchment scales (Curtis and De Lacy 1995, 1996), with implementation at the local scale. Investment in the Rivercare program triggered a boom in on-the-ground rehabilitation and conservation efforts supported by staff who are dedicated to river reach rehabilitation at the local level.

More than 4,000 Care groups exist across the country. Forty percent of Australia's rural landowners, who manage 60 percent of Australia's agricultural land and 70 percent of Australia's diverted water runoff, are involved in Landcare programs of some variety (Byron and Curtis 2002). However, the Care movement has been criticized for the extent to which it places responsibility upon individual landholders to deliver on-the-ground outcomes (Curtis and De Lacy 1996; Ewing 1999) at the expense of more strategic and sustainable management with adequate monitoring. This is seen as an exercise in shifting responsibility for action from government to local communities (Martin et al. 1992). As a result, Curtis (2000) and Byron and Curtis (2002) concluded that volunteer participation programs may be approaching their limits of voluntary action. Concern has also been expressed that Landcare and NHT are a case of "too little too late" and that although works have reversed local land and water degradation, only a few plans could be considered to have made measurable improvements in landscape-scale environmental conditions (Curtis 1998). There are also ongoing issues of state agencies and regional bodies accessing large portions of Landcare funding, agency staff playing important roles in the decision making of many groups, and government funding priorities shaping group activity or landowner participation (Curtis 1998; Ewing 1999). The dismantling of state agency extension services during the shift to regionalization has also stifled volunteer enthusiasm and engendered a sense of frustration and dissatisfaction with the process in many parts of the country (Byron and Curtis 2001a, 2001b, 2002; Hillman and Howitt, forthcoming). Nevertheless the Care movement has been successful in facilitating broader development and ownership of information and activities and produced many aware, informed, skilled, and adaptive resource managers (Curtis and De Lacy 1995, 1996).

The Care experience in Australia provides many lessons for recently formed regional organizations, such as the CMAs, in particular the need to articulate regional strategies based on the appraisal and integration of local requirements, establish priorities for allocation of government resources, provide accountability for expenditure of public funds, and link and support local groups (Curtis and Lockwood 2000; Ewing 1999). The most important roles for participants in local groups are to mobilize and maintain participation, initiate and support learning, capture resources to support local efforts, and undertake on-ground work to the extent that resources are available.

Integration and Future Challenges

Key challenges in river management in Australia include integrating institutions, undertaking cross-disciplinary science, and then using that science in river management practice. Significant progress has been made in breaking down the barriers to integration across these areas in recent years. We now document this transition in practice.

The Challenge of Institutional and Operational Integration in Australia

Since the 1980s, advocates of integrated environmental management have promoted a multiple-scale, interdisciplinary nexus of research institutions, regulatory agencies, industry, and stakeholders (Bellamy et al. 1999; Morrison et al. 2004). Although progress has been made toward this ideal, fragmentation of policy and practice continue to hinder effective management of natural resources in Australia (Morrison et al. 2004). As a result, existing frameworks and procedures for water and land-use planning are not adequately protecting Australia's freshwater ecosystems (Dunn 2003, 2004; Kingsford and Nevill 2005). This lack of integration occurs for a number of reasons based upon operational (discipline-bound), institutional, and social factors.

Operational fragmentation occurs when issues are tackled independently and along discipline-specific lines. Land, water, and vegetation programs operate independently in many organizations and are rarely integrated. The compartmentalization of these programs causes inadequate definition of problems and discipline-specific perceptions and solutions (Morrison et al. 2004). Similarly, there is a lack of guidance on how economic, environmental, and social considerations can be integrated in the development of policies and programs. Another source of fragmentation is the politically driven *institutional* compartmentalization of government into departments and agencies pursuing different and sometimes competing interests and objectives. Those with common objectives often compete for the same shrinking funding source. This fragmentation results in failure of horizontal coordination (i.e., coordination among agencies) and ultimately duplication and inefficiency. The problem is exacerbated by continual structural change of institutions and programs, which results in a lack of continuity, loss of corporate knowledge, and erosion of technical expertise, which in turn causes a loss of public confidence (Pahl-Wostl 2002). *Social* fragmentation occurs because of the multiplicity of actors who are involved in decentralized decision-making for natural resource management, often with competing philosophies, interests, and agendas (Agrawal and Gibson 1999; Broderick 2005). The challenges for the current model of regionalized catchment management are many, and the path to true integration remains long and tortuous.

Crossdisciplinary River Science

Rutherfurd and Bartley (1999) and Rutherfurd (2000) asked whether we know enough about Australian rivers to start managing them effectively. While we have a solid understanding of how rivers operate and have changed since European settlement in southeastern Australia, significant gaps remain in the coverage of baseline information. We still have very little understanding of how rivers in the tropics and arid areas of the country behave and have evolved, and what their condition is (Hamilton and Gehrke 2005). We know even less about the aquatic ecology of these systems (Douglas et al. 2005). Across the majority of the country we have little appreciation of how our rivers and their associated ecosystems are likely to adjust in the future, particularly in the face of climate change. These limitations are a consequence of the small pool of scientists, the spatial area of the continent, and to some degree the lack of integrative river science. This last limitation is due in large part to the history of scientific research on Australian rivers.

Until the 1970s, river science (and management) in Australia was heavily dominated by civil engineering, with a focus on navigation, flood mitigation, water supply, and the generation of hydroelectricity. Although generally effective in meeting these objectives, this single-discipline approach ignored other values of rivers such as fish production and nature conservation. As side effects on such values became more evident, significant sections of the public grew increasingly critical of river development schemes, as in the widespread opposition in the late 1970s and early 1980s to the Lake Pedder and Gordon-below-Franklin proposals for hydroelectric developments in Tasmania. This development in public attitudes coincided with the growth of freshwater ecology (or limnology) as a discipline in Australian universities, and a consequently increasing voice of ecologists within public sector agencies. Following the massive cyanobacterial bloom in the Darling River in 1991

(Bowling and Baker 1996), and release of the National Water Quality Management Strategy in 1992, the concept of "river condition" or "river health" was increasingly promoted, and became the basis of various scientific programs, such as the National River Health Program (Davies 2000). The "river health" field has been dominated by aquatic ecologists and biologists, who generally equate "health" with the preservation of natural values, in direct contradiction to the older engineering approach (see chapter 7).

The gap between engineers and biologists has often been bridged by the geographical sciences, and particularly by geomorphologists. From about the 1970s, geomorphologists increasingly questioned the basis of hard engineering solutions to perceived problems of riparian flooding, and described the unintended physical consequences of poorly designed works. Geomorphologists worked with progressive engineers to design more sensitive and effective works, and also explored the ecological implications of geomorphology through collaborations with aquatic biologists.

With the broadening in emphasis from engineering to include a range of ecosystem sciences, integration among the sciences has improved, and integrative projects have emerged with the support of industry, state agencies, and research groups (e.g., the Upper Hunter River Rehabilitation Initiative (www.hcr.cma.nsw.gov.au/uhrri), Granite Creeks (http://www.ewater.canberra.edu.au/domino/html/Site-CRCFE/CRCFE_WebSite.nsf; www.gbcma.vic.gov.au), and Moreton Bay (Dennison and Abal 1999). These projects aim to improve understanding of Australian river systems, ecosystem dynamics, and river condition through examination of geomorphology, ecology, hydrology, and chemistry. This science is underpinning the development of management programs and the design of rehabilitation techniques for Australian conditions (Brierley et al. 2006).

While considerable progress has been made in linkages among the biophysical sciences and civil engineering, a major challenge remains in the integration of resource economics with the natural and social sciences (i.e., transdisciplinary practices that address the "triple-bottom line").

Use of Science in River Management

Recent decades have seen an almost exponential increase in the use of Australian science in river management. Prior to this, science and techniques were imported from overseas (largely from North America and Europe) where conditions and the history of human disturbance are quite different (Rutherfurd and Gippel 2001; Brierley et al. 2005). As a result, misapplication of scientific principles occurred (and still occurs). However, the increase in place-based knowledge of Australian river systems has brought about a shift in management practice, promoting the development of techniques that are designed for Australian conditions (e.g., Rutherfurd et al. 2000; Brierley et al. 2006).

Increasingly, scientists work directly with staff in state and regional agencies in productive partnerships in research and development. For example, the development of Land and Water Australia (LWA) and its programs in riparian lands and river health (among other programs) fostered collaboration among scientists, state agencies, and volunteer groups. Co-operative Research Centres for Catchment Hydrology and Freshwater Ecology (now e-Water), with their extensive research in hydrology, geomorphology, ecology, and planning/management, have successfully linked the physical and biological sciences as well as science and management. The development of a range of tools and approaches for river management has not only engendered fruitful changes in on-the-ground implementation and planning procedures, but has also changed the mindsets of people involved in the process (i.e., scientists, managers, and the broader community).

One way to track the development of science and its use in river management in Australia is to examine the themes of the four national stream management conferences held since 1996 (Rutherfurd and Walker 1996; Rutherfurd and Bartley 1999; Rutherfurd et al. 2001; Rutherfurd et al. 2005). These biennial or triennial conferences are the largest gatherings of river practitioners in the country. Their success attests to increased integration and transfer of knowledge between scientists and the river management community. Some important trends emerged over these four conferences (Fryirs et al. 2007a). First, the key themes of the conferences have evolved from single issues such as erosion, sedimentation, and human impact toward a more holistic examination of rivers in their landscape context (table 9.3). The later conferences have also featured community and education themes in river management that are absent from earlier ones. Nevertheless, by far the majority of papers have presented ecosystem science. However, it is clear that river management still relies on a significant proportion of discipline-bound research. Crossdisciplinary science is emerging, but the use of transdisciplinary research remains poor (Fryirs et al. 2007a).

The catchment was the favored scale of application across all conferences and reach-scale applications have increased slightly since 1996. National- and regional-scale applications have remained few. A more disturbing trend is the lack of monitoring or auditing papers that address success or failure in river management. Despite calls for greater evaluation and appraisal of rehabilitation projects, and supposed adoption of adaptive management frameworks, reporting of rehabilitation success is scarce. Documentation of rehabilitation failure, supporting the ability to "learn by doing" that is a mantra of adaptive management, is almost nonexistent.

Reconciling Biophysical, Economic, and Social Knowledge and Priorities: Transdisciplinary Challenges

A central issue in new approaches to river management in Australia is to integrate biophysical, economic, and social information for deciding where rehabilitation efforts should be focused. While

TABLE 9.3

Themes of the four national stream management conferences held since 1996

1996—First stream management conference	1999—Second stream management conference	2001—Third stream management conference	2004—Fourth stream management conference
Conference title			
n/a	The challenge of rehabilitating Australia's streams	The value of healthy rivers	Linking rivers to their landscapes
Conference themes			
Sediment sources and sinks	Challenges of rehabilitation	Ecosystem services provided by streams	Landscape processes that influence rivers
Stream processes and human impacts	Integration of issues between science and community	Biophysical interactions in streams	Identifying/ managing values associated with river landscapes
Organizing management	Crossdisciplinary science	Hydrological connectivity	Education and change
Vegetation and streams	Evaluation of rehabilitation projects	Tools/techniques for management	River management at the grass roots level

there are apparent parallels between biophysical and social criteria in the setting of priorities, these can be misleading and it is quite possible for these two sets of priorities to clash. For instance, it may be scientifically more rational to focus on conservation of sections of rivers in relatively good condition, whereas stakeholders' desire to remedy basket cases pushes priorities in the opposite direction (see chapter 15). The need for such cross-dimensional understanding is reinforced by the wide range of motivations—ethical, social, educational, and so on—for volunteer involvement in rehabilitation (Gooch 2005; Ewing 1999). For rehabilitation practice, such involvement is crucial, and the most inclusive and engaging agenda will be one that openly recognizes and balances this range of priorities and motivations. The development of holistic rehabilitation programs to achieve social, economic, and biophysical gains will depend on bringing together measures of ecosystem health, modeling of the economic value of rehabilitation investment, and indicators of social capital and community capacity in an integrated assessment process (Haila 1998; Thompson and Pepperdine 2003). From this perspective "socioeconomic well-being is not only consistent with enhanced ecological conditions, but also the process of restoring ecological processes and functions itself contributes toward socioeconomic well-being" (Baker 2004, 6).

There is also increasing recognition in Australian river management practice of uncertainty and the need to adopt adaptive approaches (chapter 14). However, the philosophy of adaptive management does not often align with command-and-control attitudes that still permeate some institutions charged with natural resource management (Allan and Curtis 2005). A shift toward more effective prioritization of activities and the emergence of a range of frameworks that advocate a conservation-first approach has also occurred (see Rutherfurd et al. 1999; Koehn et al. 2001; Brierley and Fryirs 2005). A similar recognition of the need to work within a catchment context and facilitate natural recovery has engendered methods to examine river condition and recovery potential (e.g., Brierley and Fryirs 2005; Brierley et al. 2002; Fryirs 2003). Approaches to visioning and application of adaptive management principles have also evolved (Gippel 1999; Hillman and Brierley 2002). The monitoring and evaluation of river rehabilitation projects have started to receive long-term funding from regional authorities and scientific support groups such as state government departments, universities, and the CSIRO. Information from these programs can be used to assess prospects for river recovery, promoting targeted management actions to enhance these prospects.

Conclusion

It is only in recent decades that scientific understanding has started to come to terms with the unique physical and environmental conditions that have shaped Australian rivers, the impacts of colonization around 220 years ago, and the conditions under which our rivers now operate. Increasingly, effective information transfer is promoting the use of this integrative knowledge base as a platform for management actions. River management prospects are bright if these insights are used effectively, we invest adequately and wisely in initiatives, and we get our governance structures right. Ultimately, the key to this process is effective engagement of society in the process of river repair—many initiatives in Australia prompt hope in this regard.

Acknowledgments

We would like to thank Robyn Watts and Brian Finlayson for constructive review comments on this manuscript. Kirstie A. Fryirs acknowledges the Australian Research Council for their financial support to a number of the authors.

References

Agrawal, A., and C. G. Gibson. 1999. Enchantment and disenchantment: the role of community in natural resource conservation. *World Development* 27:629–49.

Allan, C., and A. Curtis. 2005. Nipped in the bud: Why regional scale adaptive management is not blooming. *Environmental Management* 36:414–25.

Allan, J., and S. Lovett. 1997. Managing water for the environment: Impediments and challenges. *Australian Journal of Environmental Management* 4:200–210.

Australian Government. 2006. *National water initiative implementation plan*. Canberra, Australia: Australian Government.

Australian National Audit Office. 2001. *Performance information for Commonwealth financial assistance under the Natural Heritage Trust*. Audit Report No. 43. Commonwealth of Australia.

Australian National Audit Office. 2004. *The administration of the National Action Plan for Salinity and Water Quality*. Canberra, Australia: Australian Government.

Baker, M. 2004. *Socioeconomic characteristics of the Natural Resources Restoration System in Humboldt County, California*. Taylorsville, CA: Forest Community Research.

Baker, R. 1997. Landcare: Policy, practice and partnerships. *Australian Geographical Studies* 35:61–73.

Bandler, H. 1995. Water resources exploitation in Australian prehistory environment. *The Environmentalist* 15:97–107.

Barmuta, L. A. 2003. Imperilled rivers of Australia: Challenges for assessment and conservation. *Aquatic Ecosystem Health and Management* 6:55–68.

Bellamy, J. A., G. T. McDonald, G. J. Syme, and J. E. Butterworth. 1999. Policy review: Evaluating integrated resource management. *Society and Natural Resources* 12:337–53.

Black, A. W., and I. Reeve. 1993. Participation in Landcare groups: The relative importance of attitudinal and situational factors. *Journal of Environmental Management* 39:51–71.

Bowling, L. C., and P. D. Baker. 1996. Major cyanobacterial bloom in the Barwon-Darling River, Australia in 1991, and underlying limnological conditions. *Marine and Freshwater Research* 47:643–57.

Bradby, K. 2004. The baby and the bathwater. *Ecological Management and Restoration* 5:3–4.

Brierley, G. J., A. P. Brooks, K. Fryirs, and M. P. Taylor. 2005. Did humid-temperate rivers in the Old and New Worlds respond differently to clearance of riparian vegetation and removal of woody debris? *Progress in Physical Geography* 29:27–49.

Brierley, G. J., and K. A. Fryirs. 2005. *Geomorphology and river management: Applications of the River Styles framework*. Oxford, UK: Blackwell.

Brierley, G., K. Fryirs, D. Outhet, and C. Massey. 2002. Application of the River Styles framework as a basis for river management in New South Wales, Australia. *Applied Geography* 22:91–122.

Brierley, G. J., M. Hillman, and K. Fryirs. 2006. Knowing your place: An Australasian perspective on catchment-framed approaches to river repair. *Australian Geographer* 37:131–45.

Broderick, K. 2005. Communities in catchments: Implications for natural resource management. *Geographical Research* 43:286–96.

Brodie, J. E., and A. W. Mitchell. 2005. Nutrients in Australian tropical rivers: Changes with agricultural development and implications for receiving environments. *Marine and Freshwater Research* 56:279–302.

Brooks, A. P., and G. J. Brierley. 2004. Framing realistic river rehabilitation programs in light of altered sediment transfer relationships: Lessons from East Gippsland, Australia. *Geomorphology* 58:107–23.

Brooks, A. P., G. J. Brierley, and R. G. Millar. 2003. The long-term control of vegetation and woody debris on channel and floodplain evolution: Insights from a paired catchment study in southeastern Australia. *Geomorphology* 51:7–29.

Byron, I., and A. Curtis. 2001a. Landcare in Australia: Burned out and browned off. *Local Environment* 6:311–26.

Byron, I., and A. Curtis. 2001b. Exploring burnout in Australia's Landcare program: A case study in the Shepparton region. *Society and Natural Resources* 14:901–10.

Byron, I., and A. Curtis. 2002. Maintaining volunteer commitment to local watershed initiatives. *Environmental Management* 30(1):59–67.

Crisp, M. D., S. Laffan, H. P. Linder, and A. Monro. 2001. Endemism in the Australian flora. *Journal of Biogeography* 28:183–98.

Curtis, A. 1998. Agency-community partnership in Landcare: Lessons for state-sponsored citizen resource management. *Environmental Management* 22:563–74.

Curtis, A. 2000. Landcare: Approaching the limits of voluntary action. *Australian Journal of Environmental Management* 7:19–28.

Curtis, A., and T. De Lacy. 1995. Evaluating Landcare groups in Australia: How they facilitate partnerships between agencies, community groups, and researchers. *Journal of Soil and Water Conservation* 50:15–20.

Curtis, A., and T. De Lacy. 1996. Landcare in Australia: Does it make a difference? *Journal of Environmental Management* 46:119–37.

Curtis, A., and M. Lockwood. 2000. Landcare and catchment management in Australia: Lessons for state-sponsored community participation. *Society and Natural Resources* 13:61–73.

Curtis, A., B. Schindler, and A. Wright. 2002. Sustaining local watershed initiatives: Lessons from Landcare and watershed councils. *Journal of the American Water Resources Association* 38:1207–15.

Curtis, A., and M. Van Nouhuys. 1999. Landcare participation in Australia: The volunteer perspective. *Sustainable Development* 7:98–111.

Davies, P. E. 2000. Development of a national river bioassessment system (AUSRIVAS) in Australia. In *Assessing the biological quality of fresh waters: RIVPACS and other techniques*, ed. J. F. Wright, D. W. Sutcliffe, and M. T. Furse, 113–24. Ambleside, UK: Freshwater Biological Association.

Dennison, W. C., and E. G. Abal. 1999. *Moreton Bay study: A scientific basis for the Healthy Waterways Campaign*. Brisbane, Australia: South East Queensland Regional Water Quality Management Strategy.

Department of the Environment and Heritage. 2006. *Identified natural rivers lists*. http://www.heritage.gov.au/anlr/code/idlists.html (accessed January 8, 2006).

Douglas, M. M., S. E. Bunn, and P. M. Davies. 2005. River and wetland food webs in Australia's wet-dry tropics: General principles and implications for management. *Marine and Freshwater Research* 56:329–42.

Dunn, H. 2003. Can conservation assessment criteria developed for terrestrial systems be applied to riverine systems? *Aquatic Ecosystem Health and Assessment* 6:81–95.

Dunn, H. 2004. Defining the ecological values of rivers: The views of Australian river scientists and managers. *Aquatic Conservation: Marine and Freshwater Ecosystems* 14:413–33.

Ewing, S. 1999. Landcare and community-level watershed management in Victoria, Australia. *Journal of the American Water Resources Association* 35:663–73.

Farrelly, M. 2005. Regionalisation of environmental management: A case study of the Natural Heritage Trust, South Australia. *Geographical Research* 43:393–405.

Finlayson, B. L., and T. A. McMahon. 1988. Australia *v* the world: A comparative analysis of streamflow characteristics. In *Fluvial geomorphology of Australia*, ed. R. F. Warner, 17–40. Sydney, Australia: Academic Press.

Fryirs, K. 2003. Guiding principles for assessing geomorphic river condition: Application of a framework in the Bega catchment, South Coast, New South Wales, Australia. *Catena* 53:17–52.

Fryirs, K., and G. J. Brierley. 2001. Variability in sediment delivery and storage along river courses in Bega catchment, NSW, Australia: Implications for geomorphic river recovery. *Geomorphology* 38:237–65.

Fryirs, K., G. J. Brierley, N. J. Preston, and M. Kasai. 2007a. Buffers, barriers and blankets: The (dis)connectivity of catchment-scale sediment cascades. *Catena* 70:49–67.

Fryirs, K., G. J. Brierley, N. J. Preston, and J. Spencer. 2007b. Landscape (dis)connectivity in the upper Hunter catchment, New South Wales, Australia. *Geomorphology* 84:297–316.

Fryirs, K., M. Hillman, and A. Spink. 2007c. Challenges faced in the integration of science in river management in Australia. In *Proceedings of the 5th Australian Stream Management Conference: Australian rivers, making a difference*, ed. A. L. Wilson, R. L. Dehaan, R. J. Watts, K. J. Page, K. H. Bowner, and A. Curtis, 103–8. Albury, Australia: Charles Sturt University.

Gippel, C. 1999. Developing a focussed vision for river rehabilitation: The lower Snowy River, Victoria. In *Proceedings of the 2nd Stream Management Conference, Adelaide*, ed. I. Rutherfurd and R. Bartley, 299–306. Melbourne, Australia: Cooperative Research Centre for Catchment Hydrology.

Gippel, C. J., B. L. Finlayson, and I. C. O'Neill. 1994. Distribution and hydraulic significance of large woody debris in a lowland Australian river. *Hydrobiologia* 318:179–94.

Gooch, M. 2004. Volunteering in catchment management groups: Empowering the volunteer. *Australian Geographer* 35:193–208.

Gooch, M. 2005. Voices of the volunteers: An exploration of the experiences of catchment volunteers in coastal Queensland, Australia. *Local Environment* 10:5–19.

Haila, Y. 1998. Assessing ecosystem health across spatial scales. In *Ecosystem health*, ed. D. Rapport, R. Costanza, P.R. Epstein, C. Gaudet, and R. Levins, 81–102. Malden, MA: Blackwell.

Hamilton, S. K., and P. C. Gehrke. 2005. Australia's tropical river systems: Current scientific understanding and critical knowledge gaps for sustainable management. *Marine and Freshwater Research* 56:243–52.

Harris, G. P. 2001. Biogeochemistry of nitrogen and phosphorus in Australian catchments, rivers and estuaries: Effects of land use and flow regulation and comparisons with global patterns. *Marine and Freshwater Research* 52:139–49.

Hassall and Associates. 2005. *Natural Heritage Trust phase 1 final evaluation*. Prepared for Australian Government, Dept. Agriculture, Fisheries and Forestry and Dept. of the Environment and Heritage. Canberra, Australia.

Hawke, R. J. L. 1989. *Our country: Our future*. Statement on the environment by the Prime Minister of Australia. Canberra, Australia: Australian Government Publishing Service.

Hillman, M., G. Aplin, and G. J. Brierley. 2003. The importance of process in ecosystem management: Lessons from the Lachlan Catchment, New South Wales, Australia. *Journal of Environmental Planning and Management* 46:219–37.

Hillman, M., and G. J. Brierley. 2002. Information needs for environmental-flow allocation: A case study from the Lachlan River, New South Wales. *Annals of the Association of American Geographers* 92:617–30.

Hillman, M., and R. Howitt. Forthcoming. Institutional change in natural resource management in New South

Wales, Australia: Sustaining capacity and justice. *Local Environment*.

Jennings, J. N., and J. A. Mabbutt. 1967. *Landform studies from Australia and New Guinea*. Canberra, Australia: Australian National University Press.

Jennings, S. F., and S. A. Moore. 2000. The rhetoric behind regionalization in Australian natural resource management: Myth, reality and moving forward. *Journal of Environmental Policy & Planning* 2:177–91.

Kingsford, R. T. 2007. *Heritage rivers: New directions for the protection of Australia's rivers, wetlands and estuaries*. Sydney, Australia: University of New South Wales.

Kingsford, R. T., and J. Nevill. 2005. Scientists urge expansion of freshwater protected areas. *Ecological Management and Restoration* 6:161–62.

Koehn, J. D., G. J. Brierley, B. L. Cant, and A. M. Lucas. 2001. *River restoration framework*. Land and Water Australia Occasional Paper 01/01. Canberra, Australia.

Lake, P. S. 1995. Of floods and droughts: River and stream ecosystems of Australia. In *River and stream ecosystems: Ecosystems of the world vol.* 22, ed. C. E. Cushing, K. W. Cummins, and G. W. Minshall, 669–94. Amsterdam, Netherlands: Elsevier.

Lane, M. B., and G. McDonald. 2005. Community-based environmental planning: Operational dilemmas, planning principles and possible remedies. *Journal of Environmental Planning and Management* 48:709–31.

Lane, M., G. Mcdonald, and T. Morrison. 2004. Decentralisation and environmental management in Australia: A comment on the prescriptions of the Wentworth Group. *Australian Geographical Studies* 42:103–15.

Martin, P., S. Tarr, and S. Lockie. 1992. Participatory environmental management in New South Wales: Policy and practice. In *Agriculture, environment and society*, ed. G. Lawrence, F. Vanclay, and B. Furze, 184–208. Melbourne, Australia: Macmillan.

McCool, S. F., and K. Guthrie. 2001. Mapping the dimensions of successful public participation in messy natural resources management situations. *Society and Natural Resources* 14:309–23.

McMahon, T. A., B. L. Finlayson, A. T. Haines, and R. Srikanthan. 1991. *Global runoff: Continental comparisons of annual flows and peak discharges*. Cremlingen-Destedt, Germany: Catena Paperback.

Morgan, M., L. Streilein, and J. Weir. 2004. *Indigenous rights to water in the Murray Darling Basin: In support of the indigenous final report to the Living Murray Initiative*. Canberra, Australia: Australian Institute of Aboriginal and Torres Strait Islander Studies.

Morrison, T. H., G. T. MacDonald, and M. B. Lane. 2004. Integrating natural resource management for better environmental outcomes. *Australian Geographer* 35:243–58.

Nielsen, D. L., M. A. Brock, G. N. Rees, and D. S. Baldwin. 2003. Effects of increasing salinity on freshwater ecosystems in Australia. *Australian Journal of Botany* 51:655–65.

NLWRA. 2001. *National Land and Water Resources Audit: An initiative of the Natural Heritage Trust*. http://www.nlwra.gov.au/default.asp (accessed November 17, 2006).

Olive, L. J., and W. A. Rieger. 1986. Low Australian sediment yields: A question of inefficient sediment delivery. In *Drainage basin sediment delivery*, ed. R. F. Hadley, 305–22. International Association of Hydrological Sciences, Publication Number 159.

Outhet, D., G. Adamson, J. Armstrong, J. Graice, W. Hader, K. Lorimer-Ward, C. Massey, F. Nagel, S. Pengelly, A. Raine, M. Shepheard, T. Smith, P. Wem, and R. Young. 1999. Evolution of stream protection and rehabilitation in NSW. In *Proceedings of the Second Australian Stream Management Conference, Adelaide*, ed. I. Rutherfurd and R. Bartley, 487–92. Melbourne, Australia: Cooperative Research Centre for Catchment Hydrology.

Pahl-Wostl, C. 2002. Towards sustainability in the water sector: The importance of human actors and processes of social learning. *Aquatic Sciences* 64:394–411.

Pannell, D. J. 2006. Seeking more effective NRM policies. *Pannell Discussions* 82:1–3. http//cyllene.uwa.edu.au/~dpannell/pd/pd0082.htm (accessed August 9, 2006).

Pannell, D. J., A. Ridley, E. Seymour, P. Regan, and G. Gale. 2007. *Regional natural resource management arrangements for Australian states: Structures, legislation and relationships to government agencies (April 2007)*. SIF3 Working Paper 0701, CRC for Plant-Based Management of Dryland Salinity, Perth. http://cyllene.uwa.edu.au/~dpannell/cmbs3.pdf (accessed March 20, 2007).

Pigram, J. J 2006. *Australia's water resources: From use to management*. Collingwood, Australia: CSIRO Publishing.

Poff, N. L., J. Olden, D. Pepin, and B. Bledsoe. 2006. Placing stream flow variability in geographic and geomorphic contexts. *River Research and Application* 22:149–66.

Prosser, I. P., I. D. Rutherfurd, J. M. Olley, W. J. Young, P. J. Wallbrink, and C. J. Moran. 2001. Large-scale patterns of erosion and sediment transport in river networks, with examples from Australia. *Marine and Freshwater Research* 52:81–99.

Rogers, K. 2006. The real river management challenge: Integrating scientists, stakeholders and service agencies. *River Research and Applications* 22:269–80.

Rutherfurd, I. D. 2000. Some human impacts on Australian stream channel morphology. In *River manage-*

ment: The Australasian experience, ed. S. Brizga and B. Finlayson, 11–49. Chichester, UK: Wiley.

Rutherfurd, I., and R. Bartley, eds. 1999. *Proceedings of the 2nd Stream Management Conference, Adelaide*, Volumes 1 and 2. Melbourne, Australia: Cooperative Research Centre for Catchment Hydrology.

Rutherfurd, I. D., and C. Gippel. 2001. Australia versus the world: Do we face special opportunities and challenges in restoring Australian streams? *Water Science and Technology* 43:165–74.

Rutherfurd, I. D., K. Jerie, and N. Marsh. 2000. *A rehabilitation manual for Australian streams, volumes 1 and 2*. Canberra, Australia: Cooperative Research Centre for Catchment Hydrology, and the Land and Water Resources Research and Development Corporation.

Rutherfurd, I., K. Jerie, M. Walker, and N. Marsh. 1999. Don't raise the titanic: How to set priorities for stream rehabilitation. In *Proceedings of the Second Australian Stream Management Conference, Adelaide*, ed. I. Rutherfurd, and R. Bartley, 627–532 Melbourne, Australia: Cooperative Research Centre for Catchment Hydrology.

Rutherfurd, I., F. Sheldon, G. J. Brierley, and C. Kenyon, eds. 2001. *Third Australian Stream Management Conference Proceedings: The value of healthy rivers, 27–29 August, 2001, Brisbane, volumes 1 and 2*. Melbourne, Australia: Cooperative Research Centre for Catchment Hydrology.

Rutherfurd, I., and M. Walker, eds. 1996. *Proceedings of the First National Conference on Stream Management in Australia, Stream Management '96, Merrijig*. Melbourne, Australia: Cooperative Research Centre for Catchment Hydrology.

Rutherfurd, I. D., I. Wiszniewski, M. J. Askey-Doran, and R. Glazik, eds. 2005. *Proceedings of the 4th Australian Stream Management Conference: Linking rivers to landscapes*. Hobart, Australia: Department of Primary Industries, Water and Environment.

Tan, P.-L. 2002. Legal issues relating to water use. In *Property: Rights and responsibilities: Current Australian thinking*, ed. C. Mobbs and K. Moore, 13–42. Canberra, Australia: Land and Water Australia.

Thompson, D., and S. Pepperdine. 2003. *Assessing community capacity for riparian restoration*. Canberra, Australia: Land and Water Australia.

Unmack, P. J. 2001. Biogeography of Australian freshwater fishes. *Journal of Biogeography* 28:1053–89.

Wasson, R. J. 1994. Annual and decadal variation in sediment yield in Australia, and some global comparisons. In *Variability in stream erosion and sediment transport*, ed. L. J. Olive, R. J. Loughran, and J. A. Kesby, 269–79. International Association of Hydrological Sciences, Publication Number 224.

Wentworth Group of Concerned Scientists. 2003. *Blueprint for a national water plan*. Sydney, Australia: WWF Australia.

Young, W. J., F. M. Marston, and J. R. Davis. 1996. Nutrient export and land use in Australian catchments. *Journal of Environmental Management* 47:165–83.

Chapter 10

River Management in the United States

ELLEN WOHL, MARGARET PALMER, AND G. MATHIAS KONDOLF

> Clean water is not an expenditure of federal funds; clean water is an investment in the future of our country.
>
> U.S. representative Bob Shuster, quoted in the *Washington Post*, January 9, 1987

This chapter discusses the contemporary conditions of rivers in the United States, as well as efforts to better manage rivers. The United States covers a large geographic area with diverse ecological conditions, and this chapter can provide only a brief overview of a broad and complex topic. After briefly summarizing contemporary river health, we review the policy and legal framework that influences river management in the United States, contemporary pressures and social constraints on water resources, and issues that will likely continue to influence river management in the future. We then briefly discuss several broad categories of activities applied to river protection and rehabilitation, and end with a summary of what has been achieved through these activities. This final section focuses on four case studies that examine multidisciplinary teams employing adaptive management to restore medium- or large-sized river systems. These case studies from the Colorado River in Arizona, the Savannah River in Georgia, the Cosumnes River in California, and the Kissimmee River in Florida reflect the diversity of river systems and challenges to rehabilitating rivers within the United States.

How Healthy Are Rivers in the United States?

The United States covers a vast area; including Alaska and Hawaii, 9.2 million square kilometers, from approximately 20 degrees to 70 degrees North and 60 degrees to 170 degrees West. This geographic extent includes a corresponding diversity of ecosystems and physical environments. With 1,200 fish species, the continent of North America (including Canada, the United States, and Mexico) is second only to Asia, which has 1,500 species. North America also has the largest number of margaritiferid and unioniid mussel species (281 species and 16 subspecies) (Williams et al. 1993); 77 percent of all crayfish species, including 99 percent of all species in the family Cambaridae (Hobbs 1988); and more than 10,000 species of freshwater invertebrates (Abell et al. 2000). Approximately 5.3 million kilometers of rivers drain the diverse lands on which these organisms live, and the health of these rivers varies substantially. Seventy-nine percent of these river kilometers are somehow affected by human activities, and another 19 percent are drowned by reservoirs associated with dams, leaving only 2 percent of river kilometers relatively unimpacted (Graf 2001). A quick review of numerical summaries suggests that the majority of rivers in the United States are not especially healthy.

- In the United States, 75,187 large dams and more than 2.5 million small dams have altered river flows, and all watersheds larger than about 2000 square kilometers, excluding those in Alaska, have some dams (Graf 1999; Abell et al. 2000).
- The U.S. Geological Survey's 1991–1995 National Water Quality Assessment (NAWQA) program found that streams and ground water in basins with significant agricultural and/or urban development almost always contain complex mixtures of nutrients and pesticides, although the concentrations of these contaminants relative to drinking and other water-quality standards vary widely (U.S. Geological Survey 1999); more than a third of the rivers in the country are listed as impaired or polluted (EPA 2000).
- Almost half the nation's waters fail to meet biological water-quality standards (Doppelt et al. 1993).
- The conservation status of fifteen of the forty major watersheds or regions in the United States is rated 'critical,' the highest category for a ranking based on degree of land-cover alteration within the catchment, degradation of water quality, alteration of hydrologic integrity, degree of habitat fragmentation, effects of introduced species, and impacts of direct species exploitation (Abell et al. 2000).
- An estimated 70 to 90 percent of riparian forests have been lost nationally, from 30 to 40 million hectares historically to less than 10 million hectares at present (Doppelt et al. 1993; Innis et al. 2000).
- Wetland losses have exceeded 50 percent, and are close to 90 percent in some portions of the country (Innis et al. 2000).
- Anywhere from one third to three quarters of aquatic species nationwide are rare to extinct; 67 percent of freshwater mussels and 65 percent of crayfish are rare or imperiled; 37 percent of freshwater fish species are at risk of extinction; 35 percent of amphibians that depend on aquatic habitats are rare or imperiled; twenty-seven species of freshwater fish and ten mussel species have gone extinct during the past century (Doppelt et al. 1993; Abell et al. 2000).
- Extinction rates of freshwater fauna are five times those for terrestrial biota (Ricciardi and Rasmussen 1999).

These statistical descriptors of rivers in the United States suggest that, despite a long tradition of protecting public lands and legislating minimum standards for water quality and environmental protection, river health has been steadily eroded by a variety of human activities. Many land-use activities occur throughout the country at varying intensities, although the history of land use and the intensity and type of land use differ by geographic region within the United States. The portion of the continental United States between 100 degrees West and the coastal ranges at approximately 122 degrees West is arid or semi-arid, for example, and was among the last areas of the country to be settled by people of European descent. However, this same area currently has the highest ratio of reservoir storage/mean annual runoff (Graf 1999) (figure 10.1a). This primarily reflects the number of very large dams built on the major arid-region rivers, including the Colorado, Missouri, and Rio Grande river systems. In contrast, the eastern and southeastern margins of the United States have the greatest number of dams per unit area (Graf 1999) (figure 10.1b), reflecting the numerous smaller dams built on the rivers of these regions.

Until recent decades, it was reasonable to generalize that water quality and loss of habitat were the major limitations to river health in the wetter, more densely settled portions of the country,

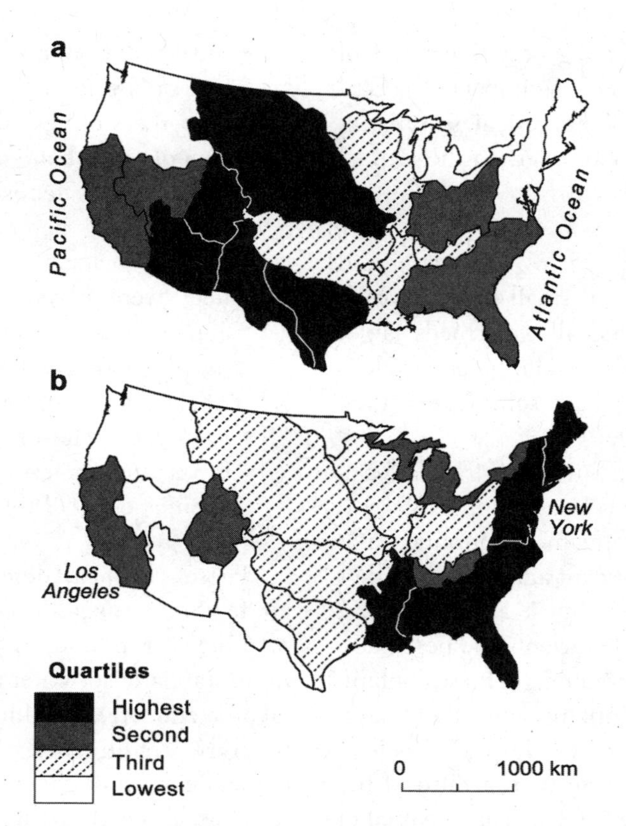

FIGURE 10.1. Geographic distribution of dams in the continental United States. (a) Ratio of reservoir storage divided by mean annual runoff, and (b) dams per unit area (dams per square kilometer) (after Graf 1999, figure 1).

whereas water quantity was the major limitation in the arid and semi-arid regions. However, this generalization no longer adequately describes a nation in which even areas with average annual precipitation exceeding 50 centimeters have effectively become more arid because of wetland drainage and floodplain settlement, lowered water tables, loss of forest cover, and urbanization. All of these alterations to the landscape decrease infiltration and water storage capacity, and increase the speed with which precipitation moves into and through drainage networks (Wohl 2004). At the same time that water quantity has become more limiting in wetter areas of the United States, contaminants in surface and shallow ground waters have become nearly ubiquitous (U.S. Geological Survey 1999). The results of these trends are that water quality, riverine habitat, and water quantity are now major limitations to river health throughout the continental United States.

Policy and Legal Framework

Governmental organization in the United States proceeds from the federal government and its legislation, which applies to the entire country, down to the level of each of the fifty states, and from there to more local entities such as counties and cities. Federal legislation that influences river health in the United States includes the Clean Water Act of 1977, which established the basic structure for regulating discharges of pollutants into the waters of the United States. The 1977 act, which was preceded by several Federal Water Pollution Control acts, gave the Environmental Protection Agency the authority to implement pollution control programs such as setting wastewater standards for industry. The act contains many additional provisions, such as funding for the construction of sewage treatment plants. Modifications to the act have continued since 1977, but the basic framework that the act established remains in place.

Other critical federal legislation that affects rivers includes the 1969 National Environmental Policy Act, which requires that all federal agencies prepare detailed environmental impact statements for "every recommendation or report on proposals for legislation and other major federal actions significantly affecting the quality of the human environment"; the 1973 Endangered Species Act, which is designed to protect critically imperiled species from extinction; the 1977 Surface Mining Control and Reclamation Act, which strengthened the standards of most state surface mining programs and created a program for reclaiming previously mined and inadequately reclaimed land; and the 1980 Comprehensive Environmental Response, Compensation, and Liability Act, better known as Superfund, which requires responsible parties to clean up environmental contamination. Enforcement of these acts has varied widely among the states, and through time, depending on the priorities of each presidential and state governor's administration. The Superfund act was rendered ineffective by funding cuts during the first term of the G. W. Bush administration, and it remains to be seen whether future administrations will resurrect the act.

Even when administrators intend to protect and improve river health, fragmentation among numerous agencies at the national, state, and local levels leads to difficulties in assessing river conditions, and in enforcing or implementing existing legislation. Among the national governmental agencies that influence river conditions in the United States are the Army Corps of Engineers, the Bureau of Land Management, the Bureau of Reclamation, the Department of Transportation, the Environmental Protection Agency, the Federal Emergency Management Agency, the Fish and Wildlife Service, the Forest Service, the National Park Service, the Natural Resources Conservation Service, and the Office of Surface Mining. Each state, county, and city also has a plethora of agencies that implement regulations or make decisions influencing river health. Rivers by their nature integrate landscapes, transporting water, sediment, nutrients, organisms, and contaminants from one political jurisdiction to another, but many of the governmental structures existing in the United States are ill-suited to deal with this integration.

Contemporary Pressures and Constraints on Water Resources

Attempts to improve river health in the United States must consider a variety of physical and social constraints present across the country, as well as the political divisions mentioned above. Physical and social constraints include an aging water infrastructure rated substandard and at risk by the American Society of Civil Engineers (ASCE 2005). Constraints also include the realities of population growth and distribution in a nation with a current population of over 300 million. The national population growth rate is less than 1 percent, but the U.S. Census Bureau estimates that the country experiences a net gain of one person every 12 seconds. Furthermore, the rate of conversion of forest and agricultural land to urban or residential land is far higher than the rate of population growth and this is having a devastating effect on river networks (NAS 2001). Headwater streams (first through third order) are disappearing at an alarming rate (Meyer and Wallace 2001). The great majority of the population lives in the eastern half of the country and along the western coast, and in urban areas. Rivers in these and other urbanized regions are commonly severely constrained because their tendency to move laterally or flood over the banks creates hazards for urban infrastructure and people. Urban rivers are also likely to receive large volumes of contaminants from storm drainage, wastewater treatment plants, and industrial effluents. Because of the large amount of impervious cover, urban watersheds often exhibit flashier peak flows, which can more readily erode stream banks and mobilize streambed substrate and kill benthic organisms (Moore and Palmer 2005). In some cities, urban stormwater drainage systems exacerbate the problem of high peak flows, with piped storm drainage networks efficiently routing stormwaters into stream channels (Bernhardt and Palmer 2007). These storm/sewer networks effectively bypass the river floodplain and route contaminants directly from roads and buildings into surface waters (Paul and Meyer 2001).

Irrigation for agriculture remains the single largest consumptive use of surface and ground water in the United States, particularly in the western half of the country, accounting for an estimated 40 percent of total offstream use in 1990 (Agricultural Research Service 2001). Both the water use and the political clout of agricultural users constrain efforts to improve either water quality in rivers or the amount of water reserved from consumptive use. In the eastern United States, for example, coastal plain streams are polluted by excess nutrients and pesticides yet are immune from provisions of the Clean Water Act in all but a few states.

Conflicts over water use and allocation commonly arise during extreme conditions that affect both agricultural and urban areas. Extremes that affect river environments include floods and droughts. Measures of total flood damage and flood damage per capita have substantial interannual variability over the period of systematic records (1926 to the present), but each metric shows a statistically significant increase over this period (Pielke et al. 2002). Overbank floods facilitate processes such as access to nursery habitat for young fish, which is critical to the health of many river ecosystems. Increasing flood damage to human communities generally results in attempts to restrict overbank flooding, which limits or destroys the geomorphic and ecological connectivity of the river and floodplain.

Droughts occur throughout the United States, and at least one region of the country experiences drought conditions in any given year. Particularly severe droughts occurred throughout the country during the 1930s and the late 1980s, and in the southwestern United States and southern Great Plains during the 1950s. Oregon's Klamath River illustrates the polarization that can result during drought in a region where water supplies are already fully allocated for normal flow conditions. Two endemic fish species living in the Klamath River, the shortnose sucker (*Chasmistes brevirostris*) and the lost river sucker (*Deltistes luxatus*), as well as the coho salmon (*Oncorhynchus kisutch*), are listed as endangered species. A Bureau of Reclamation decision to cut off irrigation water to 90 percent of the bureau's Klamath Project croplands in order to aid salmon listed under the Endangered Species Act touched off massive protests from farmers affected by the change. When irrigation flows were reinstituted the following year, an estimated 40,000 to 50,000 salmon were killed in September 2002, touching off equally massive protests from fishermen and environmentalists. The bureau has now created a water bank to meet the needs of the three endangered fish species in the river (Burke et al. 2004).

Social constraints on river health can arise from fiscal or economic conditions, where society does not consider river health of sufficient importance to devote the sums of money necessary to preserve or improve health, or from aesthetic preferences and perceptions of rivers. Many riverine "restoration" projects nationally, and most in urban areas, are actually infrastructure protection projects (Bernhardt and Palmer 2007). The normally healthy migration of rivers within their floodplains is viewed as an assault on homes and sewer lines. In the 1950s through the 1970s this was often dealt with by simply paving streams and river beds—essentially turning them into lifeless pipes whose ecological functioning was lost. As downstream and coastal areas became increasingly degraded, there was recognition that this concretization of river channels removed their natural capacity to cleanse water, but there is a distinct tension between protecting existing infrastructure and restoring river function (Palmer et al. 2005). In most cases, the infrastructure wins out and restoration involves moving the channel and recreating engineered channels. Although less concrete is being used in channels today, huge granite boulders and bundles of rocks ("rip-rap") are often brought in to keep relocated channels static (Brown 2000; Bernhardt and Palmer 2007).

Another social constraint on river health is related to recreation on or near rivers. Humans have chosen to live and work near water bodies since antiquity and entire civilizations have developed along waterways. All major rivers in the continental United States have cities somewhere along their path. Most of these rivers are highly degraded as a result of their use for transporting goods and/or industrial activities. With economic and social change has come the desire to revitalize city centers, particularly along waterways; attractive storefronts and restaurants along rivers are quite popular. Although riverfront revitalization projects may be successful in increasing economic and social activity near a river, they almost always constrain natural processes of the river and floodplain (Johansson and Nilsson 2002).

Away from city centers, recreational uses of rivers, particularly fishing and boating, can be extremely important to local economies, and can produce politically active recreational lobbies. This can be beneficial to river health where recreationalists desire a more natural flow regime or preservation of the riverine corridor, but recreational use can also be very detrimental to river health where recreationalists desire conditions that favor introduced species over native species, or desire an artificial flow regime (including reservoirs). The Colorado River downstream from Glen Canyon Dam in Arizona provides an example of a situation where some recreational users strongly favor retaining cold, clear-water releases from a dam in order to support introduced species of trout at the expense of endangered native fish that evolved to use the warm, turbid waters naturally present on the Colorado River (Collier et al. 1997).

Small dam removal is increasing rapidly in some portions of the United States. Most of these dams no longer served an economic function and needed significant repairs. Dam removal commonly increased longitudinal connectivity of the river and enhanced fish passage, but dam removal was contentious in most communities because of public concerns that flooding would increase, the reservoir sediments would become a permanently unsightly mudflat, property values would decline, fishing would decline, or reservoir-related recreation would be lost (Sarakinos and Johnson 2002). Larger dams may not be removed because they still provide services such as water storage or hydroelectric power, but the concept of changing dam operations to allow habitat maintenance flows is often strongly opposed by the entity operating the dam, or by water users such as irrigators or barge operators who depend on water storage and uniform flows throughout the year, as demonstrated by the Grand Canyon (Collier et al. 1996), the Black Canyon of the Gunnison River in Colorado (Elliott and Parker 1997; Dubinski and Wohl 2007), or the Snake River of Idaho (Benjamin and Van Kirk 1999). In some cases, however, dam operation can be changed for environmental benefit, while retaining the benefits from the dam. Recent work suggests that river basins impacted by dams will experience disproportionate ecological impacts under future climates and this will require more attention to dam operation and land management (Palmer et al. 2007).

Likely Future Influences on River Management

Each of the issues described in the preceding section will continue to influence management of rivers in the United States during the coming decades. Rapid population growth and urban/suburban sprawl, particularly in the southeastern and the arid western United States, will continue to reduce the spatial extent and function of wetlands and floodplains, as well as altering infiltration, runoff, and water quality for many river systems. Ongoing construction of dams and other structures for flow regulation in arid regions with rapid population growth (e.g., the Colorado Front Range) will contrast with continued removal of obsolete small dams in the eastern half of the country. Although the conversion of agricultural lands into conservation easements and natural areas along valley bottoms in the eastern and midwestern portions of the country has slowed within the past few years, as has the clean-up of sites with toxic contamination, these conversions and clean-ups will continue and likely increase in the future. Increasing scientific and public recognition of the environmental costs of loss of river function—including water and soil contamination and associated hazards to human health, declines in fisheries, and increased hazards from flooding—is progressively building a strong constituency for developing and applying improved methods of river management that demonstrably enhance the ecological functioning of rivers (Palmer 2004).

Growing public awareness of the consequences of global warming for water supplies in arid regions is prompting individual businesses, communities, and government entities from the city to state level to take actions to decrease the emission of greenhouse gases. The rapid loss of alpine glaciers and snowfields (Meier et al. 2006), however, as well as changes in precipitation characteristics, will likely substantially impact water supplies and river flow regimes in many parts of the United States before greenhouse gases are effectively reduced. Under warming climate in the western United States, for example, more precipitation will fall as rain rather than snow. Combined with more rapid melting of the snowpack, this will fill reservoirs to capacity earlier than normal each year, creating the potential for water shortages where reservoir capacity cannot hold the winter's precipitation (Barnett et al. 2005).

Strategies for River Protection and Rehabilitation

Given the constraints on water resources summarized previously, people interested in protecting or rehabilitating river health in the United States focus primarily on eight activities: assessment methods; protection of threatened and endangered species, habitats, or ecosystems, and regions of particularly high biodiversity; best management practices and riparian buffers; instream flow; aquifer recharge; dam operation and removal; wetland restoration; and prioritization of conservation and rehabilitation sites. All of these activities implicitly improve river form and function. After briefly discussing each activity, we then focus on programs explicitly aimed at river rehabilitation.

Assessment methods. Various qualitative or numerical measures have been developed to allow comparison of ecological integrity, biological diversity, channel stability, riparian functional condition, and other characteristics of rivers and riparian corridors in different regions. At present there is only limited consensus on which measures are most appropriate, and there are thus few standards that are consistently applied throughout the United States. Innis et al. (2000) provide a detailed review of existing methods of riparian inventory and assessment used in the United States. Different methods use invertebrates, birds and mammals, amphibians, vegetation, and hydrologic patterns as indicators (e.g., Davis and Simon 1995; Jackson et al. 2000). Where riverine environments have been extensively and severely altered, it can be difficult to define reference or expected conditions that can serve as goals for management activities.

The U.S. Geological Survey's National Water Quality Assessment (NAWQA) program, which assessed conditions at fifty-nine study drainages throughout the country over a period of a few years, provides one of the more consistent comparisons. NAWQA data for many of the study sites include a habitat degradation index, based on amount of stream modification, bank erosion, and bank vegetative stability and indices of biotic integrity, based on invertebrates, algae, or the number and types of fish, fish feeding habits, and fish abundance and health (Gilliom et al. 1998).

An emerging issue is the need to shift from static measures of river ecosystem condition to more dynamic measures; that is, instead of assessing structural attributes such as the size and shape of the channel, the diversity of habitat types, or the presence of desirable species, the focus is shifting toward assessing the functioning and resilience of river systems. Measurements of ecological functioning evaluate dynamic properties of ecosystems that underlie an ecosystem's ability to provide vital goods and services. The units of an ecological function are a process *rate* and *direction*. Functions reflect system performance and their measurement requires quantification of ecological processes such as primary production or nutrient uptake.

Healthy streams are, after all, living, functional systems. Essential ecological functions of healthy streams include: the purification of water, the removal of excessive levels of nutrients and sediments before they reach downstream waters, the processing of organic material (decomposition or biological utilization), and primary and secondary productivity (growth of photosynthetic organisms and consumers) (table 10.1). These functions are supported by ecological processes including: the flux of water that is similar to unimpaired rivers, the processing of nutrients at the same rate and form as unimpacted rivers, the decomposition of organic matter at rates typical of nearby unimpacted systems, and, microbial, primary, and secondary production the same as nearby healthy rivers (Palmer et al. 1997; Naiman et al. 2005). To determine whether these processes are brought back to the right levels and direction through rehabilitation requires that they be measured. The methods for functional measures such as nutrient processing and organic matter decomposition have been developed and are now routinely used by stream ecologists. The movement now in the United States is toward using such measures in management and mitigation contexts, particularly because the Clean Water Act makes specific reference to ecological function (see chapter 6).

Protection of threatened and endangered species/habitats, and biodiversity hot spots. Endangered and threatened species are formally listed by the U.S. Fish and Wildlife Service following a strict legal process that is often lengthy to implement. Nongovernmental organizations also establish priority rankings of species, ecosystems, or places that they wish to protect (e.g., The Nature Conservancy's Great Places Network, the Sierra Club's Critical Ecoregions Program, or American Rivers' annual Most Endangered Rivers list). These rankings are based on the same types of criteria described in the section on conservation and rehabilitation priorities below. With the exception of the American Rivers program, most of these rankings do not explicitly focus on rivers. Governmental agencies such as the National Forest Service and the National Park Service are currently working to develop methods for assessing habitat diversity and habitat sensitivity to land uses within their land-management units (e.g., Wohl et al. 2007), but much of the information used by these agencies to date has come from nongovernmental research by organizations or academic scientists. Faced with increasing threats to river health and

TABLE 10.1

Rivers and streams provide a number of goods and services, called ecosystem functions, which are critical to their health and provide benefits to society. The major functions are outlined along with the ecological processes that support the function, how it is measured, and why it is important.

Ecosystem function	Ecological Process that supports this	Measurements required	Without it what happens
Water Purification (a) Nutrient Processing	Biological uptake and transformation of nitrogen, phosphorus	Direct measures of rates of transformation of nutrients; for example: microbial denitrification, conversion of nitrate to the more useable N form of ammonium	Excess nutrients can build up in the water, making it unsuitable for drinking or unable to support life
Water Purification (b) Processing of contaminants	Biological removal by plants and microbes of materials such as excess sediments, heavy metals, contaminants, and so on	Direct measures of contaminant uptake or changes in contaminant flux	Toxic contaminants kill biota; excess sediments smother invertebrates, foul the gills of fish, and so on; water not potable
Decomposition of organic matter	The biological (mostly by microbes and fungi) degradation of organic matter such as leaf material or organic wastes	Decomposition is measured as a rate. Usually expresses as the slope of a line showing weight loss over time of organic matter that is heated to high temperatures to convert all the particulate carbon to gas (CO_2)	Without this, excess organic material builds up in streams; this leads to low oxygen levels which in turn leads to death of invertebrates and fish and water that is not drinkable
Primary Production (growth of algae and aquatic plants) Secondary production (growth and reproduction of consumers such as insects and fish)	Measured as a rate of new plant or animal tissue produced over time	For primary production, measure the rate of photosynthesis in the stream; for secondary, measure growth rate of organisms	Primary production supports the food web; secondary production supports fish
Temperature Regulation	Water temperature is "buffered" if there is sufficient infiltration in the watershed and riparian zone *and* shading of the stream by riparian vegetation keeps the water cool	Measure the rate of change in water temperature as air temperature changes or as increases in discharge occur	If the infiltration or shading are reduced (due to clearing of vegetation along stream) then the stream water heats up beyond what the biota are capable of tolerating
Flood Control	Slowing of water flow from the land to streams or rivers so that flood frequency and magnitude are reduced; intact floodplains and riparian vegetation help buffer increases in discharge	Measure the rate of infiltration of water into soils *or* discharge in stream in response to rain events	Without the benefits of floodplains, healthy stream corridor, and watershed vegetation, increased flood frequency and flood magnitude occur

limited budgets, personnel, and time, most private and governmental organizations that include river management are adopting some type of prioritization scheme in order to enhance their efficiency and effectiveness. Prioritization may be based on EPA 303d listings (impaired waters), but this approach is problematic if applied across states because the listing of streams as impaired is based on criteria and standards that are unique to each state (Hassett et al. 2005).

Best management practices and riparian buffers. The phrase "best management practices" is applied to a wide variety of commercial activities. In the context of this chapter, the phrase refers to a set of guidelines used primarily during timber harvest, farming, and residential, commercial, and industrial construction to minimize disruption to adjacent rivers. Timber best management practices (BMPs) specify guidelines for actual harvest, new road construction, and haul road maintenance, and generally focus on the maintenance of streamside buffers composed of some minimum width of undisturbed vegetation (MacDonald et al. 1991; Rashin et al. 1999). Road BMPs include guidelines for road drainage (specifically culverts), and construction and stabilization of cutslopes and fillslopes on road segments draining to streams. Measures of BMP effectiveness include erosion and sediment delivery to rivers, physical disturbance of stream channels, and condition of aquatic habitats and biological communities (Rashin et al. 1999).

Agricultural BMPs include guidelines to minimize the application of pesticides through integrated pest management (a management strategy that includes an understanding of the target pest and use of a combination of physical, chemical, biological, and cultural controls); control erosion and runoff through practices such as strip-cropping, shelterbelts, use of cover crops, and riparian buffers; reduce the use of nutrients such as fertilizers and manure; and reduce nonpoint source pollution from animal feeding operations (EPA 2003a, 2003b, 2003c).

Although some aspects of timber and agricultural BMPs are regulated through federal and state legislation, use and enforcement of these practices varies widely in the United States. Further, it has been extremely difficult to determine the effectiveness of specific agricultural BMPs in improving the ecological condition of streams and rivers because obtaining accurate information on farming practices for entire watersheds that drain to a particular stream is extremely difficult (issues of privacy limit widespread release of this information by USDA and local NRCS district offices) (Moore and Palmer 2005).

Construction BMPs focus on erosion and sediment control and are ultimately overseen by the U.S. Environmental Protection Agency as part of its enforcement of the Clean Water Act (U.S. Environmental Protection Agency 2000). All construction activities that disturb greater than 0.4 hectare (1 acre) of land must be covered by a permit under the National Pollutant Discharge Elimination System program. States, counties, and towns also enforce their own regulations, which may be stricter than regulations at the national level. All of these regulations specify allowable sediment discharge from construction point sources, and recommend methods such as use of brush barriers, check dams, chemical stabilization, filter berms, geotextiles, gradient terraces, land grading, mulching, and silt fences to limit erosion and sediment discharge.

Instream flow. Instream flow refers to the quantity of water actually in the stream channel, rather than diverted for off-channel uses. Hydrologists and biologists working for government agencies having regulatory responsibility related to water development attempted to quantify instream flow needs of species and aquatic and riparian communities starting in the late 1960s. Most of the work thus far has focused on fish, macroinvertebrates, and riparian vegetation, although attention to recreational instream flow needs has recently increased. Initial approaches specified a fixed minimum flow that was not necessarily linked to a specific benefit to aquatic or riparian habitats. Incremental methods developed to quantify changes in aquatic and riparian habitat as a function of discharge are often based on habitat suitability curves (Orth and Maughan 1982). Critics quickly noted that the assumption of a positive linear relationship between potential available habitat and biomass of organisms was not justified, and that significant interactions among habitat variables can affect habitat selection by fish (Mathur et al. 1985). Moreover, fish use habitat at a much finer scale (e.g., low velocity zones downstream from boulders) than most hydraulic models can predict (Kondolf et al. 2000). The basic concepts underlying such an approach remain the most widely used because of their simplicity (e.g., Rood et al. 2003; Strakosh et al. 2003), but there is little evidence to link model outputs with actual fish populations (Castleberry et al. 1996). Models of interactions among hydrology and physical habitat parameters have, however, grown much more sophisticated (e.g. Stewart et al. 2005). Scientists have also increasingly emphasized that a range of flow levels are critical to maintain channel processes, habitat, and riverine ecosystems, as articulated in the concept of the natural flow regime (Poff et al. 1997).

Aquifer recharge. Aquifer recharge occurs through natural processes of infiltration, but can be artificially enhanced through infiltration and/or pumping of water into the subsurface in order to restore subsurface water supplies and associated stream flow. Where changes in land cover have altered infiltration capacity and runoff processes, and pumping of groundwater has lowered the water table, aquifer recharge becomes particularly important to prevent land subsidence and to maintain stream flows for water supply, water quality, and riverine habitat. Natural aquifer recharge can be enhanced with reclaimed water—wastewater that has received at least secondary treatment and basic disinfection or better, and that is reused after leaving a municipal wastewater treatment facility. This water can then be used to facilitate infiltration from irrigated lands, unlined storage ponds, wetlands, septic tanks, or injection into the subsurface (Cook 1997). Management activities also focus on preserving natural recharge areas such as floodplains and wetlands from land uses that impede infiltration. Many management agencies responsible for individual river basins such as the South Platte River of Colorado (Aikin and Turner 1989), the rivers fed by the Edwards Aquifer in south-central Texas (Ferrill et al. 2004), or rivers in Florida (McKim et al. 1995), are actively pursuing the protection of natural recharge areas and the enhancement of recharge through artificial methods.

Dam operation and removal. As unforeseen environmental changes and costs associated with dams became more apparent in the late twentieth century, the idea of removing dams entirely or at least changing the manners in which they operated gained momentum. Discussions of dam removal and operational changes remain contentious because of opposing perceptions of the cost-benefit ratio of dams, in which costs are commonly environmental and borne by the entire human and nonhuman community, whereas benefits are usually specific to users such as power or water consumers. Dam removal can also be contentious because of uncertainties that arise from the slow pace of scientific research and understanding relative to bureaucratic decisions and public expectations (Graf 2002). The nonprofit organization American Rivers lists more than 200 dams removed in the United States from 1999 through 2004. The great majority of these were small dams (less than 10 meters in height), but they were spread across the country and they provide useful case studies on river response to dam removal (e.g., Stanley et al. 2002; Stanley and Doyle 2003) from which more generalized conceptual models can be developed (e.g., Hart et al. 2002; Poff and Hart 2002). The first large dams are scheduled to be removed in the states of Washington and California within the next five years, and ongoing research on the geomorphic effects of dam removal includes field studies, laboratory experiments, and numerical simulations (NCED 2005).

Wetland restoration. In 1977 President Jimmy

Carter granted protection to wetlands. A national wetlands policy forum convened by the EPA in 1987 promoted the theme of no net loss of wetlands, implying that if some wetlands were destroyed by land use, then new wetlands must be constructed elsewhere. However, basic questions such as how to define a wetland and no net loss have kept the United States from attaining anything close to no net loss of wetlands (Goldstein 1991; Committee on Mitigating Wetlands Losses 2001), despite a great interest in wetland restoration.

Wetland restoration can take the form of reclaiming naturally occurring wetlands that have been filled and/or drained, or constructing new wetlands. These activities can be undertaken to restore aquifer recharge, as described previously, to provide habitat for wetland plants and animals, or to provide water filtration with the aim of improving the quality of surface and ground waters (Hook 1988; Hammer 1989). The Environmental Protection Agency website (http://www.epa.gov/owow/wetlands/restore/) lists hundreds of past and current projects, including twelve national showcase watersheds (ranging from less than 5 square kilometers to nearly 28,000 square kilometers) in which wetland and river rehabilitation are linked to land uses occurring throughout the watershed. The spread of mosquito-transmitted West Nile virus during the first years of the twenty-first century has locally constrained some wetland restoration efforts as public perception of wetlands as mosquito-breeding grounds diminishes support for their protection.

Conservation and rehabilitation priorities for rivers in the United States. No single national authority sets conservation and rehabilitation priorities for rivers in the United States. A plethora of federal, state, and local agencies with overlapping and sometimes contradictory mandates are involved in prioritizing different types of related programs that conserve and rehabilitate species, ecosystems, and different types of protected lands such as national parks and forests. Nongovernmental agencies also develop their own priorities. Each year the environmental organization American Rivers, for example, issues a list of the ten most endangered rivers in the country. These rivers are deliberately chosen to reflect the geographic span of the country. A review of the rivers designated during the first years of the twenty-first century indicates recurring threats regardless of geographic location: dams, diversions, and flow regulation; contaminants from human sewage to factory farm and industrial effluents; and watershed development, whether for coalbed methane drilling or urbanization. Under a political climate that facilitates ignoring or even dismantling existing legislation and programs for environmental protection, much of the energy of managers, activists, and scientists is directed toward simply enforcing existing measures, in part by raising public awareness both of specific threats and more broadly of the critical importance of healthy rivers.

Priorities for river conservation and rehabilitation can be discussed in terms of concepts or geography. Conceptual priorities include helping agencies and the public to focus on process rather than form when rehabilitating or conserving rivers. A focus on process facilitates protection and recovery of physical and ecological integrity. A river with physical and ecological integrity is a functioning ecosystem that continually responds to changes in water, sediment, nutrients, contaminants, or organisms entering the river (Graf 2001). It is characterized by connectivity in the longitudinal, lateral, and vertical dimensions, and a dynamic flow regime and channel (Kondolf et al. 2006). Such a river is not defined by a constant, stable form, but because a stable form is relatively simple to conceptualize and impose (although not to maintain), it presently is too often a goal of river conservation and rehabilitation.

Another, very basic conceptual priority of river conservation and rehabilitation is to view a river as an entire watershed with atmospheric, surface, and subsurface links to the rest of the region and, ultimately, the world. Conservation and rehabilitation are too often undertaken at a very local scale with larger-scale constraints being ignored. For example, river rehabilitation that seeks to reinstate a former meandering configuration to a segment of river that is now braided will be unlikely to succeed if the changes in slope, bank stability, and sediment supply that initiated braiding are not addressed. Changes in sediment supply in particular are likely to reflect processes occurring either upstream from the river segment under consideration, or on adjacent valley side slopes. Similarly, protection of a river segment used for spawning by an endangered fish species is unlikely to be successful by itself if the fish is migratory and its access to the reach is blocked by dams or diversions, or its marine habitat is threatened by pollution. Thinking at a watershed scale is much more difficult than focusing on a limited segment of a river, because the larger scale of the watershed incorporates processes and exchanges seemingly far removed from the river itself, and of much greater complexity. Although difficult, this kind of thinking is essential to rehabilitation efforts. Rehabilitation projects in river reaches that are downstream from highly degraded waters typically fail. In some cases, the watershed-scale constraints are so severe that rehabilitation of single reaches is not a wise use of limited resources. Yet, with sufficient watershed planning, the cumulative effects of multiple projects may yield great benefits.

Geographical priorities for river conservation and rehabilitation can be organized by ecoregions, specific types of river ecosystems or communities, or individual species. Such subdivisions can also be prioritized to favor high biodiversity, high rates of endemism or uniqueness, or highly stressed rivers where permanent loss of species or ecosystems is imminent. Existing priority lists often incorporate all of these considerations. Watersheds assigned a critical conservation status during a study undertaken by the World Wildlife Fund cluster predominantly in the Great Lakes region, the southeastern United States, and the southwestern United States (Abell et al. 2000) (figure 10.2).

In general, the southeastern United States has the greatest species richness and the greatest total number of endemic species for both fish and other aquatic organisms. The Colorado River basin in the southwestern United States has the lowest species richness and the western Great Plains have the lowest number of endemic species. Compilation of these and other criteria results in the southeastern portion of the country receiving the highest ranking for biological distinctiveness, whereas the southwestern region and the Great Lakes are in the second ranking, and the western Great Plains are at the lowest (fourth) ranking (Abell et al. 2000). The conservation status rankings of the three most endangered regions reflects these biological metrics, as well as extensive alteration of land cover, degradation of surface water quality, alteration of hydrographic integrity, fragmentation of habitat, and likelihood of future threats. This comprehensive ranking thus implies that rivers in the Great Lakes region, and the southeastern and southwestern United States, should receive the highest priority for river conservation and rehabilitation. There has also been a recent discussion among managers, scientists, and activists of identifying criteria for establishing freshwater protected areas, similar in scope to the marine protected areas concept.

Examples of River Rehabilitation

During the past two decades, river rehabilitation has become highly popular and widely practiced among government agencies and public organizations. Many small- to medium-sized projects are contracted to private consulting firms with little to no expertise in fluvial geomorphology or aquatic ecology. Rehabilitation projects are commonly undertaken without a conceptual model of river ecosystems or a clearly articulated understanding of ecosystem processes, without recognition of the multiple, interacting temporal and spatial scales of river response, and without long-term monitoring of success or failure in meeting project objectives following completion (Pedroli et al. 2002; Bernhardt et al. 2005; Hassett et al. 2005; Wohl et al.

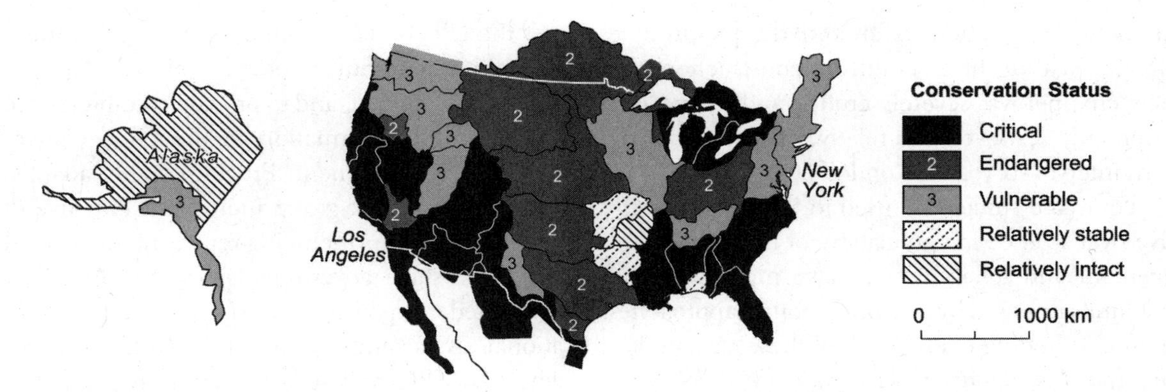

FIGURE 10.2. Relative ranking of conservation status for major watersheds in the continental United States and Alaska (after Abell et al. 2000, figure 4.10).

2005; Simon et al., forthcoming). In some cases, rehabilitation projects attempt to fix a static form on a dynamic system, rather than restoring process; a common example is the engineered meander bend in which the rip-rap prevents the normal bend migration that is an integral part of a meandering channel planform. In other cases rehabilitation attempts to superimpose a culturally preferred channel planform in a setting where that type of planform does not naturally exist, and cannot exist without continual intensive maintenance (e.g., Kondolf et al. 2001). Geomorphologists and ecologists are now actively responding to the numerous rehabilitation projects that have failed because practitioners ignored or misapplied scientific understanding of riverine systems (Smith and Prestegaard 2005). Position statements by professional organizations, publications in peer-reviewed journals (Kondolf et al. 2001; Malakoff 2004; Wohl et al. 2005), and the compilation of a database on rehabilitation projects (National River Restoration Science Synthesis, http://www.restoringrivers.org) are among the responses to date. However, it is clear that research scientists will have to become more proactive in the arena of public policy in order to correct misconceptions that are now common and widespread. Palmer et al. (2005) have argued that progress will not be made in rehabilitating rivers and streams until criteria for judging ecological success in rehabilitation have been agreed upon and adopted by federal and local authorities. On behalf of more than twenty U.S. scientists, Palmer and Allan (2006) have proposed specific federal-level regulatory and legislative policy reforms to improve the effectiveness of river rehabilitation and the health of the nation's waterways.

Large-scale rehabilitation projects undertaken by federal or state agencies and involving numerous groups are much more likely to receive scientific input, have adopted some criteria for evaluation and careful peer review, and thus to have a stronger conceptual framework and appropriate monitoring. Rehabilitation efforts on the Colorado River in the Grand Canyon of Arizona, the Kissimmee River in Florida, and several rivers in California exemplify this approach, and are discussed in more detail later in this chapter.

River rehabilitation project design is driven in many cases by public perception and aesthetic preferences more than by scientific concepts of river health (Bernhardt et al. 2005; Wohl et al. 2005). An illustrative example is Uvas Creek, which drains forested mountains along the central coast of California. The region is seasonally dry, with high interannual variability in floods that produces temporal variability in channel planform, from braiding shortly after flooding to gradual narrowing and increase in sinuosity with time since flooding (Kondolf et al. 2001). A 1995 river rehabilitation project attempted to impose a narrow, symmetrically meandering channel stabilized with boulders and rootwads, reflecting a common aesthetic expectation of an orderly, picturesque stream. Within three months of construction, the new channel was severely eroded and the project essentially destroyed by a relatively small flood (return interval six years) (Kondolf et al. 2001).

Four case studies are used to illustrate integrative river science in application, or how river management and rehabilitation have moved beyond the limits imposed by discipline-bound approaches to understanding rivers. Each of these case studies incorporates adaptive management (Walters 1997) and monitoring, which facilitate learning from experience.

Grand Canyon of the Colorado River in Arizona

Glen Canyon Dam was completed just upstream from the Grand Canyon in 1963. Designed for water storage and hydroelectric power generation, the 216-meter-high dam changed downstream river segments by effectively trapping sediment; altering the pre-dam flow regime that was characterized by large snowmelt peaks during late spring and early summer to a more uniform flow throughout the year, with short-term fluctuations imposed by hydroelectric power generation; releasing cold, clear, oxygenated water, rather than the historically warm, turbid water flowing in the river; and blocking fish migration along the river (Collier et al. 1996, 1997). Partly as a result of these changes, the endemic fish species of the upper Colorado River—the humpback chub (*Gila cypha*), the Colorado pikeminnow (*Ptychocheilus lucius*), and the razorback sucker (*Xyrauchen texanus*)—are endangered, and their geographic distribution is drastically curtailed. Other changes along the river downstream from the dam include severe beach erosion, erosion of Native American archeological sites, and invasion of exotic riparian (e.g., tamarisk, *Tamarix chinensis*) and aquatic (e.g., rainbow trout, *Oncorhynchus mykiss*) species (figure 10.3).

The Grand Canyon Monitoring and Research Center (GCMRC) was established in 1995 to assess the effects of the dam on biological, cultural, and physical resources of the Colorado River in the Grand Canyon, and to provide credible, objective scientific information to the Glen Canyon Adaptive Management Program. The adaptive management work group includes twenty-five diverse stakeholders including representatives of federal and state governments, Native American tribes, federal power purchase contractors, recreational users, and environmental organizations. The GCMRC and stakeholders utilize information generated by research scientists, and reviewed by independent panels of peers, to make recommendations on river management to the U.S. secretary of the interior.

Lack of sand storage along the channel margins has been identified as one of the key management issues for the Colorado River in Grand Canyon. Channel-margin sand deposits provide: camping areas for river rafters; lateral support that reduces erosion of older, higher sediment deposits which contain archeological materials; riparian habitat; and channel-margin diversity that supports shallow-water habitat for endemic fish species (Collier et al. 1997).

An experimental week-long flood release of 1,270 m^3s^{-1} from Glen Canyon Dam in March 1996 was instituted to test the idea that larger flows might mobilize sand stored on the streambed and deposit this sand along the channel margins. Pre-dam flood peaks were on the order of 3,500 m^3s^{-1}, with a gradual rise and fall that lasted for several weeks (Collier et al. 1997). Despite the much lower magnitude of the 1996 experimental flood, researchers used the flood to test hypotheses and to design requests for further experimental floods and for changes in how Glen Canyon Dam is operated. Continuing experimental flow releases are particularly important in providing opportunities to actively test conceptual and numerical models of river system response to perturbations in flow and sediment supply. At present, such releases are requested when floods along the Paria River, the most important source of sediment between Glen Canyon Dam and the Grand Canyon, provide

Figure 10.3. Matching photographs at river level in the Grand Canyon, just downstream from tributary Three Springs Canyon. The upper photograph was taken by Robert Stanton on March 1, 1890. The lower photograph was taken by Jane Bernard on February 26, 1991. Note how xeric shrubs now occupy what were the high-water zone and beach sands in the historical view. (From Webb 1996, figures 6.9 A and B).

sufficient sediment for redistribution along the river corridor in Grand Canyon (figure 10.4). Despite the potentially integrative framework provided by the concepts of adaptive management, however, it has in practice proven very difficult to obtain additional experimental flow releases because of concerns with revenue lost when hydroelectric power generation is not maximized, and because of a severe drought in the Colorado River basin during the period from 2000 to 2006. Current proposals to reduce the negative downstream effects of Glen Canyon Dam include dredging sediment from the reservoir and reintroducing the sediment downstream from the dam; altering the magnitude of daily flow fluctuations, ramping rates, and base flows; and stocking and/or translocating native fish populations (U.S. Geological Survey 2006).

Cosumnes River, California

The Cosumnes is a small river (3,000 square kilometers drainage) that flows 130 kilometers from headwaters in the Sierra Nevada to the Sacramento-San Joaquin Delta. Historically, the river

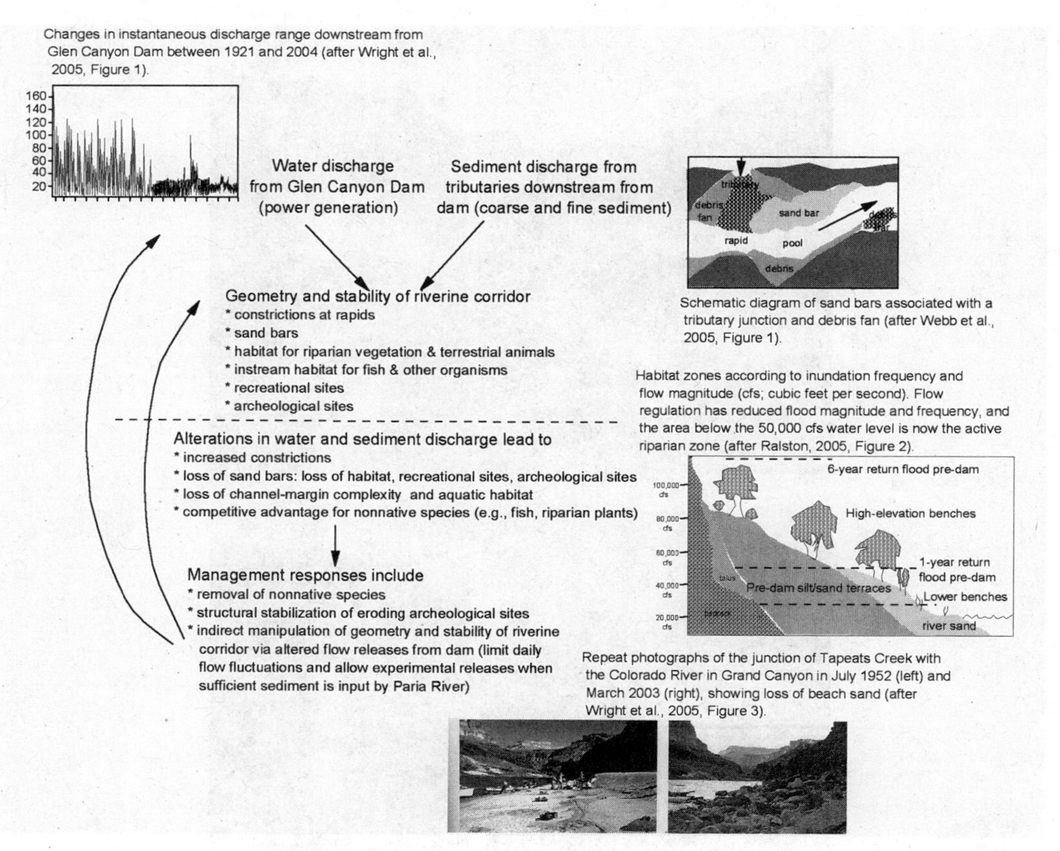

FIGURE 10.4. Simplified schematic flow chart showing relations among factors being considered in adaptive management for the Colorado River ecosystem in Grand Canyon. Alterations in water and sediment moving downstream along the Colorado River have caused changes in the geometry and stability of the river corridor downstream from Glen Canyon Dam. Changes in the river corridor have altered aquatic and riparian habitat, the stability of archaeological sites, and the availability of recreational sites (beaches for camping). Geologists and hydrologists work with ecologists and archaeologists to understand the processes involved in these changes. Management responses include altering flow releases from the dam, and potentially dredging sediment from the reservoir and reintroducing the dredged sediment downstream from the dam. Ongoing research focuses on quantifying the magnitude, duration, frequency, and seasonal timing of flow releases that optimize desired changes in downstream river geometry and stability.

beyond the mountain front had an anastomosing planform with multiple channels, seasonal marshes, and small floodplain lakes. The Cosumnes inundated broad floodplains during winter overbank flooding, and channel avulsion was an important process in maintaining floodplain heterogeneity (Swenson et al. 2003). Although it remains one of the few undammed rivers in this portion of California, the Cosumnes has been heavily altered by floodplain deforestation, filling and drainage of floodplain lakes and wetlands, agricultural uses of the floodplain, and construction of levees (Florsheim and Mount 2002). These alterations created a single main channel that was disconnected from adjacent floodplain wetlands and former secondary channels. In 1978 The Nature Conservancy established a preserve along the Cosumnes, which has subsequently grown to 5,260 hectares. From 1995 to 1997, preserve managers and scientists, working with support from the CALFED Bay-Delta watershed program, created two breaches each 30 meters in length in the levees. Levee breaching was designed to facilitate overbank flooding in order to encourage regeneration of cottonwood (*Populus fremontii*), willow (*Salix* spp.), and oak species in the riparian forest, as well as provide habitat for salmon and the rare Sacramento splittail (*Pogonichthys macrolepidotus*). As in the Grand Canyon example, ecologists and geomorphologists at Cosumnes work together to specify habitat needs and the physical processes that create habitats, using an adaptive management approach where observations of environmental response to levee breaching are used to design the overbank flood regimes necessary to sustain the channel-floodplain ecosystem (figure 10.5).

Florsheim et al. (2006) identified discharge thresholds of 23 to 25.5 m^3s^{-1} for establishing hydrologic connectivity between the channel and adjacent floodplain, and 100 m^3s^{-1} for sediment transfer from the main Cosumnes River through the levee breaches onto floodplain sand-splay complexes (figure 10.6). They then used the volume of sand deposited on floodplain splay complexes as an indicator of floodplain habitat, developed a model to aid in predicting volume of sand deposition as a function of discharge magnitude and duration, and specified a procedure by which sand deposition can be used as a tool to assess progress toward restoration goals developed at the start of the project. Assessment of progress toward restoration goals then feeds back via the adaptive management process into decisions about whether to modify active manipulations such as artificial levee breaches and reforestation with native plants.

The Cosumnes River project is one of many supported by the CALFED Bay-Delta program, which focuses on the Sacramento and San Joaquin river drainages, and the San Francisco Bay into which they drain. The region encompassed by the program supports 750 plant and animal species, and provides drinking water to more than 22 million people, supports 80 percent of California's commercial salmon fisheries, and supports a $27 billion agricultural industry (CALFED 2005). Prior to development of the CALFED program in 1998, competing interests were deadlocked, fish species and water quality were declining, and water supplies were becoming increasingly unreliable. In response, twenty-five state and federal agencies developed a framework in which they could work cooperatively with each other and with local and regional interests. One of the largest water management and ecosystem rehabilitation programs in the nation, CALFED has invested more than $850 million in ecosystem rehabilitation across several hundred projects. Some fish populations increased in the late 1990s and early years of this century, probably due in large part to a series of wet years. The Sacramento splittail has been removed from the endangered species list (CALFED 2005).

Collaborative Projects between the Army Corps of Engineers and The Nature Conservancy

The Army Corps of Engineers was established by Congress when the nation was founded, but its involvement in levees and other forms of river engineering dates to the first decades of the nineteenth century (Wohl 2004). The agency has a long history of water management, including construction of dams, diversions, and levees, channelization and dredging, and oversight of permits under the Clean Water Act. The Corps presently operates 630 dams across the country. The Nature Conservancy is a private, nonprofit organization founded in 1951 for the purpose of preserving habitat and species. In 2002, the Corps and the Conservancy developed the Sustainable Rivers Project to establish a process for planning, implementing, and evaluating proposed rehabilitation and water-management projects and programs, and for coordinated review and alteration of dam operations. At present, fourteen candidate sites on ten rivers are being examined under the program. The Savannah River basin of Georgia provides an example of joint activities.

The Savannah River drains more than 29,000 square kilometers from its headwaters in the Blue Ridge Mountains through the piedmont and

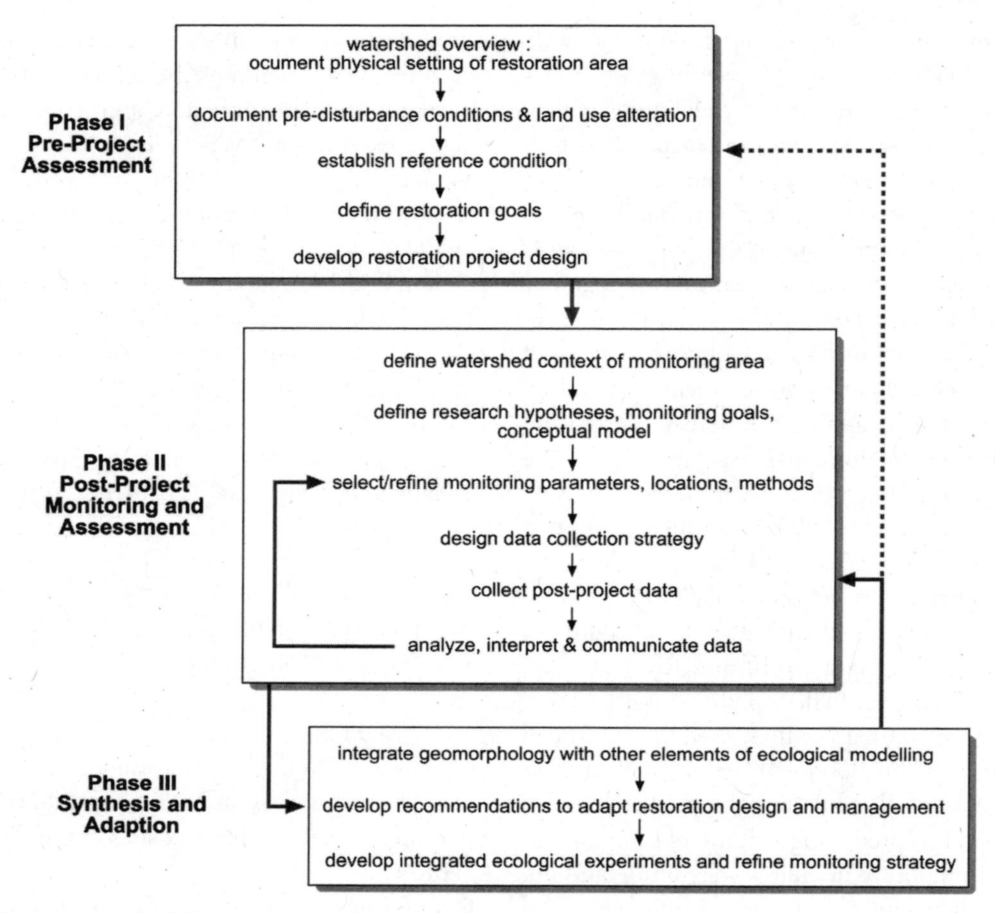

FIGURE 10.5. Schematic diagram of three-phase adaptive monitoring, assessment, and management framework used at the Cosumnes River, Califorina. (After Florsheim et al. 2006, figure 2)

coastal plains regions to the Atlantic Ocean. The basin supports more than seventy-five species of rare plants and animals, including eighteen rare fish species (out of a total of more than one hundred and ten fish species in the basin). Blackwater tributaries in particular support highly diverse vertebrate and invertebrate communities. Flow on the river is regulated by four major dams, including the J. Strom Thurmond Dam constructed by the Corps of Engineers from 1946 to 1954 to provide flood control, hydropower, and navigation. This dam backfloods more than 150 kilometers along four rivers. Scientists in the Sustainable Rivers Project are now attempting to quantify the flow regime necessary to sustain ecosystems along the Savannah River, which is predicated on being able to change operations of the dams within the watershed.

Of particular concern is rehabilitation of some portion of the historical peak flows, which inundated an estimated 28,300 hectares of riverine floodplain covered by mixed bottomland hardwood forest (Postel 2006). Eighty species of fish, as well as migratory neotropical song birds, rely on the floodplains for some portion of their lifecycle. Restored peak flows would also help endangered Atlantic sturgeon (*Acipenser oxyrhynchus*) to pass through some of the lock and dam systems along the river that currently limit sturgeon migration. The Corps conducted a series of controlled floods in 2004 and 2005, with scientists from the

(a)

(b)

FIGURE 10.6. View of (a) splay channel and (b) sand splay deposit on the Cosumnes River floodplain, California (photographs courtesy of Joan Florsheim).

Conservancy and the University of Georgia monitoring fish movements, as well as responses of floodplain invertebrate and forest communities to the higher flows. As for the Grand Canyon, the physical and biological responses of the river ecosystem to the experimental releases will be monitored and used to adjust the timing, magnitude, and duration of future experimental floods.

Kissimmee River, Florida

The Kissimmee River drains a low-relief portion of central Florida. Historically, the river meandered across a 90–kilometer length of floodplain 2 to 3 kilometers wide that dropped from 15.5 meters above sea level at Lake Kissimmee to 4.6 meters at Lake Okeechobee. Flows exceeded 7 m^3s^{-1} about 95 percent of the time, and overbank flooding occurred 35 to 50 percent of the time. The slowly flowing water averaged less than 0.6 ms^{-1} in velocity. Lateral migration of the channel along with overbank flooding maintained a complex mosaic of habitats, including more than 14,000 hectares of wetlands, which supported numerous species of plants, invertebrates, fish, birds, reptiles, and mammals (Pierce et al. 1982). Maintenance of this ecosystem depended on the interactions among subtle topographic differences associated with natural levees, oxbows, and the active channel, and prolonged inundation and seasonally fluctuating water levels (Perrin et al. 1982; Toth 1991). As late as the mid-1950s, the Kissimmee River supported several highly productive commercial and recreational fisheries as well as extensive wading bird and wintering waterfowl populations (U.S. Fish and Wildlife Service 1959, 1979).

Human habitation in the Kissimmee River basin was sparse prior to 1940, although farming and cattle ranching were present. Rapid growth and development following World War II, combined with severe flooding induced by a hurricane in 1947, created political pressure to mitigate flood damages. The Corps of Engineers channelized the Kissimmee between 1962 and 1971 as part of a larger flood-control project in central and southern Florida. The Corps changed the Kissimmee to a series of reservoirs impounded by six water control structures, blocked off meanders with weirs, and built canals to regulate water levels within and between lakes in the upper basin. These activities changed the flow regime in the main channel, severed the river-floodplain connections, and degraded floodplain wetlands (Toth 1993, 1995). The diverse, meandering main channel became a canal 90 kilometers long, 9 meters deep, and 91 meters wide. Excavation of this canal and deposition of the resulting spoil destroyed 58 kilometers of river channel and 2,510 hectares of floodplain wetlands. Another 10,500 to 12,500 hectares of floodplain wetlands were drained or converted into canal. Stable water levels allowed vegetation including exotic water lettuce (*Pistia stratiotes*) and water hyacinth (*Eichhornia crassipes*) to encroach on the main channel. As these plants died and decayed, the formerly sandy channels filled with up to a meter of fine organic matter that greatly increased the biological oxygen demand of the system (Toth 1990).

Floodplain utilization by wintering waterfowl declined by 92 percent shortly after the project was completed (Perrin et al. 1982), and the wading bird population also declined (Toland 1990). Chronically low levels of dissolved oxygen favored species of invertebrates and fish that could tolerate these conditions, resulting in massive changes in the species present in the river (Toth 1993; Trexler 1995). Plans to restore the channelized river began immediately after channelization was completed.

In 1992 Congress authorized the Kissimmee River Restoration Project, which is designed to restore over 110 square kilometers of river and floodplain habitat, including 72 kilometers of meandering river channel and 10,900 hectares of wetlands. The Corps and the South Florida Water Management District are jointly overseeing rehabilitation, which is designed to recreate pre-channelization attributes to the extent possible. The estimated total cost of $414 million is being split by the national government and the state of Florida. The first phase of canal backfilling and installing weirs across the canal in order to force flow into relict meander bends began in 1999, and the project is scheduled to be completed in 2010. Scientists have actively field-tested various rehabilitation measures including water-level manipulations, floodplain impoundments, weir construction, and canal backfilling. Results to date indicate that levels of dissolved oxygen in the river have increased and fine organic material has been flushed downstream, restoring the sandy streambed. Flows have recontoured the uniformly flat, shallow channel into pools and riffles. Benthic invertebrates have become more numerous and diverse, and game fish are now more abundant. As in the example of the Grand Canyon, the primary objectives of adaptive management in rehabilitating the Kissimmee ecosystem are to (1) understand the interactions among flow regime, sediment supply, channel and floodplain morphology/habitat, and biological communities, and (2) use observations of the initial responses of the river ecosystem to continually modify flow regime and sediment supply as necessary to optimize desired biological recovery. Agricultural and urban water demands remain high in central Florida, and the Kissimmee will never be fully restored to its pre-channelization state, but the careful monitoring and explicit attempt at adaptive management associated with this rehabilitation project illustrate a shift in approach likely to influence other, future projects (Wohl 2004).

The four case studies briefly summarized here illustrate some of the regional differences and the common constraints that influence large-scale river rehabilitation projects in the United States. In each of the rivers, disruption of historical patterns of water and sediment movement downstream and across floodplains has resulted in loss of habitat diversity and abundance, and associated declines in diverse aquatic and riparian species. Continuing demands for hydroelectric power generation (Colorado and Savannah Rivers), flood control (Cosumnes, Savannah, and Kissimmee Rivers), navigation (Savannah River), and water supply (Colorado and Kissimmee Rivers) limit the options for altering existing dams and levees in these watersheds. Increasing societal demands for species and habitat protection, however, has encouraged the agencies operating structures such as levees and dams to explore alternative operating procedures that limit the negative impacts of these structures. In each of these case studies, adaptive management is based on a conceptual model that incorporates seasonal peak flows and the effects of these flows on habitat quality and abundance. The "adaptive" portion of the management comes from monitoring ecosystem response to initial experimental alterations of levees or of flow releases from dams, and then adjusting subsequent levee alteration or dam releases based on observed ecosystem response.

Conclusion

It seems appropriate to paraphrase Charles Dickens's famous words, "It was the best of times, and the worst of times." On the one hand, the United States appears to be in the midst of a period of growing disparities in personal incomes, lax enforcement of protective legislation, and extreme laissez-faire capitalism. Such phases have occurred before in American history, such as at the end of the nineteenth and start of the twentieth centuries, and they correspond to poorly regulated development that results in severe environmental deterioration (Morris 2001). On the other hand, national polls consistently indicate that most citizens are concerned about the environment. This concern is reflected in the number of local watershed organizations such as Friends of the Poudre, the Anacostia (Potomac) Watershed Society, and Friends of Five Creeks, where the authors of this chapter live. A century of systematic, government-sponsored research has given scientists and activists concerned with river health in the United States a solid base of information and understanding, although of course questions always remain to be answered. National networks including the Long-Term Ecological Research sites, the National Ecological Observatory Network, and the NAWQA study sites facilitate comparisons among the diverse river environments present in the United States. The country's major environmental organizations recognize the fundamental importance of healthy river ecosystems, and the organization American Rivers is devoted solely to rivers. Further, increasing attention is being given to the environmental causes of rapidly rising rates of cancer and other severe health effects (e.g., Colborn et al. 1996; Steingraber 1997), and how these in turn relate to river health (e.g., Wohl 2004).

Contemporary conditions provide an interesting contrast to those of a century ago. During the nineteenth and early twentieth centuries, the release of pollutants into rivers in the United States was unregulated and routine. However, because populations and extent of urbanization were smaller, many kilometers of rivers remained largely unaffected. Human pressure on the landscape is much greater today, but this is balanced by greater awareness of the importance of river ecosystems, and by more regulations to protect river integrity. In many cases these regulations originated in attempts to protect fishery stocks. Regulations against overfishing for salmon were in place on the Sacramento River by the 1870s, for example, largely to prevent practices such as stretching nets across the entire channel. No regulations governed discharges from the multiple canneries operating in the valley, however. Most of the rivers in the Sacramento Valley were still passable by spring run chinook salmon (*Onchorhynchus tshawytscha*), and flow regimes were largely unaltered, so the fish spawned in the upper reaches of the watersheds. Today, most of the flow in these rivers is diverted from the channels, and the banks and floodplains have been extensively modified, but stronger regulations on the discharge of waste and taking of the few remaining fish help to protect what is left of the spring runs (Botsford and Brittnacher 1998). In other words, the United States at least has the legal mechanisms and scientific understanding to protect and recover river health.

Recovery of rivers from past land uses also provides evidence of the resiliency of river ecosystems. Watersheds in coastal northern California that underwent widespread clearcutting during the 1950s, resulting in massive aggradation, have subsequently stabilized and recovered much of their ecological function (House 1992). Migratory fish returned to upstream habitats that had been inaccessible for decades within a year or two of dam removals on the Kennebec River in Maine, Bear Creek in Oregon, the Mad River in California, and the Neuse River in North Carolina (Hart et al. 2002). Although dam removal can create problems associated with excess sediment mobilization, downstream spread of contaminated sediments, and spread of exotic species (e.g., Stanley and Doyle 2003; Thomson et al. 2005), some impounded rivers prove resilient to the impacts of impoundments once the dams are removed.

How resilient rivers will be in the face of climate change remains to be seen. Given the extent to which most major river basins are already impacted by dams and nearby infrastructure, rivers have reduced or removed natural ability to adjust and absorb disturbances such as higher magnitude or more frequent flows that are expected in some regions of the United States by the middle of the century (Lettenmaier et al. 1999; Poff et al. 2002). The inability of rivers to adjust and buffer the impacts of new flows outside the range previously experienced may create serious problems and recent analyses suggest that the area of large river basins that will need proactive or reactive management intervention is much higher for basins impacted by dams than for basins with free-flowing rivers (Palmer et al. 2007). But we do have the knowledge that could guide proactive measures; it is a matter of will.

Because of the existing framework of environmental regulations, increasing public awareness of the importance of river health and the dangers of environmental contamination, and the resiliency of river ecosystems, those who care deeply about river health have many reasons to be optimistic. There is no lack of scientific understanding of the causes of river degradation or of methods that can be used to protect and rehabilitate rivers. If the citizens of the United States care sufficiently about their rivers to think carefully about how their daily lifestyle choices and their votes influence river health, they can change the list of dismaying statistics cited at the start of this chapter.

Acknowledgments

We would like to thank two reviewers for constructive comments on this manuscript.

References

Abell, R. A., D. M. Olson, E. Dinerstein, P. T. Hurley, J. T. Diggs, W. Eichbaum, S. Walters, W. Wettengel, T. Allnutt, C. J. Loucks, and P. Hedao. 2000. *Freshwater ecoregions of North America: A conservation assessment.* Washington, DC: Island Press.

Aikin, A. R., and A. K. Turner. 1989. *Geochemical assessment of aquifer recharge effects in the southwest Denver Basin.* Completion Report. Fort Collins, CO: Colorado Water Resources Research Institute.

Agricultural Research Service. 2001. *Action plan: Compo-*

nent I: Agricultural watershed management. Washington, DC: USDA Agricultural Research Service.

ASCE (American Society of Civil Engineers). 2005. *2005 report card for America's infrastructure*. http://www.asce.org/reportcard/2005/index.cfm (accessed December 8, 2007).

Barnett, T. P., J. C. Adam, and D. P. Lettenmaier. 2005. Potential impacts of a warming climate on water availability in snow-dominated regions. *Nature* 438:303–9.

Benjamin, L., and R. W. Van Kirk. 1999. Assessing instream flows and reservoir operations on an eastern Idaho river. *Journal of the American Water Resources Association* 35:899–909.

Bernhardt, E. S., and M. A. Palmer. 2007. Restoring streams in an urbanizing world. *Freshwater Biology* DOI:10.1111/j.1365-2427.2006.01718.x.

Bernhardt, E. S., M. A. Palmer, J. D. Allan, and the National River Restoration Science Synthesis Working Group. 2005. Synthesizing U.S. river restoration efforts. *Science* 308:636–37.

Botsford, L. W., and J. G. Brittnacher. 1998. Viability of Sacramento River winter-run chinook salmon. *Conservation Biology* 12:65–79.

Brown, K. 2000. *Urban stream restoration practices: An initial assessment*. Ellicot City, MD: The Center for Watershed Protection.

Burke, S. M., R. M. Adams, and W. W. Wallender. 2004. Water banks and environmental water demands: Case of the Klamath Project. *Water Resources Research* 40:W09S02.

CALFED Bay Delta Program. 2000. *Ecosystem restoration program plan, strategic plan for ecosystem restoration*. Final programmatic EIS/EIR technical appendix. Sacramento, CA: CALFED.

CALFED. 2005. CALFED standard presentation. http://calwater.ca.gov/index.aspx (accessed 25 May 2005).

Castleberry, D. T., J. J. Czech, D. C. Erman, D. Hankin, M. Healey, G. M. Kondolf, M. Mangel, M. Mohr, P. B. Moyle, J. Nielsen, T. P. Speed, and J. G. Williams. 1996. Uncertainty and instream flow standards. *Fisheries* 21:20–21.

Colborn, T., D. Dumanoski, and J. P. Myers. 1996. *Our stolen future: Are we threatening our fertility, intelligence, and survival? A scientific detective story*. New York, NY: Dutton.

Collier, M. P., R. H. Webb, and E. D. Andrews. 1997. Experimental flooding in Grand Canyon. *Scientific American* 276:66–73.

Collier, M., R. H. Webb, and J. C. Schmidt. 1996. *Dams and rivers: Primer on the downstream effects of dams*. U.S. Geological Survey Circular 1126. Tucson, AZ.

Committee on Mitigating Wetland Losses. 2001. *Compensating for wetland losses under the Clean Water Act*. Washington, DC: National Academies Press.

Cook, K. 1997. *Guidance document for the establishment of critical aquifer recharge area ordinances*. Washington State Department of Ecology Water Quality Program, Publ. no. 97-030.

Davis, W. S., and T. P. Simon, eds. 1995. *Biological assessment and criteria*. Boca Raton, FL: Lewis Publishers.

Doppelt, B., M. Scurlock, C. Frissell, and J. Karr. 1993. *Entering the watershed: A new approach to save America's river ecosystems*. Washington, DC: Island Press.

Dubinski, I. M., and E. Wohl. 2007. Assessment of coarse sediment mobility in the Black Canyon of the Gunnison River, Colorado. *Environmental Management* 40:147–60.

Elliott, J. G., and R. S. Parker. 1997. Altered streamflow and sediment entrainment in the Gunnison Gorge. *Water Resources Bulletin* 33:1041–54.

EPA. 2000. *National water quality inventory*. EPA Publ. 841-R-02-001. Washington, DC.

EPA. 2003a. *Beef cattle and environmental stewardship*. Environmental Protection Agency 305-F-03-004, Washington, DC.

EPA. 2003b. *Dairies and environmental stewardship*. Environmental Protection Agency 305-F-03-003, Washington, DC.

EPA. 2003c. *Poultry production and environmental stewardship*. Environmental Protection Agency 305-F-03-002, Washington, DC.

Ferrill, D. A., D. W. Sims, D. J. Waiting, A. P. Morris, N. M. Franklin, and A. L. Schultz. 2004. Structural framework of the Edwards Aquifer recharge zone in south-central Texas. *Geological Society of America Bulletin* 116:407–18.

Florsheim, J. L., and J. F. Mount. 2002. Restoration of floodplain topography by sand-splay complex formation in response to intentional levee breaches, Lower Cosumnes River, California. *Geomorphology* 44:67–94.

Florsheim, J. L., J. F. Mount, and C. R. Constantine. 2006. A geomorphic monitoring and adaptive assessment framework to assess the effects of lowland floodplain river restoration on channel-floodplain sediment continuity. *River Research and Applications* 22:353–75.

Gilliom, R. J., D. K. Mueller, and L. H. Nowell. 1998. *Methods for comparing water-quality conditions among National Water-Quality Assessment Study Units, 1992–1995*. U.S. Geological Survey Open-File Report 97-589.

Goldstein, J. H. 1991. The road to no-net-loss of wetlands. *Forum for Applied Research and Public Policy* 6:73–77.

Graf, W. L. 1999. Dam nation: A geographic census of American dams and their large-scale hydrologic impacts. *Water Resources Research* 35:1305–11.

Graf, W. L. 2001. Damage control: Restoring the physical integrity of America's rivers. *Annals of the Association of American Geographers* 91:1–27.

Graf, W. L. 2002. Summary and perspective. In *Dam removal research: Status and prospects*, ed. W. L. Graf, 1–21. Proceedings of the Heinz Center's dam removal workshop, October 23–24, 2002. Washington, DC: The Heinz Center.

Hammer, D. A., ed. 1989. *Constructed wetlands for wastewater treatment: Municipal, industrial, and agricultural*. Chelsea, MI: Lewis Publishers.

Hart, D. D., T. E. Johnson, K. L. Bushaw-Newton, R. J. Horwitz, A. T. Bednarek, D. F. Charles, D. A. Kreeger, and D. J. Velinsky. 2002. Dam removal: Challenges and opportunities for ecological research and river restoration. *BioScience* 52:669–81.

Hassett, B., M. A. Palmer, E. S. Bernhardt, S. Smith, J. Carr, and D. Hart. 2005. Status and trends of river and stream restoration in the Chesapeake Bay watershed. *Frontiers in Ecology and the Environment* 3:259–67.

Hobbs, H. H. 1988. Crayfish distribution, adaptive radiation, and evolution. In *Freshwater crayfish biology, management, and exploitation*, ed. D. M. Holdich and R. S. Lowery, 52–82. London, UK: Croom Helm.

Hook, D. D., ed. 1988. *The ecology and management of wetlands*. Portland, OR: Timber Press.

House, F. 1992. Dreaming indigenous. *Restoration Management News* 10:60–63.

Innis, S. A., R. J. Naiman, and S. R. Elliott. 2000. Indicators and assessment methods for measuring the ecological integrity of semi-aquatic terrestrial environments. *Hydrobiologia* 422/423:111–31.

Jackson, L. E., J. C. Kurtz, and W. S. Fisher. 2000. *Evaluation guidelines for ecological indicators*. EPA/620/R-99/005. Research Triangle Park, NC: U.S. Environmental Protection Agency, Office of Research and Development.

Johannson, M. E., and C. Nilsson. 2002. Responses of riparian plants to flooding in free-flowing and regulated boreal rivers: An experimental study. *Journal of Applied Ecology* 39:971–86.

Kondolf, G. M., A. Boulton, S. O'Daniel, G. Poole, F. Rahel, E. Stanley, E. Wohl, Å. Bång, J. Carlstrom, C. Cristoni, H. Huber, S. Koljonen, P. Louhi, and K. Nakamura. 2006. Process-based ecological river restoration: Visualising three-dimensional connectivity and dynamic vectors to recover lost linkages. *Ecology and Society* 11(2):Article 5. http://www.ecologyandsociety.org/vol11/iss2/art5/ (accessed December 8, 2007).

Kondolf, G. M., E. W. Larsen, and J. G. Williams. 2000. Measuring and monitoring the hydraulic environment for assessing instream flows. *North American Journal of Fisheries Management* 20:1016–28.

Kondolf, G. M., M. W. Smeltzer, and S. F. Railsback. 2001. Design and performance of a channel reconstruction project in a coastal California gravel-bed stream. *Environmental Management* 28:761–76.

Lettenmaier D. P., A. W. Wood, R. N. Palmer, E. F. Wood, and E. G. Stakhiv. 1999. Water resources implications of global warming: a U.S. regional perspective. *Climate Change* 43:537–79.

MacDonald, L. H., A. W. Smart, and R. C. Wissmar. 1991. *Monitoring guidelines to evaluate effects of forestry activities on streams in the Pacific Northwest and Alaska*. Seattle, WA: Environmental Protection Agency EPA 910/9-91-001.

Malakoff, D. 2004. The river doctor. *Science* 305:937–39.

Mathur, D., W. H. Bason, E. J. Purdy, and C. A. Silver. 1985. A critique of the instream flow incremental methodology. *Canadian Journal of Fisheries and Aquatic Sciences* 42:825–31.

McKim, T. W., G. Harkness, and A. W. Aikens. 1995. Aquifer recharge modeling for the Reedy Creek Improvement District. In *Artificial recharge of groundwater, II: Proceedings of the second international symposium*, ed. A. I. Johnson and R. D. G. Pyne, 888–97. New York, NY: ASCE.

Meier, M. F., M. B. Dyurgerov, U. K. Rick, S. O'Neel, W. T. Pfeffer, R. S. Anderson, and S. P. Anderson. 2006. Disappearing glacial ice: A global synthesis. *Eos, Transactions, American Geophysical Union* 87(52), Fall Meeting Suppl., Abstract C14B-08.

Meyer, J. L., and J. B. Wallace. 2001. Lost linkages and lotic ecology: Rediscovering small streams. In *Ecology: Achievement and challenge*, ed. M. C. Press, N. J. Huntly, and S. Levin, 295–317. Oxford, UK: Blackwell.

Moore, A. M., and M. A. Palmer. 2005. Agricultural watersheds in urbanizing landscapes: Implications for conservation of biodiversity of stream invertebrates. *Ecological Applications* 15:1169–77.

Morris, E. 2001. *Theodore rex*. New York, NY: Random House.

Naiman, R. J., N. Décamps, M. E. McClain. 2005. *Riparia: Ecology, conservation, and management of streamside communities*. Amsterdam, Netherlands: Elsevier/Academic Press.

NAS (National Academy of Science). 2001. *Growing populations, changing landscapes: Studies from India, China, and the United States*. Washington, DC: National Academies Press.

NCED. 2005. *National center for earth-surface dynamics workshop on sediment remobilization and channel morphodynamics in active and abandoned reservoirs*. http://www.nced.umn.edu (accessed January 14, 2008).

Orth, D. J., and O. E. Maughan. 1982. Evaluation of the incremental methodology for recommending instream flows for fishes. *Transactions of the American Fisheries Society* 111:413–45.

Palmer, M. A., and J. D. Allan. 2006. River restoration: as the need for river restoration grows, supporting federal policies should follow. *Issues in Science and Technology* 22:40–48.

Palmer, M. A., E. S. Bernhardt, J. D. Allan, P. S. Lake, G. Alexander, S. Brooks, J. Carr, S. Clayton, C. Dahm, J. Follstad Shah, D. J. Galat, S. Gloss, P. Goodwin, D. H. Hart, B. Hassett, R. Jenkinson, G. M. Kondolf, R. Lave, J. L. Meyer, T. K. O'Donnell, L. Pagano, P. Srivastava, and E. Sudduth. 2005. Standards for ecologically successful river restoration. *Journal of Applied Ecology* 42:208–17.

Palmer M. A., A. P. Covich, B. J. Finlay, J. Gibert, K. D. Hyde, R. K. Johnson, T. Kairesalo, P. S. Lake, C. R. Lovell, R. J. Naiman, C. Ricci, F. Sabater, and D. Strayer. 1997. Biodiversity and ecosystem processes in freshwater sediments. *Ambio* 26:571–77.

Palmer, M. A., C. A. Reidy, C. Nilsson, M. Flörke, J. Alacmo, P. S. Lake, and N. Bond. 2007. Climate change and the world's river basins: Anticipating response options. *Frontiers in Ecology and the Environment* 6, DOI: 10.1890/060148.

Palmer, T. 2004. *Lifelines: The case for river conservation, second edition*. Lanham, MD: Rowman & Littlefield.

Paul, M. J., and J. L. Meyer. 2001. Streams in the urban landscape. *Annual Review of Ecology and Systematics* 32:333–65.

Pedroli, B., G. de Blust, K. van Looy, and S. van Rooij. 2002. Setting targets in strategies for river restoration. *Landscape Ecology* 17(Suppl. 1): 5–18.

Perrin, L. S., M. J. Allen, L. A. Rowse, F. Montalbano, K. J. Foote, and M. W. Olinde. 1982. *A report on fish and wildlife studies in the Kissimmee River basin and recommendations for restoration*. Okeechobee, FL: Florida Game and Fresh Water Fish Commission, Office of Environmental Services.

Pielke, R. A., M. W. Downton, and J. Z. B. Miller. 2002. *Flood damage in the United States, 1926–2003: A reanalysis of National Weather Service estimates*. Boulder, CO: University Corporation for Atmospheric Research.

Pierce, G. J., A. G. Amerson, and L. R. Becker. 1982. *Pre-1960 floodplain vegetation of the lower Kissimmee River valley, Florida*. Biological Service Report 82-3. Dallas, TX: Final Report Environmental Consulting.

Poff, N. L., J. D. Allan, M. B. Bain, J. R. Karr, K. L. Prestegaard, B. D. Richter, R. E. Sparks, and J. C. Stromberg. 1997. The natural flow regime. *BioScience* 47:769–84.

Poff N. L., M. M. Brinson, and J. W. Day. 2002. *Aquatic ecosystems and global climate change: Potential impacts on inland freshwater and coastal wetland ecosystems*. Prepared for the Pew Center on Global Climate Change. Arlington, VA. http://www.pewclimate.org/global-warming-in-depth/all_reports/aquatic_ecosystems (accessed January 14, 2008).

Poff, N. L., and D. D. Hart. 2002. How dams vary and why it matters for the emerging science of dam removal. *BioScience* 52:659–68.

Postel, S. 2006. *To save a sturgeon*. Arlington, VA: The Nature Conservancy.

Ralston, B. E. 2005. Riparian vegetation and associated wildlife. In *The state of the Colorado River ecosystem in Grand Canyon*, ed. S. P. Gloss, J. E. Lovich, and T. S. Melis, 103–22. U.S. Geological Survey Circular 1282. Reston, VA.

Rashin, E., C. Clishe, A. Loch, and J. Bell. 1999. *Effectiveness of forest road and timber harvest best management practices with respect to sediment-related water quality impacts*. Olympia, WA: Timber/Fish/Wildlife Cooperative Management, Evaluation, and Research Committee, Washington State Department of Ecology, Publication No. 99-317.

Ricciardi, A., and J. B. Rasmussen. 1999. Extinction rates of North American freshwater fauna. *Conservation Biology* 13:1220–22.

Rood, S. B., C. R. Gourley, E. M. Ammon, L. G. Heki, J. R. Klotz, M. L. Morrison, D. Mosley, G. G. Scoppettone, S. Swanson, and P. L. Wagner. 2003. Flows for floodplain forests: A successful riparian restoration. *BioScience* 53:647–56.

Sarakinos, H., and S. E. Johnson. 2002. Social perspectives on dam removal. In *Dam removal research: Status and prospects*, ed. W. L. Graf, 40–55. Proceedings of the Heinz Center's dam removal workshop, Oct. 23–24, 2002. Washington, DC: The Heinz Center.

Shuster, B. 1987. Quoted in E. Walsh, Reagan's First Hill Showdown Already Brewing: Pocket-Vetoed $20 Billion Clean-Water Bill Revised in House by Lopsided 406–8. *Washington Post*, January 9.

Simon, A., M. Doyle, G. M. Kondolf, F. D. Shields, B. Rhoads, G. Grant, F. Fitzpatrick, K. Juracek, M. McPhillips, and J. MacBroom. Forthcoming. Do the Rosgen Classification and Associated Natural Channel Design Methods integrate and quantify fluvial processes and channel response? *Journal of the American Water Resources Association*.

Smith, S. M., and K. L. Prestegaard. 2005. Hydraulic performance of a morphology-based stream channel design. *Water Resources Research* 41(11): W11413 10.1029/2004WR003926.

Stanley, E. H., and M. W. Doyle. 2003. Trading off: The ecological effects of dam removal. *Frontiers in Ecology* 1:15–22.

Stanley, E. H., M. A. Luebke, M. W. Doyle, and D. W. Marshall. 2002. Short-term changes in channel form and macroinvertebrate communities following low-head dam removal. *Journal of the North American Benthological Society* 21:172–87.

Steingraber, S. 1997. *Living downstream: An ecologist looks at cancer and the environment*. Reading, MA: Addison-Wesley Publishing.

Stewart, G., R. Anderson, and E. Wohl. 2005. Two-dimensional modeling of habitat suitability as a function of discharge on two Colorado rivers. *River Research and Applications* 21:1–14.

Strakosh, T. R., R. M. Neumann, and R. A. Jacobson. 2003. Development and assessment of habitat suitability criteria for adult brown trout in southern New England streams. *Ecology of Freshwater Fish* 12:265–74.

Swenson, R. O., K. Whitener, and M. Eaton. 2003. Restoring floods to floodplains: Riparian and floodplain restoration at the Cosumnes River Preserve. In *California riparian systems: Processes and floodplain management, ecology, restoration, 2001 riparian habitat and floodplains conference proceedings*, ed. P. M. Faber, 224–29. Sacramento, CA: Riparian Habitat Joint Venture.

Thomson, J. R., D. D. Hart, D. F. Charles, T. L. Nightengale, and D. M. Winter. 2005. Effects of removal of a small dam on downstream macroinvertebrate and algal assemblages in a Pennsylvania stream. *Journal of the North American Benthological Society* 24:192–207.

Toland, B. R. 1990. Effects of the Kissimmee River pool B restoration demonstration project on Ciconiiformes and Anseriformes. In *Proceedings of the Kissimmee River Restoration Symposium, Orlando, Florida, Oct. 1988*, ed. M. K. Loftin, L. A. Toth, and J. Obeysekera, 83–92. West Palm Beach, FL: South Florida Water Management District.

Toth, L. A. 1990. Impacts of channelization on the Kissimmee River ecosystem. In *Proceedings of the Kissimmee River Restoration Symposium, Orlando, Florida, Oct. 1988*, ed. M. K. Loftin, L. A. Toth, and J. Obeysekera, 47–56. West Palm Beach, FL: South Florida Water Management District.

Toth, L. A. 1991. *Environmental responses to the Kissimmee River Demonstration Project*. Tech. Publ. no. 91-2. West Palm Beach, FL: South Florida Water Management District.

Toth, L. A. 1993. The ecological basis of the Kissimmee River restoration plan. *Florida Scientist* 56:25–51.

Toth, L. A. 1995. Principles and guidelines for restoration of river/floodplain ecosystems: Kissimmee River, Florida. In *Rehabilitating damaged ecosystems, 2nd edition*, ed. J. Cairns, 49–73. Boca Raton, FL: Lewis Publishers/CRC Press.

Trexler, J. C. 1995. Restoration of the Kissimmee River: A conceptual model of past and present fish communities and its consequences for evaluating restoration success. *Restoration Ecology* 3:195–210.

U.S. Environmental Protection Agency. 2000. *Storm water phase II compliance assistance guide*. EPA 833-R-00-002.

U.S. Fish and Wildlife Service. 1959. A detailed report of the fish and wildlife resources in relation to the Corps of Engineers' plan of development, Kissimmee River basin, Florida, Appendix A. In *Central and Southern Florida project for flood control and other purposes: Part II. Kissimmee River basin and related areas, supplement* 5. Jacksonville, FL: U.S. Army Corps of Engineers, Jacksonville District.

U.S. Fish and Wildlife Service. 1979. *Impacts of channelization on ecological resources*. U.S. Fish and Wildlife Service Report.

U.S. Geological Survey. 1999. *The quality of our nation's waters: nutrients and pesticides*. U.S. Geological Survey Circular 1225. Reston, VA.

U.S. Geological Survey. 2006. *Assessment of the estimated influence of four experimental options on resources below Glen Canyon Dam*. Draft for review. Washington, DC: Glen Canyon Adaptive Management Program Science Advisors, U.S Department of the Interior.

Walters, C. 1997. Challenges in adaptive management of riparian and coastal systems, *Conservation Ecology* 1(2): Article 1. http://www.consecol.org/vol1/iss2/art1/ (accessed December 8, 2007).

Webb, R. H., P. G. Griffiths, C. S. Magirl, and T. C. Hanks. 2005. Debris flows in Grand Canyon and the rapids of the Colorado River. In *The state of the Colorado River ecosystem in Grand Canyon*, ed. S. P. Gloss, J. E. Lovich, and T. S. Melis, 139–52. U.S. Geological Survey Circular 1282. Reston, VA.

Williams, J. D., M. L. Warren, K. S. Cummings, J. L. Harris, and R. J. Neves. 1993. Conservation status of freshwater mussels of the United States and Canada. *Fisheries* 18:6–22.

Wohl, E. 2004. *Disconnected rivers: Linking rivers to landscapes*. New Haven, CT: Yale University Press.

Wohl, E., P. L. Angermeier, B. Bledsoe, G. M. Kondolf, L. MacDonnell, D. M. Merritt, M. A. Palmer, N. L. Poff, and D. Tarboton. D. 2005. River restoration. *Water Resources Research* 41:W10301.

Wohl, E., D. Cooper, L. Poff, F. Rahel, D. Staley, and D. Winters. 2007. Assessment of stream ecosystem function and sensitivity in the Bighorn National Forest, Wyoming. *Environmental Management* 40:284–302.

Wright, S. A., T. S. Melis, D. J. Topping, and D. M. Rubin. 2005. Influence of Glen Canyon Dam operations on downstream sand resources of the Colorado River in Grand Canyon. In *The state of the Colorado River ecosystem in Grand Canyon*, ed. S. P. Gloss, J. E. Lovich, and T. S. Melis, 17–31. U.S. Geological Survey Circular 1282. Reston, VA.

Chapter 11

Integrative River Science and Rehabilitation: European Experiences

Hervé Piégay, Larissa A. Naylor, Gertrud Haidvogl, Jochem Kail, Laurent Schmitt, and Laurent Bourdin

> The challenge for the early twenty-first century of European legislation in the form of a Water Framework Directive . . . requires interdisciplinarity with a need to emphasize the integrated roles of geomorphology, hydrology, hydraulics and freshwater ecology in informing the "vision" of riparian human communities.
>
> Newson and Large 2006, 1608

Following the United Nations conference in Rio de Janiero in June 1992, politicians agreed to promote the concept of sustainable development as a means for environmental management. Stakes underpinning this concept are well demonstrated, at least in terms of public health and technologies. Increasingly, rehabilitation initiatives are applied to freshwater systems. Recent rehabilitation activities in Europe have stemmed from the pioneer work of Danish and German scientists and managers in the 1980s (Brookes 1987; Landesumweltamt Nordrhein-Westfalen 1989). Since this period, official guidelines for stream rehabilitation have been published and local experimental trials have been conducted in many parts of Europe. In October 2000, the European Parliament legislated a broader-scale strategy for water ecosystem improvement called the Water Framework Directive (WFD).

Two complementary tasks have been identified in the WFD: (1) to define a program of measures to achieve a good ecological status for waterbodies by 2015, considering biological, physico-chemical, hydromorphological, and economic criteria in application of sustainable development principles, and (2) engaging the public in the process. In this context, water resource management is now part of a long-term planning process with evidence-based management actions, cost-benefit analysis, and new governance arrangements.

The WFD involves three main spatial scales of management: (1) river basin districts, (2) waterbodies, and (3) individual monitoring sites. Within each river basin district, designated authorities (e.g., water authorities, state representatives) are charged with designing a program of measures to be applied in 2009 that will allow waterbodies to reach a good ecological status by 2015. River basin districts are spatially delimited hydrographic areas that enable countries to design holistic strategies to prioritize catchment-scale management. At the smallest scale, scientific tools are used to assess the ecological status of individual sampling areas within each waterbody unit. These data are then used to assess, monitor, and report on the current ecological status of each waterbody on a six-year cycle. This state is then compared to the "reference condition" for each scientific parameter measured. The status of individual waterbodies is considered as part of the larger scale river basin planning process, to prioritize which waterbodies are in most need of mitigation and rehabilitation.

As part of the river basin management planning process, scientific evidence, expert judgement, and local knowledge are used to select appropriate mitigation and rehabilitation measures (known as programs of measures in the WFD) for each river basin district. Scientific evidence is used to demonstrate the direct benefit of a given rehabilitation or mitigation measure and to monitor whether the measures used actually achieve the management objectives set during the river basin management planning process. For this process to be successful, integrative science is essential, along with clear communication and encouragement of participation by the wide array of disciplines and organizations involved in the WFD process (figure 11.1). Many scientific questions have been posed as part of the process (e.g., how to define good ecological status, what are the reference conditions to consider, what is a physical alteration, what is the effectiveness of the rehabilitation measures promoted?). Most of these questions cannot be answered without considering the problems in a crossdisciplinary fashion and with strong

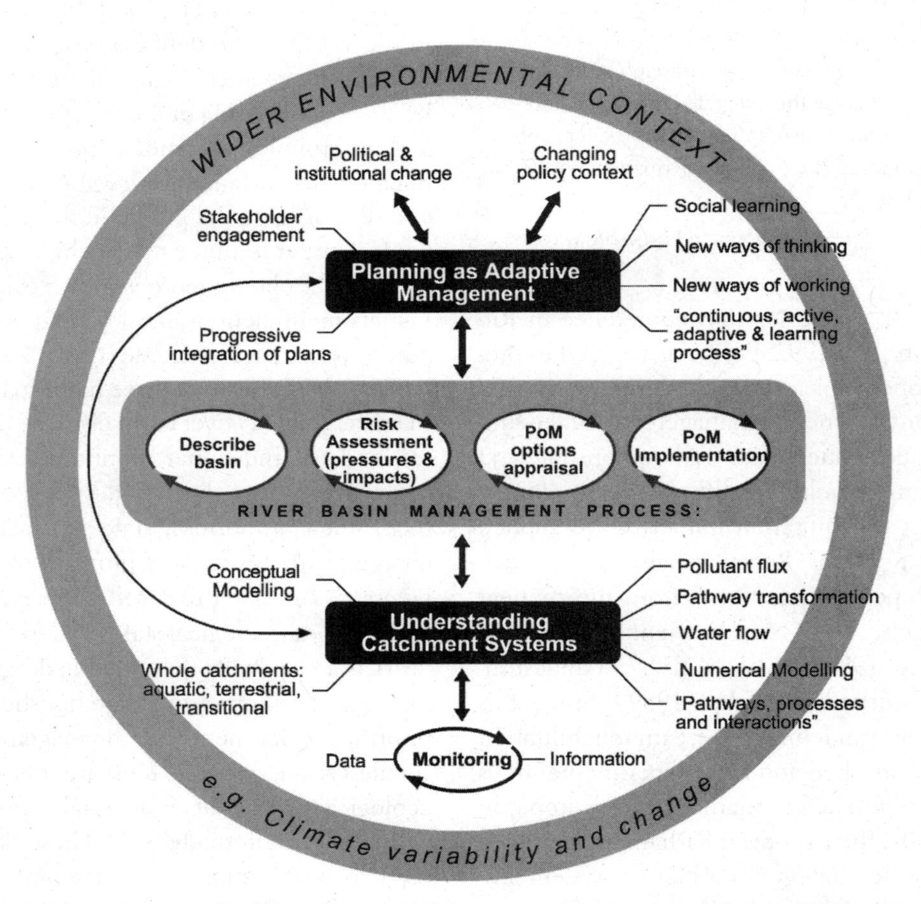

FIGURE 11.1. Strategic framework of the science and management tasks required to enable WFD implementation. Designed by L.A. Naylor, based on results of a workshop held in June 2004 that brought together Environment Agency Research Scientists and WFD Policy and Implementation staff (Colvin et al. 2005).

interactions between scientists and managers (i.e., a transdisciplinary approach; see chapters 1 and 3). Such approaches are not new in European river research, as exemplified by pioneering work in the Rhône Basin, France, by Roux (1982). What is now emerging is better integration between scientists and managers. Such interactions are not yet commonplace in each country, but WFD implementation requirements are encouraging the two communities to work more closely (Collins et al. 2006).

In this chapter, the progressive emergence of integrative river science is examined in the European context. We first provide a historical context of management and the evolution of integrative scientific perspectives. We then examine four case studies from European countries to illustrate how the scientific community has applied integrative river science. Finally, given that the WFD now drives river management practice in Europe, we examine some of the scientific and practical challenges faced in its implementation and how it has improved planning and macro-management strategies.

The Emergence of Integrative River Science in European Countries

Across Europe, integrative river science has emerged in different ways at different times. In this section we examine the evolution of integrative river science, how it is used by managers, and how public participation has enhanced this integration.

Emergence of Integrative River Science

Early integrative river science experiences occurred in France in the early 1980s with PIREN (Programme Interdisciplinaire de Recherches sur l'Environnement) interdisciplinary programs on the environment. Some of these focused on large rivers, emphasizing interactions between ecological, chemical, and physical factors in aquatic ecosystems. This led to discussion of "hydrosystem" dynamics (Amoros and Petts 1993). More recently (1998–2002), the PEVS (Program Environment, Life, and Society) program began with a clear aim to improve interactions among earth, life, and social sciences. Scientists involved in such programs moved from the hydrosystem concept to the anthroposystem, examining the interactions between natural and social systems (Lévêque et al. 2000). The recognition of human impacts on ecosystems has led to new questions being posed, such as quantification of ecosystem services and the function and variability of natural ecosystems from an ecological perspective. Since 2000, programs such as *Zone Ateliers* (long-term ecological research sites) consider social aspects as well as scientific integration, and interactions with managers are strongly encouraged.

The need for crossdisciplinary cooperation and the necessity of scaling up from local to regional stages began in Austria in the late 1980s. Consideration of the ecological integrity of rivers was a consequence of the amendment of the Austrian Water Law in 1985. Early work focused on restructuring riverine habitats at a local scale, usually to meet flood risk management objectives (e.g., *Restructuring River Melk Project*; Jungwirth and Waidbacher 1990; Jungwirth et al. 1993). During this period, water management institutions integrated crossdisciplinary cooperation into their management and planning tools (Redl 1990; RIWA-t 1994). Transdisciplinary river projects in Austria started with the introduction of the *Gewässerbetreuungskonzepte* (river management concepts). In these planning tools, the cooperation of natural scientists, ecologists, and water engineers was seen as essential to ensure adequate protection of human life as well as the environment. Nevertheless, it has to be recognized that these river management concepts were initiated by authorities responsible for flood protection. Hence, specific issues such as residual flow and ecological impacts were often excluded from detailed analysis. Also, economic and social aspects were only occasionally considered in this early phase of "integrated" river projects (e.g., river Salzach: Land Salzburg, Ref. Landesplanung, and SAGIS 1997; river Lech: Lang et al. 1996).

Large summer floods in 2002 in Switzerland, Liechtenstein, and Austria prompted greater awareness of the need to integrate concerns for water management, flood protection, and ecological problems related to channel incision, a consequence of intense gravel mining, but also reduced sediment transport from tributaries due to weirs (IRKA and IRR 2005). However, progress in cross- and transdisciplinary river science is limited to a few research projects including the development of an integrative river management plan for the river Möll, a tributary of the river Drau in Carinthia (Muhar et al. 2003) and the Riverine Landscape Planning Project for the river Kamp (Preis et al. 2006). The latter is one of the first examples where representatives of stakeholders and interest groups have been integrated in the planning process from the beginning.

Managers and Integrative River Science

Since World War II (1945–1990), managers have often viewed rivers as constraints upon agriculture, transportation, and urban development. Locally posed problems such as bank erosion and flooding were "solved" by civil engineering works. In general terms, river scientists and managers emphasized different spatial and temporal scales, with very different objectives. Geologists or physical geographers investigated long-term processes over large spatial scales (e.g., changes in erosion and sedimentation caused by climate change) whereas river managers and biologists mainly considered short-term processes typically at the local scale (e.g., preventing local bank erosion and habitat availability for invertebrates). Despite examining similar time-space scales, biologists did not always work closely with managers, as environmental issues and ecosystem conservation were not central to management agendas. The two spheres, scientists and managers, did not start to interact closely until the 1980s, or later in many cases.

As managers have changed their focus, their questions have been progressively posed at larger spatial and longer temporal scales. The need to adopt integrated and comprehensive analyses is increasingly recognized. The idea of working with nature rather than fighting against nature emerged in the 1980s, when managers and elected representatives began to understand the limits of engineering-based solutions. The evidence was compelling. Nearly 50 percent of bank protection works created between 1970 and 1995 have been undermined along the Galaure River, France, demonstrating the shorter-than-expected life span of these structures (Piégay et al. 1997). Moreover, increasing the number and size of dikes did not solve flooding problems. Instead, the dikes often created problems such as increasing flood frequencies downstream (Dister et al. 1990). Gravel mining was conducted without considering other stakeholders. These activities often increased incision and flood risk, undermining dikes and destroying bridges (Bravard et al. 1999). Many flood defenses protecting human settlements were degraded and the long-term maintenance of such works is now a major concern. These examples illustrate some of the problems that arise when decisions are made locally, without being part of a catchment-scale management strategy.

With the recent emergence of integrative river science and management in Central Europe, river managers are increasingly contacting scientists for advice and support. Water authorities increasingly involve or employ biologists and geomorphologists. New study courses, such as the transnational ecosystem-based water management program at the University of Nijmegen (Netherlands) and Duisburg-Essen (Germany), adopt an integrative approach that spans hydraulic, biological, geomorphological, and social aspects in the curriculum.

Public Participation and Integrative Science

Since the 1980s, more people in Europe have become concerned about the environment and their relationship to it. This paradigm shift has paralleled the emergence of the "sustainable development" concept that stemmed from the Brundtland report in 1987 (WCED 1987). More recently, this shift has led to greater public participation in the decision-making process and new governance rules, notably in river management.

In the 1990s, interactions between managers and scientists increased as ideas such as ecosystem-based management and public participation became more mainstream. Although integrated approaches are often now used by scientists and managers, the public, in a wider sense, still has limited involvement. For example, public involvement in the 1980s and 1990s in Austria was limited to public relations activities such as press conferences, press articles, and information campaigns. Representation by stakeholders and interest groups in planning processes has only recently begun, with a few exceptions. In France, public participation in management began with the Water Law of 1964, but more intensely after 1992 with the Local Water Management and Development plans (e.g., SAGE; Piégay et al. 2002). Nevertheless, most of these initiatives did not involve scientists, whose concerns are sometimes perceived by local elected people as disconnected from their day-to-day priorities. Scientists must be aware of this history and be open to wider debates with civil society and politicians. Implicit in this is accepting that the scientific basis of different management options could be discussed and subject to scrutiny by nonscientists.

Risk prevention and precautionary approaches are increasingly being put forward as innovative scientific solutions and/or management options. However, landowners, politicians, or the public often find such approaches difficult to grasp without sufficient engagement in the process (see chapter 14). These groups are also typically more focused on immediate to short-term concerns, rather than longer-term impacts of issues such as how climate change may affect rivers and, ultimately, the human activities derived from them. To implement sustainable development, environmental education becomes important and challenging (figure 11.2). Scientists can promote environmental education when participating in management schemes. However, such local-scale approaches to education and awareness may not be sufficient. Participatory decision-making may necessitate a large-scale environmental education program as well as exploration of how communities and institutions need to adapt to effectively implement emerging legislation.

Integrative River Science in Pioneer Rehabilitation Programs

Case studies are used to highlight the kind of rehabilitation measures that have been promoted in Europe during the last two decades and the level of integration that they have achieved (figures 11.3 and 11.4). These case studies illustrate how rehabilitation concepts have evolved and how scientists, managers, and the public interact.

Wood Reintroduction in Central Europe

Many studies, mainly from North America, demonstrate that large wood can have a strong influence on physical features and biota in river systems (Maser and Sedell 1994; Gregory et al. 2003). Therefore, river managers in Central Europe increasingly use large wood in rehabilitation projects (e.g. Gerhard and Reich 2001; Siemens et al. 2005) (figure 11.4a). As we move from an engineering-dominated to an ecosystem-based approach to rehabilitation (see chapter 1), there is a strong need for methods that will create or accelerate natural processes.

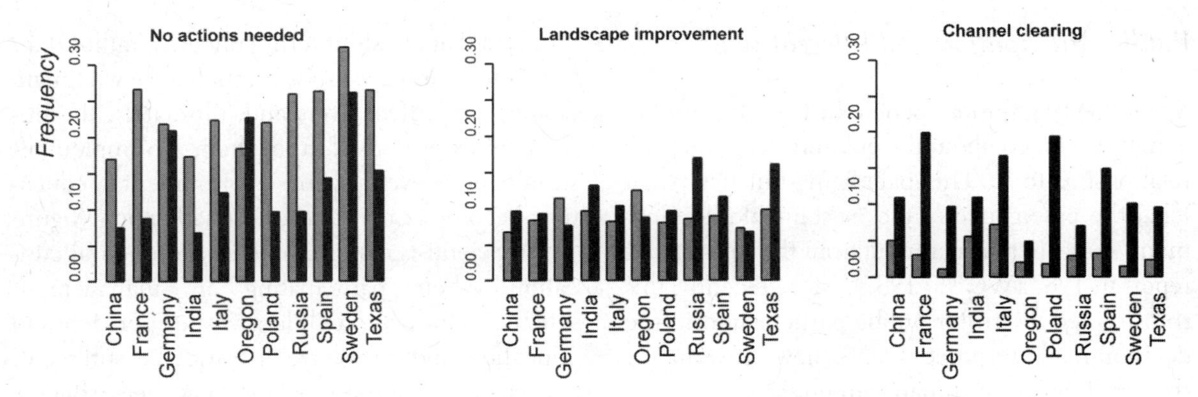

FIGURE 11.2. Differences in riverscape perception among groups of students from different geographic areas (see details in Piégay et al. 2005b). Students were preented with photos of river channels and were asked to decide how they would improve the riverscapes they saw (possible responses: no improvement needed, landscape improvement, channel clearing). The black bars represent students' responses when presented with photos of a river channel with wood, and the gray bars represent their responses when the river channel contained no wood. The variation in answers from one region to another clearly illustrates that rehabilitation objectives differ from one cultural setting to another. The idea of working with nature is not accepted by all, nor is sustainable development always discussed. Most students from Sweden, Germany, and Oregon responded that no action should be taken when the riverscape images contained wood deposits. However, the majority of students from Russia and Texas recommended landscape improvement when presented with the same images, and the French, Italian, Polish, and Spanish students strongly recommended channel clearing for the wooded riverscapes.

Because the use of wood is socially controversial, this topic illustrates some of the challenges of practicing integrative river science. There are potential hazards to reintroduction of wood in streams—both perceived and actual (e.g., increasing water tables, downstream transport of large wood, and damage to infrastructure such as bridges). Experiences from some projects show that if the residents are not involved in the planning process, they may oppose the project, even if the project does not pose any hazard to them. These examples emphasize the need to also consider socioeconomic aspects (e.g., perception of hazards, analysis of cost-effectiveness). Unfortunately, in most Central European rehabilitation projects that involve large wood, only stream managers and hydraulic engineers are involved. Large wood is also typically engineered (e.g., cylindrical logs heavily fixed with boulders or cables) as trial-and-error management solutions, with limited monitoring of their effects. Of course, it is neither possible nor necessary to carry out extensive monitoring for every small project. However, detailed monitoring of "pilot projects" is necessary to quantify the effects of large wood and its potential use in rehabilitation projects under different conditions (e.g., stream type and size, land-use pressure).

Channel Widening in Austria and Switzerland

In the 1990s, integrated river management projects in Austria were primarily initiated by the authorities responsible for flood protection. The amendment of the national water law in 1985 to include ecological issues alongside increasing problems associated with traditional flood protection measures (channel incision, increases in flood discharge, and so on), demanded new management approaches. River widening became an important tool to link ecological objectives with flood protection and

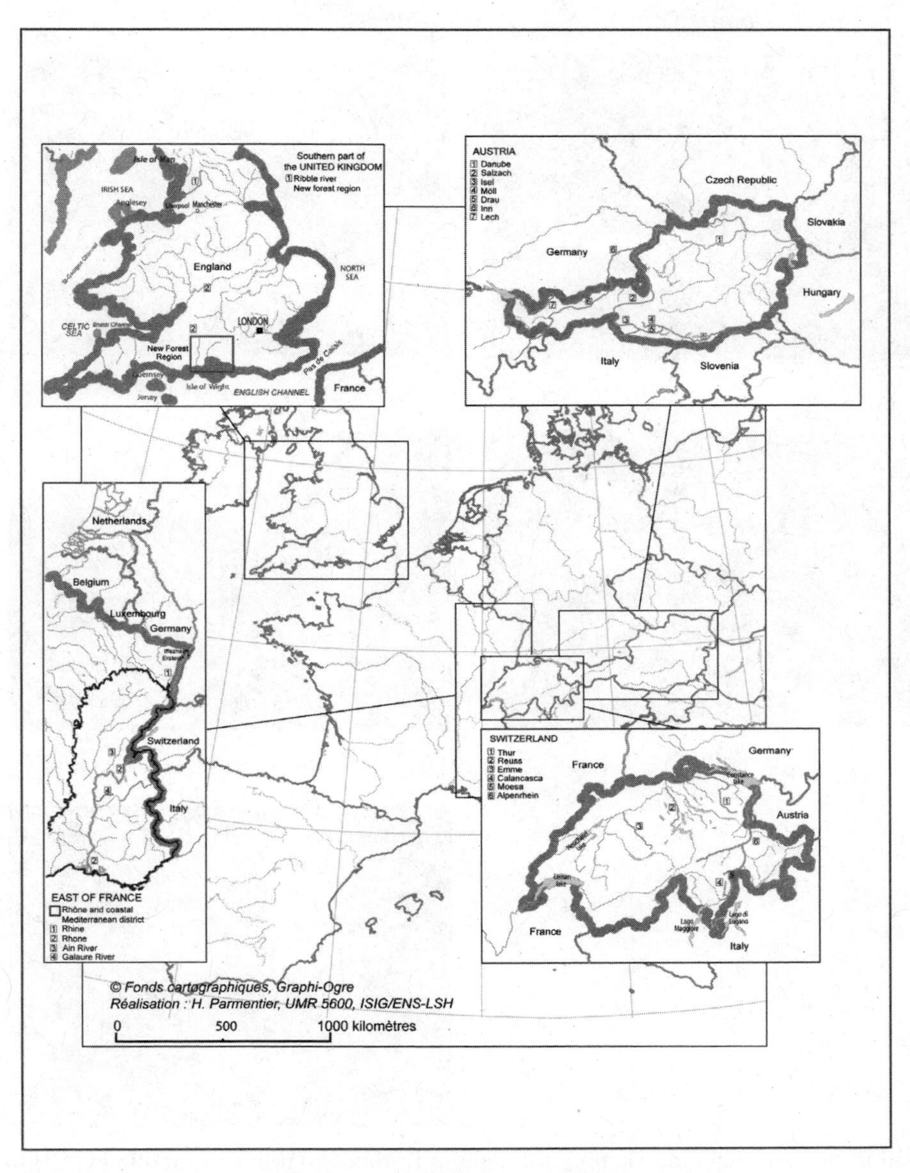

Figure 11.3. Location of the river reaches and areas exemplified in this chapter.

water engineering requirements, recreating multichanneled networks in previously channelized systems. This approach also fulfils the commitment for "specific ecological reference conditions" required by WFD. This is especially important because wide, branching river channels and a structured floodplain were typical features of alpine braided rivers before their channelization for flood protection from the nineteenth century to the 1980s (Jungwirth et al. 2003).

River widening has become a common rehabilitation measure in Austria and Switzerland (Jäggi and Zarn 1999; Habersack et al. 2000) (figure 11.4b). The first examples were applied in Austria about thirty years ago (e.g., River Isel in East Tirol; figure 11.3). In the 1990s, thirty-six river widening

FIGURE 11.4. Illustration of restoration actions throughout Europe: (a) wood reintroduction in Germany (Wolfgang Klein), (b) channel widening in Austria (Jürgen Petutschnig), (c) reflooding of the Rhine at the German-France border (Service de la Navigation), (d) recreation of perifluvial aquatic zones along the Rhône, France, (e) Channel remeandering in the New Forest in the UK.

projects were accomplished in Austria in rivers with a catchment area greater than 500 square kilometers (Habersack 2003). In Switzerland the first project was accomplished on the river Emme in 1992. Several projects followed on different rivers (Rohde 2004) (figure 11.3).

The largest channel widening projects in Austria occurred on the rivers Drau and Lech. Seventy kilometers of the river Drau and about 40 kilometers of the river Lech as well as three tributaries were included in these projects. For both projects, specific measures were set within a regional scale integrated river management framework. They were based on a detailed analysis of the current condition by geomorphologists, hydrologists, ecologists, and river engineers and on a crossdisciplinary target concept or guiding image (*leitbild*). Both rehabilitation programs were put into practice in close cooperation with water management and nature conservation agencies. Additionally, the European Commission supported their implementation as LIFE projects. As public acceptance of these initial rehabilitation efforts grew, projects became more comprehensive, creating different types of specific habitats and enhancing a range of morphodynamic processes. Together, this led to lateral erosion and sedimentation, pioneer site development, and flooding of riverbanks as well as adjacent wetlands (Muhar et al. 2004). On the river Lech the EU-life Nature project and Wild River Landscape Tyrolean Lech aim to restore the physical environment in order to maintain endangered riverine species. For both projects, ecological success is evaluated by crossdisciplinary monitoring programs. In general, however, long-term, detailed monitoring of river response to rehabilitation is often lacking.

Rehabilitation Plans of the Rhine and Rhône Rivers, France

Channelization of a 150-kilometer section of the Rhine River from Kembs to Iffezheim between 1928 and 1977 resulted in the disconnection of 130 square kilometers of floodplain (figure 11.3). This resulted in increased flood risk downstream of the channelized section, groundwater table lowering, almost total disconnection between anabranching channels and the main channel on the Rhine's left bank, ecological alteration (no rejuvenation of ecological successions, and so on), and loss of water purification processes in the flooding areas (Dister et al. 1990). One of the technical solutions to decreased flood risk, devised by the Commission of the Study of the Rhine Floods, was the construction of eighteen *polders* (i.e., retention basins in the former flooding area) in France and Germany. Following criticism of this action, an alternative integrated approach has been promoted (Dister 1992). Polders and the reconnection of anabranching channels to the main channel are created to enable "retention flooding" as well as "ecological flooding" (with running water when possible) for frequent high flows (Dister 1992). In France, this concept has been recently applied in the construction of the polder of Erstein (figure 11.4c) after an important negotiation between hydraulic managers (French Service of Fluvial Navigation), and biological scientists and associations of nature protection. After the construction of this polder, a six-year scientific monitoring program was initiated. Rehabilitation objectives are based on local opportunities (e.g., polders), rather than a holistic and planned view of the problem.

This case study also demonstrates how national boundaries can cause barriers to integrative river science. A French rehabilitation project, which consists of channel reconnections on the left side of the river, is partially funded by the European Union (LIFE Nature Program), but it is separated from rehabilitation projects on the German side. The French project seeks to meet public and political demand for rehabilitation of the Rhine floodplain (sixty-six towns participate), through public education in the rehabilitation and monitoring program. However, no interdisciplinary scientific study of the rehabilitation possibilities (and limits) of the Upper Rhine hydrosystem was undertaken prior to the application of rehabilitation works.

In the case of the Rhône River (figure 11.3), managers have stated that the channel has been altered by the construction of groynes and embankments in the nineteenth century, and by eighteen hydroelectric dams constructed between 1948 and 1986. A decadal rehabilitation plan was set in place in 1998, focusing on fish species mobility and habitat improvement with minimum flow increase, but also recreation of former channel conditions (Lamouroux et al. 2004; Amoros et al. 2005) (figure 11.4d). Rehabilitation policy has been based on a participatory approach, with a strong involvement by communes. Rehabilitation did not have the same meaning for each stakeholder group nor each commune; the final selection is a compromise of the different objectives and desires. Local managers wanted to recreate a more effective link between the Rhône and the local population, reminiscent of their childhood regardless of the ecological basis of rehabilitation schemes based on these goals. Selecting which former channels to rehabilitate has been a political challenge, as it has been based on ecological criteria and local scale political considerations. For example, most of the communes want to have their own rehabilitated channel and/or improved ecosystem functioning of a single former channel geometry (i.e., a channel with deep clear water always connected to the main channel). Such local-scale measures may not be ecologically effective if replicated for each commune along the rivers. Scientists' involvement has been limited to the very end of the decision-making process—for advice on monitoring design, instead of helping select appropriate rehabilitation goals and/or contributing to the design of rehabilitation schemes. In this context, scientific monitoring could prove to be a very effective and efficient means of demonstrating the ecological integrity of rehabilitation choices made by the communes and help design future rehabilitation works.

Catchment-Scale Wetland and River Floodplain Rehabilitation in the United Kingdom

In the United Kingdom, there is growing recognition of the potential benefits to be gained from planning rehabilitation projects at a more strategic, catchment scale. This is evidenced by two initiatives. The Royal Society for the Protection of Birds leads the Wetland Vision Project on behalf of government agencies and nongovernmental organizations (RSPB 2006). The project is designed to generate three maps. The first is a GIS map of where wetlands occurred historically. The second maps the current location of freshwater wetlands. The third identifies where it might be possible to recreate wetlands and renew river channel-floodplain connectivity at a national scale.

The Vision Map will identify those areas where scientists and stakeholders think it is feasible, in terms of science, social, and economic issues, to focus wetland rehabilitation efforts. Thus, in addition to the scientific element of identifying suitable locations, the project is also working with stakeholders and regulators to obtain a vision of what the UK would like wetlands to look like in fifty years' time. This combination of public and stakeholder perspectives and science on suitability of wetlands is novel. It will be interesting to monitor the success of the visioning process in securing government support to deliver the vision. If uptake by policy makers is good, this process demonstrates how integrative science and social science can deliver a fresh approach to environmental management.

At a more regional-local scale, one of several EU-LIFE projects is presented to demonstrate implementation of a strategic approach to wetland rehabilitation. The New Forest EU LIFE Project was a catchment-scale rehabilitation project where strategic planning of channel rehabilitation and channel-floodplain reconnections occurred (figure 11.4e). Preliminary (less than three years into the project) results strongly suggest that this approach has been successful. First, the project was collaborative with numerous organizations, local and national, working together. The benefits of collaboration were realized during the project. For example, many people were concerned that flood risk would increase after the rehabilitation works. A crossdisciplinary team of scientists successfully demonstrated and communicated that the project would lead to a reduction in flood risk. Second, to date this is the largest scale rehabilitation scheme in the UK and it demonstrated the utility of integrative approaches. Lastly, scientists involved in the project have remarked on the change in language and understanding of integrative rehabilitation over the past fifteen years (Sear, personal communication 2006), which echoes the recent work of Newson and Large (2006).

Feedback from the Case Studies

These case studies emphasize that integrated science is performed in Europe, although the degree and scales of integration (e.g., national, regional, local levels) vary considerably. Interactions between scientists, managers, and the public take different forms and are variable in intensity.

Academic scientists are involved in a few rehabilitation projects, but real transdisciplinary projects in which natural sciences as well as socioeconomic aspects are considered are still rare. The case studies also underline that the scientific evidence-base generated by monitoring programs is typically weak and not adequate to provide a basis for evidence-based policy and management required under WFD. There are several possible reasons for this. First, project manager funds are limited; and the rehabilitation measures, which are well intended, are assumed to be inherently beneficial. Therefore, routine or integrative scientific monitoring is not considered necessary. Second, water board and local authority funding is often restricted to the planning and implementation of projects, with little or no funding provision for post-project appraisal and/or monitoring. Detailed quantitative monitoring is often considered research, and thus not the responsibility of the project or the government agencies commissioning the rehabilitation works. Third, many rehabilitation projects are planned and implemented by engineering consultants. Companies who outline the uncertainty of planned rehabilitation measures, and advise that a detailed monitoring program be established, are unlikely to receive the order due to increased cost. In addition, project specifications are scoped such that the wide range of sciences (natural and social) required to deliver an integrated science approach are not offered by companies undertaking the work. Fourth, many natural scientists and/or project managers often undervalue the need for, and expertise in, socioeconomic aspects; they often complete the socioeconomic aspects themselves, although they lack the necessary background. Where monitoring does exist, the results are often not made available beyond the monitoring organizations. This makes it difficult to compare the effectiveness of different rehabilitation strategies, many of which could still benefit from improvement. For example, scientists often only design selected parts of projects, such as the monitoring program and/or consultation on plans already made, rather than being involved in the overall project design. This limits scientific guidance of the assessment of risks and uncertainties associated with rehabilitation plans and potential measures of success. The majority of rehabilitation projects occur at the local scale, without consideration of regional or national strategic plans and priorities. This can have implications for the availability of funding, ownership of the project, and whether the local project can serve as part of a national evidence base to support further rehabilitation activities. This gap may be considered more seriously in the future with the implementation of the WFD.

Interesting insights for social, planning, and strategic aspects of river rehabilitation have also emerged in the case studies. Observations can be made about the role of the public and stakeholders in the decision-making process. The public is rarely involved in the rehabilitation programs that are decided and prepared by the manager sphere, even if new communication programs are increasingly promoted (see figure 11.5). Stakeholder viewpoints often differ, leading to compromises over the rehabilitation goals selected. Ecological improvement as a specific objective is generally not meaningful for stakeholders and politicians. Instead, projects usually try to maximize benefits, with combinations of flood management, landscape improvement for practicing leisure activities, fish resource enhancement for fishing purposes, natural heritage restoration, and ecological

FIGURE 11.5. Examples of communication documents created in LIFE—projects (Ain and Drôme rivers) to valorize the gravel in the public perception and promote ideas of artificial sediment supply for ecological and risk management purposes. (a) "l'Ain, la rivière aux pierres précieuses"—The Ain, the River with precious stones / «Rivière en fête / l'Ain un trésor à deux pas de chez vous» - River Festival / The Ain, a treasure close to your home. (b) The river Drôme. "Pierres qui roulent" / Rolling gravels.

improvement. Often, the main objective of rehabilitation activities relates to flood risk management—either to improve it or to compensate for its adverse ecological impacts, while integrating ecological, recreational, or other social benefits. Thus, managers seek to create a win-win project, balancing the needs of different stakeholders and statutory requirements within the context of traditional management operations. This approach, if properly underpinned by sound scientific evidence and as part of planned environmental management at the catchment scale, could meet the WFD as well as other objectives such as flood risk or conservation (Wharton and Gilvear 2006).

Nevertheless, feedback from recent integrative science experiences has already been positive, leading to a shift in practice. More and more integrated science and process-based operations are now promoted (Henry and Amoros 1995; Dufour and Piégay 2005; Habersack and Piégay 2007). Managers now consider the inherent natural dynamics of rivers and understand that unregulated systems can provide immediate benefits and services in terms of risk management, water resource availability, and quality of human environment. Following this evolution, flooding, bank erosion, and wood in rivers are considered more positively than previously by managers, although there is not yet widespread acceptance by local politicians or residents. National laws and the WFD provide important legal frameworks upon which such policy decisions can be based. For example, the French Ministry of ecology and sustainable development passed a risk-based law in 2003 that recognizes that the preservation and rehabilitation of floodplains and eroding river corridors can be used as tools to manage flood and bank erosion (Piégay et al. 2005).

Challenges Approaching Implementation of the European Water Framework Directive

The WFD directive is ambitious and innovative. Its implementation has led to active debates in the scientific community with some doubt about its application. Cooperation between scientists and managers has rapidly increased as WFD implementation is required by all member states. Cooperation has occurred for a number of key elements of WFD, such as identification of waterbody types and their boundaries; designing methodologies to identify highly modified waterbodies, good ecological status, and reference conditions; identifying suitable mitigation and rehabilitation measures; designing a program for implementing such measures; and, finally, development of monitoring tools to assess the efficiency of the program of measures. The latter is designed to demonstrate a causal link between rehabilitation or mitigation options selected by managers and an ecological response at six-year intervals.

The Reference Conditions Debate: From Theory to Practice

Changes in scientific and management perspectives have radically altered the way hydrologists, biologists, and geomorphologists consider rivers. No longer do they think of rivers as natural features, but instead as features with natural attributes under human influence. This has led to controversial discussions as to whether and how one could objectively define reference and target conditions for restoration required by WFD. Basically, two different approaches are used to define the reference state. The first approach has its roots in the assumption that before major human changes such as systematic channelization or the construction of hydro power plants (i.e., from the late nineteenth century onward), rivers exhibited hydromorphologically dynamic, varied habitats and consequently, a diverse ecology—even if there were human changes prior to the major engineering works (Muhar 1994). Such a historical state is used as a reference state (e.g., in Austria, where this concept was developed in the early 1990s).

Scientists who favor the second approach argue that many streams have been altered by human activity since the Mesolithic Age and irreversible changes to the natural setting cannot be reversed even if the streams are no longer affected by human activity (e.g., subsidence caused by mining). Hence, as the recreation of any historical state is impossible, the reference state is defined as the state which potentially would develop under the present natural setting (including the irreversible changes) without further human intrusion (Deutscher Verband für Wasserwirtschaft und Kulturbau 1996). This "guiding image" (*leitbild*) is comparable to the concept of the potential natural vegetation by Tüxen (1956). In such an approach, historical stages are considered not as target stages, but as knowledge helping to define the present potential natural state.

In this light, many authors consider actions for ecological improvement in Europe as rehabilitation rather than restoration, moving beyond links to a historical state as a reference target (see chapter 1). The rehabilitation concept is broader and can include actions to improve existing conditions, with the aim of maximizing ecological potential. These ideas emerged more recently and conflict slightly with the WFD. Any reference state based on historical states is difficult to define due to a lack of data and does not always represent a desirable state to attain under present conditions. Geomorphic patterns are easy to determine from old maps, but ecological, hydrological, and water-quality conditions are usually more difficult to reconstruct. Moreover, some anthropogenic features can provide interesting biodiversity (e.g., some mining sites). Also, river patterns that existed prior to the implementation of major engineering works in the nineteenth and twentieth centuries were not necessarily natural, and their higher ecological quality is far from being demonstrated (see Kondolf et al. 2007). Therefore, it seems sensible to promote rehabilitation rather than restoration strategies to set targets against which to measure improvements. In both cases, this does not change the practice of restoration/rehabilitation efforts, as both improve ecological features. It may, however, change the way of measuring the response of ecosystems following the implementation of the WFD.

The debate on reference states is still raging, but time is limited as managers are currently implementing the WFD. Managers use scientific information and technical tools to implement the WFD, taking decisions pragmatically to meet the tight timetable set by the directive. As such, scientific methods have been adapted to provide some meaningful input to managers, to meet their rigid timescales (Newson and Large 2006). In the Rhône-Mediterranean district, the existing state of waterbodies was published in March 2005. Some of the recommendations, such as how to best identify waterbodies or assess their state of alteration, were based on existing data (mainly for water quality and biology) and expert knowledge (for geomorphic indicators) provided by local stakeholders, river managers, and scientists (figure 11.6). Once waterbody states were assessed, local experts identified the rehabilitation measures deemed appropriate to reach the goal of "good ecological status" for the reaches they manage, taking into account technical and financial capabilities. In order to help local managers, district managers worked with scientific experts from different disciplines, building upon the results of previous rehabilitation studies to provide guidelines on generic measures. The toolbox suggests that four components of physical features need to be evaluated (hydrology, sediment management, biological connectivity, and geomorphology of the channel and floodplain). The measures proposed at the local scale will be compared with propositions from the state services, to agree on the most suitable (scientifically, economically, and socially) rehabilitation and/or mitigation measures to be implemented. In this context, it is clear that the rehabilitation program is

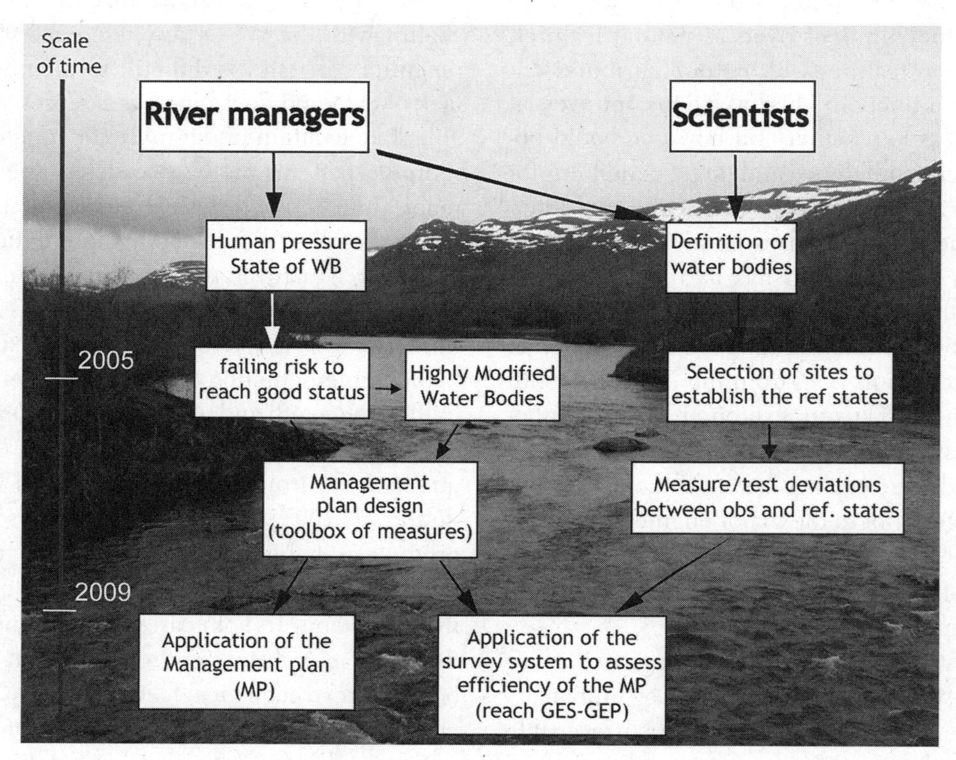

FIGURE 11.6. General schedule of the WFD implementation within the Rhone—Mediterranean district showing the interactions between the river managers (Water Agency, Representative of the Ministry of Ecology and Sustainable Development) and the scientists (expert groups). (WB refers to Water Body, GES is Good Ecological Status, and GEP is Good Ecological Potential.)

designed pragmatically by managers with the occasional involvement of scientists, and a good appraisal of reference conditions is not so important.

Assessment of Good Ecological Status

If in practice managers design their program of measures to reach a good ecological status independently of scientists, major scientific efforts are required to design tools that can be used by managers to assess the effectiveness of rehabilitation measures and determine if a given waterbody has reached good ecological status. A deviation between a reference state and an observed state must be calculated based on a large set of physico-chemical, hydromorphological, and biological metrics. The definition of a biological reference for each waterbody type has been based on expert knowledge from state managers and scientists. A reference reach is considered as being subjected to minimal human pressures (e.g., weirs, dikes) on the floodplain, channel, and banks.

In France, identification of reference reaches has comprised two steps: (1) a preliminary list of reference reaches has been generated by scientists using GIS data analysis, based on land-use conditions and physical river characteristics; (2) managers conducted field surveys to validate whether the pressure analysis from the GIS was accurate. Their assessment was based on a survey of the amount of human alteration observed (i.e., a "hydromorphological pressure," e.g., number of weirs) and physico-chemical condition of the river. This presents a good example of integrative science where scientists provide data and preliminary analysis to help managers conduct the risk assessment phase of WFD implementation. Integration also occurred across spatial scales where objective setting for each waterbody was conducted through dialog between managers in charge of the local catchment plans and the managers in charge of the implementation of the WFD at the river basin district scale.

Key scientific questions need to be considered alongside WFD implementation to provide information to improve its application. To date, there have been considerable collaborative or parallel efforts to develop monitoring tools (e.g., FAME, outlined below). WFD monitoring is mainly associated with the design, testing, and use of monitoring to validate that good ecological status is reached by 2015. This work is not based on interdisciplinary efforts, but on the challenging task of comparing national or regional methodologies and databases to derive a European calibration of the different quality indices for ecological status (known as the intercalibration exercise by managers). This is based on the development of tools at the network scales of Europe. Various programs have been funded by Europe in this context. FAME (2001–2004), a research project supported by the European Commission under FP5, focused on development, evaluation, and implementation of a standardized fish-based assessment method for the determination of ecological status of European Rivers, by twelve countries. The main aims were to determine: (A) how to define a good ecological status for rivers using fish as a biological indicator and (B) how to evaluate and quantify a deviation from the reference status to determine specific requirements at the European scale for intercalibrating biological tools developed at national or regional levels. This promoted new common biological tools at the European scale. Other projects such as Euro-Limpacs (2004–2009, twenty-two countries involved) aim to provide scientific information to evaluate the effects of future global change on Europe's freshwater ecosystems. This research can also inform the WFD, as climate variability and change was not explicit in the directive, yet it is imperative that tools, monitoring, and measures are designed within the envelope of climate variability and change (Wilby et al. 2006).

The efficacy of scientific inputs being tailored to the timescales required by EC legislation has yet to be evaluated, as it is too early to assess the costs and benefits of meeting such short time horizons.

It is postulated that this approach may require more effort by EC nations than if the time horizon was relaxed to enable a more integrated approach to the design and implementation of tools and measures to meet the directive's requirements. What is clear, however, is that future environmental legislation would benefit from being designed in close collaboration with scientists, to ensure that the timescales set by such legislation are realistic before transposing time-dependent legislative endpoints into law.

Involvement of Scientists in Local Rehabilitation Projects

In 2009 the program of measures promoted to reach good ecological status (or potential) will begin. Clear challenges exist in terms of knowledge production and tool development for selecting appropriate measures. Integrative sciences for helping river managers at this scale have largely been developed in the European LIFE programs, where funding can be used to experiment and test new management options, within which monitoring can be done and academic scientists involved. Such programs have been very important for supporting rehabilitation actions and testing their efficiency. A recent LIFE program on Forest for Water (2005–2007) highlighted how foresters in Europe could assist in the achievement of the objectives of the WFD. It provides information to help managers design their program of measures for improving water and aquatic ecosystem quality and reach good ecological status, by changing forestry practices. Expert groups of both scientists and managers are supporting WFD implementation in forests, by working together to produce a comprehensive handbook on river basin management planning. Additionally, group members from countries that are not partners in the project will deliver information on forestry and forestland in the implementation of the WFD in their countries. Moreover, the program is used to monitor and test both for effects of actions and for WFD implementation, as well as to explore how monitoring partnerships could be established.

In the United Kingdom, a pilot catchment and river basin management planning unit on the River Ribble, in northwest England, is being used to test two specific aspects of WFD implementation: (1) river basin planning, and (2) guidance on public participation (Environment Agency 2005). Preliminary work with local stakeholders enabled researchers to highlight some of the key mechanisms to overcome barriers to social learning and public participation required to implement the WFD. This spawned research and development work for other aspects of the WFD. New biological classification tools to support WFD for various riverine biota have been tested. Similarly, state-of-the-art equipment for monitoring the chemical status of water (e.g., web-linked automatic sensors for a suite of chemicals) and the biological quality of rivers (e.g., use of geonomics) have been tested.

Scientific Challenges Posed by the WFD

Future contributions by scientists to WFD applications and water ecosystem improvement should target three key themes.

1. *Conceptual background and continuous knowledge production*. In this domain, social scientists can provide significant expertise and additional knowledge to improve the decision-making process. For example, the services provided by natural ecosystems need to be better understood and estimated to provide economic arguments for rehabilitation. Further discussion between managers and scientists is required to promote interactions and improve efficiency, enhancing the design and monitoring of rehabilitation programs.

2. *Large-scale tools* are required to provide diagnosis at the catchment scale and help to prioritize rehabilitation strategies, help in the design of planning strategies, and create a mechanism by

which rehabilitation success can be evaluated. The development of a Europe-wide project database is required. Standardization of sampling designs and measurement procedures would help to populate such a large-scale dataset. The WFD demands explicitly the development of typologies of river reference status.

3. *Local scale post-rehabilitation monitoring* is required to provide feedback to identify the most efficient measures, models, and prediction tools, potentially leading to a reduction in monitoring and improvement actions. Integrated science approaches are required to link human pressures to ecological status and responses, so that measurable improvements in river functioning (or ecological status) can be made.

Conclusion

The period from 2007 to 2015 is a challenging one in the European context for developing integrative sciences for water management. The legal framework for the WFD places very tight timescales on managers, where individual countries risk infraction fines from the EC if they do not deliver components of the directive within rigid timeframes. In this deadline-driven context, scientists can provide expertise and experience to inform management decisions through improved interactions between scientists and managers and through scientific involvement in public discussions on river basin management planning decisions. Integrative sciences have clear yet challenging tasks at the continental scale, to develop decision-making tools and to identify priority reaches. Integrative sciences also have a clear role to play at the local scale, to better design rehabilitation and mitigation measures and assess their efficiency.

There are still many unanswered questions. At what spatial and time scale do we expect biological improvement? Do we need restoration, rehabilitation, and/or mitigation of human pressures? Will these satisfy long-term human needs? Which measures are socially desirable and/or acceptable? Can participatory management and education lead to environmental improvement? What landscape attributes do people want and are they willing to pay for improvement? How can the concepts of biodiversity, naturalness, and functionality be incorporated into management planning debates? What strategies can be implemented to manage an environment that has been heavily modified by humans, moving beyond creating stability to encouraging natural processes and dynamics to deliver more sustainable river systems? Over the past decade there has been increasing acceptance and uptake of the necessity of living with nature by scientists, managers, and the public. In this light, it would be very informative, both academically and practically, to assess the extent to which ongoing management programs embrace these concepts, integrating our understanding of biophysical connections, variability, and sensitivity at local and catchment scales.

The urgency of WFD implementation in EC member states has heightened the need for integrative and collaborative approaches to river science across Europe. Much work is currently being undertaken that embraces aspects of integrative river science, such as biologists working with chemists and economists working with geomorphologists. Some aspects and projects that support WFD implementation are more integrative in nature, although these are the exceptions rather than the rule. However, the number and variety of these new collaborations to grapple with implementing the same directive across Europe presents a fruitful future research opportunity for natural as well as social scientists to explore.

Acknowledgments

The French authors are grateful for the interdisciplinary discussions and experiences of the group of the Rhône Basin LTER involving also river managers in an integrative science project including hydrobiologists, geomorphologists, hydraulic engineers, and social scientists (http://www.graie.org/zabr/index.htm). We also thank the group of scientists and managers involved in the LIFE Forests for Water (2003–2007). Larissa A. Naylor values discussions at recent UK hydrogeomorphology workshops for informing her contribution to this paper. The authors also thank kindly Ken Gregory, Matt Kondolf, and the book editors for their helpful comments during the reviewing process. Y. Le Lay and H. Parmentier provided some of the figures.

References

Amoros, C., and G. E. Petts. 1993. *Hydrosystèmes fluviaux*. Paris: Masson.

Amoros, C., A. Elger, S. Dufour, L. Grosprêtre, H. Piégay, and C. Henry. 2005. Flood scouring and groundwater supply in side-channel rehabilitation of the Rhône River, France. *Archiv für Hydrobiologie* 155:147–67.

Bravard, J. P., G. M. Kondolf, and H. Piégay. 1999. Environmental and societal effects of river incision and remedial strategies. In *Incised river channels*, ed. A. Simon and S. Darby, 303–41. Chichester, UK: Wiley.

Brookes, A. 1987. Restoring the sinuosity of artificially straightened stream channels. *Environmental Geology and Water Sciences* 10:33–41.

Collins, K., R. Ison, and C. Blackmore. 2006. River basin planning project: Social learning. *Environment Agency Science Report* SC050037/SR1.

Deutscher Verband für Wasserwirtschaft und Kulturbau (DVWK). 1996. Fluß und Landschaft: Ökologische Entwicklungskonzepte. *DVWK-Merkblätter zur Wasserwirtschaft*.

Dister, E. 1992. La maîtrise des crues par la renaturation des plaines alluviales du Rhin Supérieur. *Bulletin de la Société Industrielle de Mulhouse* 824: 73–82.

Dister, E., D. Gomer, P. Obrdlik, P. Petermann, and E. Schneider. 1990. Water management and ecological perspectives of the Upper Rhine's floodplain. *Regulated Rivers: Research and Management* 5:1–15.

Dufour, S., and H. Piégay. 2005. Restoring floodplain forests. In *Forest restoration in landscapes: Beyond planting trees*, ed. S. Mansourian, D. Vallauri, and N. Dudley, 306–12. New York, NY: Springer.

Environment Agency. 2005. The Ripple Effect: Encouraging active involvement in the Water Framework Directive. *Proceedings from the Ripple Effect Conference*, Bolton, October 12–14, 2004.

Gerhard, M., and M. Reich. 2001. *Totholz in Fliessgewässern, Empfehlungen zur Gewässerentwicklung*. Mainz, Germany: GFG mbH und WBW.

Gregory, S. V., K. L. Boyer, and A. M. Gurnell. 2003. The ecology and management of wood in world rivers. Symposium 37. Bethesda, MD: American Fisheries Society.

Habersack, H. 2003. Erfahrungen mit Gerinneaufweitungen in Österreich. Communication at the Workshop Gerinneaufweitungen organized in March 2003 at EAWAG Kastanienbaum, Switzerland.

Habersack, H., M. Koch, and H. P. Nachtnebel. 2000. Flußaufweitung in Österreich: Entwicklung, Stand, Ausblick. *Österreich Wasser- und Abfallwirtschaft* 7/8:143–53.

Habersack, H. M., and H. Piégay. 2007. Challenges in river restoration in the Alps and their surrounding areas. In *Gravel-bed rivers 6: From process understanding to river restoration*, ed. M. Rinaldi, H. Habersack, and H. Piégay, 703–37. Amsterdam, Netherlands: Elsevier.

Henry, C. P., and C. Amoros. 1995. Restoration ecology of riverine wetlands. I. A scientific base. *Environmental Management* 19:891–902.

IRKA and IRR. 2005. *Entwicklungskonzept Alpenrhein*. Kurzbericht, Hg. von der Int. Regierungskommission Alpenrhein (IRKA) & Int. Rheinregulierung. Thusis and Rorschach, Switzerland.

Jäggi, M., and B. Zarn. 1999. Stream channel restoration and erosion control for incised channels in alpine environments. In *Incised river channels*, ed. S. Darby and A. Simon, 343–70. Chichester, UK: Wiley.

Jungwirth, M., O. Moog, and S. Muhar. 1993. Effects of river bed restructuring on fish and benthos of a 4th-order stream, Melk, Austria. *Regulated Rivers, Research and Management* 8:195–204.

Jungwirth, M., and H. Waidbacher. 1990. Fischökologische Zielsetzungen bei Fließgewässerregulierungen. *Wiener Mitteilungen* 88:105–20.

Jungwirth, M., G. Haidvogl, O. Moog, S. Muhar, and S. Schmutz. 2003. *Angewandte Fischökologie an Fließgewässern Facultas*. Vienna, Austria: Facultas.

Kondolf, G. M., H. Piégay, and N. Landon. 2007. Changes in the riparian zone of the lower Eygues River, France, since 1830. *Landscape Ecology* 22:367–84.

Lamouroux, N., A. Chanderis, J. F. Fruget, S. Dufour, H. Piégay, A. Citterio, C. Amoros, C. Henry, J. Barbe, C. Bedeaux, P. Joly, S. Plenet, and M. Babut. 2004. Chute de Pierre Bénite. Suivi de l'incidence de l'augmentation du débit réservé dans le vieux-Rhône. Phase II (2001–2004). Unpublished final report.

Landesumweltamt Nordrhein-Westfalen. 1989. *Richtlinie für naturnahen Ausbau und Unterhaltung der Fließgewässer in NRW*. Düsseldorf, Germany.

Land Salzburg, Ref. Landesplanung, and SAGIS. 1997. Gesamtuntersuchung Salzach. Gesamtübersicht und Kurzfassungen der Teiluntersuchungen der Teile A "Basisuntersuchung" und B "Regionale Studie." Salzburg, Austria: Kammer für Arbeiter und Angestellte Salzburg.

Lang, O., M. Fries, H. Wallnöfer, and W. Bubik. 1996. Regionales Pilotprojekt Lech-Ausserfern. Ökologisch-ökonomische Entscheidungsgrundlagen zur Erhaltung, Pflege und Enwicklung des alpinen Kulturlandschaftsraumes nach Schwerpunkten.ILU—Institut für Landschaftspflege und Umweltschutz. Hg. BMLF Wien, Amt der Tiroler Landesregierung Innsbruck, EW Reutte, Gemeinden des Lechtals.

Lévêque, C., A. Pavé, L. Abbadie, A. Weill, and F. D. Vivien. 2000. Les zones ateliers, des dispositifs pour la recherche sur l'environnement et les anthroposystèmes. Une action du programme "Environnement, vie et sociétés" du CNRS. *Natures, Sciences et Sociétés* 8:43–52.

Maser, C., and J. R. Sedell. 1994. *From the forest to the sea: The ecology of wood in streams, rivers, estuaries, and oceans*. Delray Beach, FL: St. Lucie Press.

Muhar, S. 1994. Stellung und Funktion des Leitbildes im Rahmen von Gewässerbetreuungskonzepten. *Wiener Mitteilungen* 120:136–58.

Muhar, S., S. Preis, S. Schmutz, M. Jungwirth, G. Haidvogl, and G. Egger. 2003. Integrativ-ökologisches Management von Flussgebieten (Integrative and Ecological River Basin Management). *Österreich Wasser- und Abfallwirtschaft* 55:213–20.

Muhar, S., G. Unfer, S. Schmutz, M. Jungwirth, G. Egger, and K. Angermann. 2004. Assessing river restoration programs: Habitat conditions, fish fauna and vegetation as indicators for the possibilities and constraints of river restoration. *Proceedings 5th Int. Symposium on Ecohydraulics. Aquatic Habitats: Analysis & Restoration*. Madrid September 12–14, 2004.

Newson, M. D., and A. R. G. Large. 2006. "Natural" rivers, "hydromorphological quality" and river restoration: A challenging new agenda for applied fluvial geomorphology. *Earth Surface Processes and Landforms* 31:1606–24.

Piégay, H., M. Cuaz, E. Javelle, and P. Mandier. 1997. A new approach to bank erosion management : The case of the Galaure river, France. *Regulated Rivers: Research and Management* 13:433–48.

Piégay, H., S. A. Darby, E. Mosselmann, and N. Surian. 2005. The erodible corridor concept: applicability and limitations for river management. *River Research and Applications* 21:773–89.

Piégay, H., P. Dupont, and J. A. Faby. 2002. Questions of water resources management: Feedback of the French implemented plans SAGE and SDAGE (1992–1999). *Water Policy* 4:239–62.

Preis, S., S. Muhar, H. Habersack, C. Hauer, S. Hofbauer, and M. Jungwirth. 2006. Nachhaltige Entwicklung der Flusslandschaft Kamp: Darstellung eines Managementprozesses in Hinblick auf die Vorgaben der EU-WRRL. *Österreich Wasser- und Abfallwirtschaft* 58:159–67.

Redl, G. 1990. Moderne wasserwirtschaftliche Planungsansätze. *Wiener Mitteilungen* 88:149–90.

RIWA-t 1994. Richtlinien für die Bundeswasserbauverwaltung. BM für Land- und Forstwirtschaft, Sektion IV—Wasserwirtschaft und Wasserbau. Vienna, Austria.

Rohde, S. 2004. River Restoration: Potential and limitations to re-establish riparian landscapes. Assessment & Planning. PhD diss. Swiss Federal Inst. of Technology, Zurich. Roux, A. L. 1982. Cartographie polythématique appliquée à la gestion écologique des eaux. Etude d'un hydrosystème fluvial, le haut-Rhône français. Lyon, France: CNRS.

RSPB (Royal Society for the Protection of Birds). 2006. *A 50-year vision for wetlands: A future for England's water and wetland biodiversity*. Sandy, UK: Royal Society for the Protection of Birds.

Siemens, M., S. Hanfland, W. Binder, M. Herrmann, and W. Rehklau. 2005. Totholz bringt Leben in Bäche und Flüsse. Bayrisches Landesamt für Wasserwirtschaft, Color-Offset, München, 47.

Tüxen, R. 1956. Die heutige potentielle natürliche Vegetation als Gegenstand der Vegetationskartierung. *Angewandte Pflanzensoziologie* 13:5–42.

WCCED (World Commission on Environment and Development). 1987. *Our common future*. Oxford, UK: Oxford University Press.

Wharton, G., and D. J. Gilvear. 2006. River restoration in the UK: Meeting the dual needs of the European Water Framework Directive and flood defence? *International Journal of River Basin Management* 4:1–12.

Wilby, R., H. G. Orr, M. Hedger, D. Forrow, and M. Blackmore. 2006. Risks posed by climate change to the EU Water Framework Directive. *Environment International* 32:1043–55.

Chapter 12

The Light and Dark of Sabo-Dammed Streams in Steepland Settings in Japan

TOMOMI MARUTANI, SHUN-ICHI KIKUCHI, SEIJI YANAI, AND KAORI KOCHI

> While water quality has improved remarkably during the past decades, Japanese rivers are still heavily impacted by canalization, loss of most dynamic flood plains, flow regulation, invasion by exotic species, and intensive urbanization. Currently 49 percent of the entire human population concentrates on 14 percent of the land, and the annual flood damage is the highest worldwide. As a consequence, major recent restoration initiatives aim to protect people and property against floods as well as simultaneously improving the ecological integrity of river ecosystems.
>
> Yoshimura et al. 2005, 93

Due to the topographic and climatic setting of Japan, along with its high population density, a concerted effort has been made to minimize the impacts of sediment disasters upon Japanese society. In order to protect cities and infrastructure from debris flow hazards, multiple sets of Sabo dams have been constructed along more than 80,000 streams in Japan. The construction of these dams has been especially pronounced since the 1960s, due to rapid economic growth. Since the 1990s, there has been a trend to implement more natural engineering techniques across the country. In 1997, river legislation placed specific emphasis on environmental issues, extending previous policies that focused solely on flood control and water utilization. This trend accelerated after enactment of the Law for Promotion of Nature Restoration. However, many rivers remain in poor condition, with ecological gradients that are severely disrupted by various kinds of dams. Rehabilitating these rivers and improving biological diversity are among the main technological and social challenges faced in Japan today. This chapter provides an overview of some recent modifications to the construction and operation of Sabo dams in steepland settings of Japan, indicating how environmental principles are increasingly integrated into river management. Examples highlight concerns for the management of riparian forests, instream habitat, and ecosystem functionality. Before discussing these developments, this chapter will briefly outline the environmental setting of Japan.

Why Have We Developed the Sabo Dam Country?

The ever-changing and fragile beauty of Japan's landscape reflects the country's environmental diversity. The islands of Japan are located on the northwestern edge of the Pacific Rim mobile belt and ring of fire. They are dominated by steep, tectonically active folded mountain ranges that are prone to earthquakes and volcanic hazards (Yoshikawa 1974; Tsukamoto 1973). The Asian monsoon climate yields heavy rainfall, often associated with typhoon-induced storms that move along the mountain ranges. More than ten typhoons can occur in a year, with rainfall intensities as high as 500 millimeters per day and 50 millimeters per hour. Collectively, these environmental, geomorphic, and climatic conditions have resulted in an intricate pattern of land-use and sediment disasters in Japan (figure 12.1).

Extensive modification of Japan's rivers has resulted in limited opportunity to investigate fluvial processes under pristine conditions (Oguchi et al. 2001). Although much is known about the geographical characteristics of rivers in Japan, especially the ecological status of lowland rivers (Yoshimura et al. 2005), prospects for urban/wetland restoration (Nakamura et al. 2006), and impacts of dams upon fish populations (Fukushima et al. 2007), few studies have adopted a whole-of-system approach to geoecological investigations. However, many of the problems arising in the lowland parts of Japanese rivers reflect processes operating in the steepland settings of the upper catchment. Hence, it is important to frame analyses in relation to the entire river system, from sediment source to sink. These relationships are especially important when examining storm-related sediment disasters in Japan (Marutani et al. 2001). Concerted efforts have been made to negate these impacts through construction of Sabo dams. These features, in turn, have affected the geoecology of stream environments through their onsite, direct effect on upland streams, and indirectly through offsite impacts on downstream reaches.

Storm-related sediment disasters and associated watershed management problems in Japan are accentuated by the narrowness of the land mass, the high population density, and the intense land utilization. Steep mountainous catchments comprise 70 percent of Japan's land area, occupying approximately 3,800,000 square kilometers. The Japanese population recently reached 130 million, a tenfold increase over the last century. Population growth has been accompanied by increases in food and

Figure 12.1. Local community close to a hazardous hillslope, Saru River catchment, Hokkaido, Japan.

energy production, and urban and industrial expansions. During this phase, many farmers, foresters, and tourists expanded their range of activities beyond relatively safe zones into hazardous mountainous areas. As urban growth has accelerated in recent decades, many remote areas have been abandoned without due care or rehabilitation. Abandoned landscapes have been subjected to surface erosion and sediment generation, promoting many debris and sediment flows, landslides, and cliff failures along steep channels. Such sediment disasters dominate Japan's landscape today, with significant impacts on human life (figure 12.1 and table 12.1).

Despite the recent development in preventative technologies and early warning systems, the number of sediment disasters increased during the five years prior to 2005. Debris flow damage to homes, facilities, and infrastructure can be personally tragic and economically devastating. More than 80,000 streams remain prone to debris flow hazards in Japan. These are primarily steep, low-order streams with small catchment areas.

Debris flows not only have a huge impact on human life and property, but they also affect landscape and river health. Many Japanese scientists and technicians have interpreted the physical and hydrological mechanisms of debris flows (e.g., Takahashi 1978). Typically, these processes occur on slopes above 14 degrees during periods of intense rainfall (more than 40 millimeters per hour). Debris flows are especially dangerous given their velocity (3 to 14 ms^{-1}) and difficulties faced in predicting the time and place of their occurrence. This is a prime reason why Sabo dams have been constructed along most steepland streams in Japan.

Given the high degree of landscape connectivity in these steeplands (Harvey 2002; Fryirs et al. 2007), sediment yielded from hillslope process by landslides and debris flows is rapidly transferred through drainage systems. This is especially the case in the upper reaches of drainage systems where debris flows and hyperconcentrated flows denude bedrock channels by abrasion processes. Riverbed denudation causes foot slope incision, resulting in lateral slope failure. The sediment produced from lateral sources is immediately transferred downstream, further contributing to the denudation of the riverbed. In order to manage this cycle of degradation, facilities are required to retard the downstream transfer of sediment and stabilize bed-level adjustments. System responses to the placement of Sabo dams in these mountainous landscapes have resulted in an upstream extension in the length of dammed streams.

A schematic drawing of a Sabo dam system is shown in figure 12.2. The dam is attached to both basal and lateral bedrock. The dam wall is commonly between 5 and 20 meters high. To protect the base of Sabo dams from bedrock incision, a lower or secondary (sub) dam is attached below the main structure. Although the Sabo dam system traps coarse materials and debris, fine suspended materials pass through and are not retained within the sand pocket behind the dam. Therefore, fine materials do not accumulate, whereas coarse sand is intercepted until the dam's pocket fills up. As a result, the particle size distribution on the river bed is very different in the lower stream where large

TABLE 12.1

Fatalities and missing persons for recent sediment disasters in Japan (excluding volcanic and flooding disasters) (Ministry of Land, Infrastructure and Transport, Japan, 2006).

Year	2000	2001	2002	2003	2004	2005
Cliff failure	291(6)	365(2)	275(2)	712(2)	1511(28)	483(8)
Land slide	137(0)	96(1)	218(0)	128(1)	461(7)	173(5)
Debris flow	180(0)	48(1)	46(2)	57(20)	565(27)	158(17)

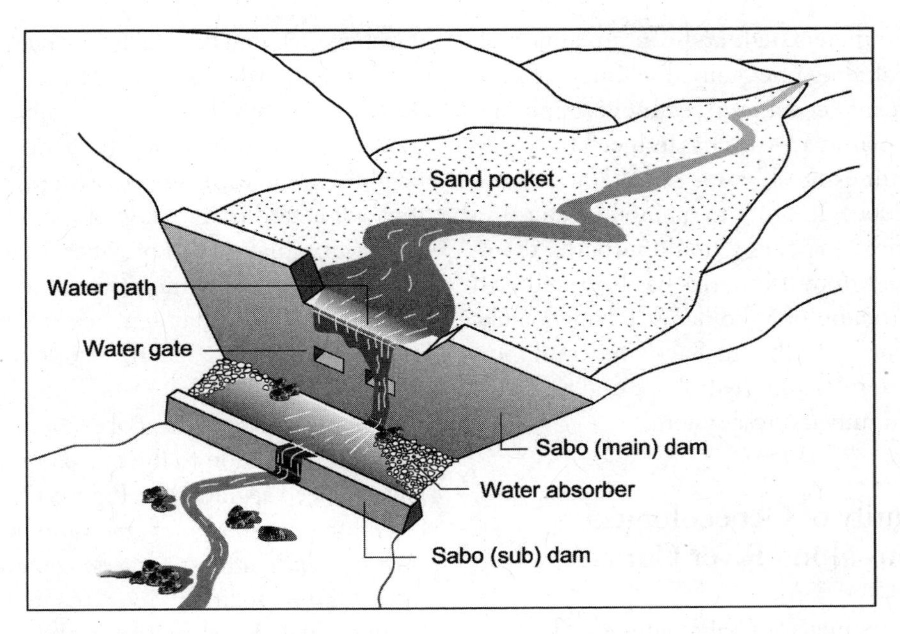

FIGURE 12.2. Schematic drawing of a Sabo dam.

rocks dominate, relative to the upper stream where sand dominates.

Decoupling of lateral and longitudinal sediment transfer and retardation of sediment in upstream parts of catchments are the primary aims of Sabo dam construction. There are several ways in which this is achieved. First, stabilizing steep slopes and/or valley heads prevents sediment release from slopes into the channel network, disconnecting the lateral transfer of sediment. Chisan dams are constructed in the uppermost reaches of the catchment to reduce erosional impacts of debris torrents (table 12.2). These check dams are small structures, typically 3 to 5 meters high. Second, storing sediment on the riverbed is an emergency measure to reduce longitudinal sediment transfer further downstream. This is achieved through the construction of Sabo dams (table 12.2). Chisan dams are generally associated with reforestation and comprise a sequence of stepwise dam structures with increasing bed friction. In contrast, Sabo dams aim to interrupt debris flows and accumulate sediment.

TABLE 12.2

Design and effect of Chisan dam and Sabo dam

	Expected effect	Design criteria
Chisan dam	Preventive measure to reduce lateral erosion caused by channel bed incision in mountainous streams, by increasing bed friction	Expected channel bed slope after infilling the sand pocket behind the dam is one half to two thirds that of the natural channel bed slope
Sabo dam	Traps debris flows and controls sediment discharge along the stream channel	Expected channel bed slope after infilling the sand pocket behind the dam is one half that of the natural channel bed slope

Industrial development and increased population pressure in Japan have required expansion to electric power and water resource facilities. In response, multipurpose dams have been constructed in mountainous areas. These dams usually range from 10 to 50 meters high and capture almost all sediment (including fine-grained sediment) resulting in significant decreases in sediment supply and transport. In many cases this has induced the recession of coastlines (Yoshimura et al. 2005; Nakamura et al. 2006). In order to maintain water storage capacities in these multipurpose dams, sediment input from tributaries has been reduced by the construction of Sabo dams. Unfortunately, the construction of Chisan, Sabo, and multipurpose dams also interrupts hydrologic and biotic cycles and continuity in these systems.

Discontinuity of Geoecological Interactions along River Courses

In this section, impacts of Sabo dams are analyzed in terms of consequences for aquatic habitat, riparian forests, fish stocks, and measures of ecosystem functionality (focusing on organic matter processing).

Impact of Sabo Dams on Geomorphic Continuity and Aquatic Habitat

The riverbeds of upland streams in Japan are characterized by step-pool sequences, especially in sediment-dominated channels. Steps form naturally from large rock and organic material jams. Pool spacing between steps reflects channel slope. Step-pool sequences tend to occur where sediment containing large amounts of rock is supplied from hillslope erosion sources such as debris flows and lateral cliff failures. Steps and pools gradually change their size, distribution, and gradient according to the sediment transport capacity and power of stream flow (Grant et al.1990; Abrahams et al. 1995).

In these streams, pools provide important aquatic habitat and ecosystems for fish, insect, and vegetation communities. Pool size is important for determining environmental capacity. For example, it is suggested that the maximum population of fish species such as the Yamame (*Oncorhynchus masou masou*) is controlled by pool capacity. Furthermore, current turbulence (dependent upon step height) produces dissolved oxygen, whereas current stagnation (depending on pool depth) allows sand accumulation that provides suitable spawning ground for fish communities (Hirose and Marutani 1993). Thus, maintenance of river health requires the maintenance of step-pool sequences at the reach scale.

The stability of step-pool sequences can be understood in relation to the relaxation path that follows riverbed aggradation. Figure 12.3 shows a representative example of relaxation in sediment storage depth after riverbed aggradation in the mountainous sedimentary cascade along the Hitotsuse River, Kyushu. Two typhoons in 1997(B) and 2003(C), following a typhoon in 1993(A), caused up to 2 meters of riverbed aggradation in response to sediment input from tributaries. The 1993 typhoon was a major triggering event that restarted the channel bed stabilization process. After the event, the creek mobilized sediment, regaining its step and pool structures. The recovery process after 1993 can be represented as an exponential relaxation curve (Marutani et al. 2003).

Even after regaining the step-pool structures, a reasonable amount of sediment continued to flow into and out of the reach, maintaining a state of dynamic equilibrium. The channel bed of lower and upper reaches attained a stable condition eight and four years after major triggering events (figure 12.3a; Kasai et al. 2004a, 2004b). The upper reach is more sensitive in its response to hillslope and fluvial processes than the lower reach. This is because downstream fining of sediment is dependent on flow power and the lower reach has a gently undulating riverbed that is less susceptible to sediment transport.

The difference in relaxation path between the upper and lower reaches represents the natural sedimentary process without any impact by Sabo dams. Each relaxation path between storms

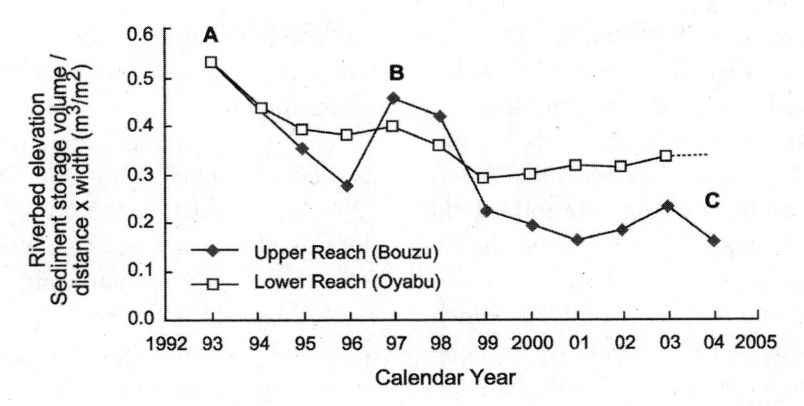

FIGURE 12.3. Changes in riverbed elevation calculated as the sediment storage volume per unit distance and width plotted against calendar year (after Marutani et al. 2003).

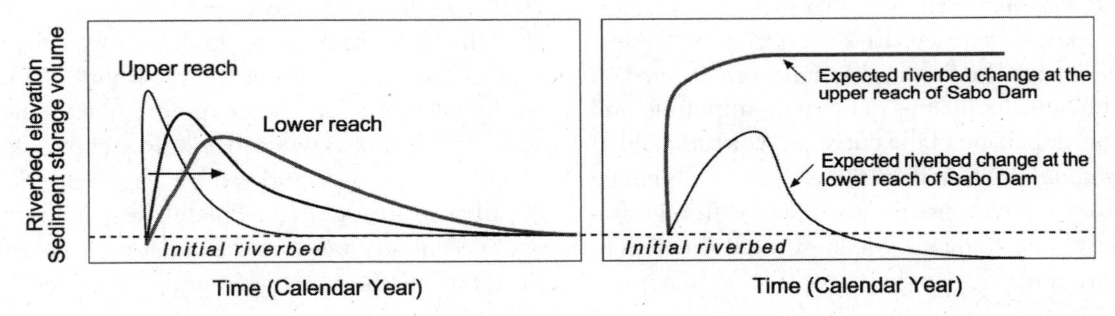

FIGURE 12.4. Idealistic model of the difference in the response of the relaxation pathway between the upper reach and the lower reach of a (a) natural river and a (b) dammed stream.

converged toward the initial riverbed elevation as determined by characteristics of riverbed materials. Exchanges of riverbed materials, with sediments freshly produced by storms, resulted in regime transition.

If a Sabo dam had been constructed between the upper and lower reaches after the 1993 storm, riverbed aggradation in the upper reach and degradation in the lower reach would have been advanced (figure 12.4a). Idealistic trajectories of expected riverbed changes shown in figure 12.4b do not converge; rather, they diverge, maximizing differences in riverbed elevation. Sabo dams interrupt the natural geomorphological and hydrological processes of river channels, making a significant impact on the availability of aquatic habitat. Hence, in order to restore step-pool sequences in systems where Sabo dams have been constructed, sediment stored behind the dam must be rapidly released, while retaining their primary function of inhibiting debris flows (i.e., a dual function is established).

Decoupling of Riparian Ecosystems by Dams, with Special Reference to Riparian Forests

Most streams and rivers in Japan have been altered for land development purposes, leaving few natural riparian forests remaining. Some areas with undisturbed riparian zones remain in Hokkaido, northern Japan. In most instances, riparian forests on floodplains and alluvial fans tend to consist of pioneer species such as *Salicaceae* (Niiyama 1990). Environmental gradients change abruptly at the boundary of forest patches. Distinct trends in species distribution are associated with soil conditions, surface elevation, and frequency of flood inundation (Nakamura et al. 1997; figure 12.5). Three broad-leaved pioneer species dominate bars and floodplains adjacent to the channel. However, their modes shift from lower to higher elevation, with the amplitude of distribution curves decreasing in the following order: *Alnus hirsuta Turcz, T. urbaniana*, and *P. maximowiczii*. The two species of conifers (*Picea jezoensis* [Sieb. Et Zucc.] Carr. and *Abies sachalinensis* [Fr. Schm.] Masters) were located on the highest floodplains. The high amplitude and narrow dispersion of the curves of *A. hirsuta* and *T. urbaniana* suggest that these species dominate high-stress environments associated with high water level and frequent flooding, without competition from other species. By contrast, *P. maximowiczii* and the two conifers have a scattered distribution pattern, suggesting a preference for low-stress environments.

The composition of pioneer species and conifers also differs with time (stand age) (figure 12.6; Nakamura et al. 1997). The dominance of the three pioneer species declined with age and fell almost to zero when stands were more than eighty years old. By contrast, the conifers became dominant with time, increasing sharply in stands more than eighty years old. Transitional stands, consisting of a mixture of pioneer and late successional species, were not found. This is because recurrent reworking of erosion-prone areas (e.g., adjacent to active channels) limits vegetation establishment and succession (Nakamura and Kikuchi 1996). In contrast, stable sites for late successional species are located away from channels on higher elevation geomorphic surface where flood disturbance is reduced. Consequently, floodplain areas are differentiated into high- and low-disturbance areas until the floodplain geomorphology is entirely reset by extraordinary events. Relationships between geomorphic surfaces and

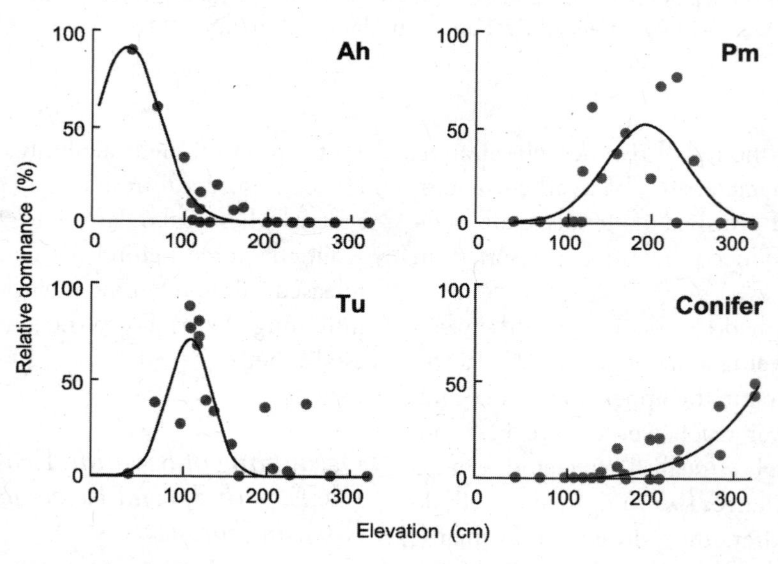

FIGURE 12.5. Changes in relative dominance with elevation gradient. Species abbreviations: Ah, *Alnus hirsuta*; Tu, *Toisusu urbaniana*; Pm, *Populus maximowiczii*; Conifer, *Picea jezoensis* and *Abies sachalinensis* (after Nakamura et al. 1997).

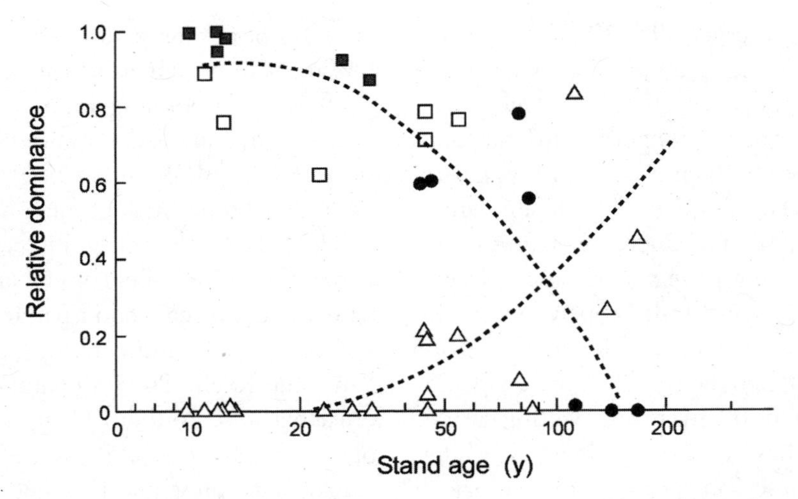

FIGURE 12.6. Changes in relative dominance of pioneer species and conifers with stand age. Black squares, white squares and black circles indicate the sum of the relative dominance of the three pioneer species that were established on active, semiactive, and stable sites, respectively. The triangles indicate the sum of the relative dominance of the two conifers (after Nakamura et al. 1997).

tree species distribution are also reported for braided and meandering channels by Shin and Nakamura (2005).

As patterns of riparian forests reflect the heterogeneity of site conditions created by frequent flood disturbances, geomorphic adjustment caused by direct alterations to flow regime can cause a substantial shift in riparian vegetation composition. Sabo dams and reservoir dams drastically change the disturbance regime and the formation of geomorphic surfaces (Takagi and Nakamura 2003). Hence, geomorphic adjustments following dam construction alter the composition of riparian forests.

Decoupling of forest ecosystems following dam construction affects other elements of aquatic ecosystems. Dam regulation alters the frequency and quantity of organic matter supply, affecting the provision of litter as a food resource and the impact of wood on habitat availability. Therefore, conservation of indigenous plant communities in riparian zones and protection of river health are strongly dependent upon efforts to preserve the dynamic nature and processes of river systems. Maintenance of flow variability in a manner that establishes and regenerates riparian plant species remains an important challenge in Japan.

The Influence of Dam Structures on Fish Associations and Their Rehabilitation

Dam construction has impacted significantly upon fish ecology in Japan. This is especially the case for salmonid species, which are mostly distributed in northern Japanese rivers. Salmonid fish in Japan are classified into three main genera: *Oncorhynchus*, *Salvelinus*, and *Hucho*. The *Oncorhynchus masu* (masu salmon or cherry salmon) depend on a freshwater habitat for at least one year after hatching. Half of the males become resident and spend their whole life in a stream. These are known as Yamame. The female salmon, having migrated from the ocean, spend a further half-year maturing before spawning in early autumn together with resident anadromous males. The masu salmon is important for commercial fisheries along Japan's sea coast, and there is an urgent need to improve freshwater habitat. *Hucho perryi* (Japanese Huchen or

Sakhalin Taimen) is especially vulnerable to poor water quality, and was recently listed as the most endangered fish species by IUCN (category IA). This fish has the greatest longevity (around fourteen years) and is the largest salmonid in Japan, growing up to 2 meters in size. They depend on riparian habitat and favor meandering channels and deep pools containing large woody debris (Fukushima 2001). They also require a spawning site in headwater streams where favorable gravel (pebbly and cobble gravel) and dense cover are provided.

Implementation of river engineering works since the 1960s has completely changed habitat availability for these fish species. The ecological linkage from headwater to ocean and/or ocean to mountain has been fragmented and lost. As a result, fish numbers have declined dramatically, resulting in extinction in some regions. Obstacles to spawning migration such as Sabo dams and check dams are among the most serious problems.

These structures impact not only the distribution of anadromous fish, but also the biological diversity of freshwater fish communities. The ability of fish to leap over a dam's wall varies among species. Most structures have falls of at least 1 meter, making it difficult for pregnant female species to cross. Furthermore, fish need deep pools to assist in jumping over an obstacle. As most riverbeds are protected by concrete structures to prevent degradation from the energy of falling water, most fish are unable to generate sufficient speed to jump over these obstacles.

Dam structures reduce propagation and intermarriage with other populations, resulting in genetic homogeneity and loss of genetic diversity (Maekawa et al. 2001), vulnerability to disease, and finally extinction of local populations. For instance, the landlocked white-spotted charr (*Salvelinus leucomaenis*) appears to have resulted as its migration has been prevented by dams.

Dams have an impact not only on anadromous fish, but also on material flows from oceans back into aquatic ecosystems. Research in North America has shown that anadromous fish bring large amounts of nutrients from the ocean to rivers and forests. Salmon carcasses increase nutrient concentrations in running waters, enhance primary production, and affect the structure of stream ecosystems (Cederholm et al. 1999). These carcasses are redistributed by large mammals such as bear and fox, ultimately contributing to the growth of riparian forests (Helfield and Naiman 2001). To date, these impacts upon riparian ecosystems have not been reported directly in Japan. Research by Yanai and Kochi (2005) in natural streams in eastern Hokkaido, northern Japan, used nitrogen stable isotope analysis to assess the effects of salmon carcasses on forest and stream ecosystems. In these rivers, numerous chum salmon (*Oncorhyhncus keta*) migrate upstream from the ocean to spawn in autumn. Although salmon carcasses did not have a significant effect on riparian vegetation, they strongly influenced the stream ecosystem by enhancing trophic levels of aquatic consumers. The movement of carcasses was traced by telemetry. Three out of five carcasses were removed and dropped within 10 meters of the stream, but one carcass was transported 500 meters to a distant hillslope ridge. Although there is currently no evidence for the effect of salmon carcasses, it is likely that the interception of material flow from the ocean by dams causes serious degradation of terrestrial ecosystems. Hence, anadromous fish provide a key indicator species with which to manage the sustainability and diversity of riparian ecosystems. Management of wild salmon, material flow, and river continuity are important challenges in Japan today.

Since the year 2000, nationwide rehabilitation of dammed streams has been underway in Japan. Fish passageways have been constructed to enhance upstream-downstream connectivity. However, these structures can cost nearly half as much as building a new dam and they often prove difficult to maintain as the approach to fish ways becomes buried by transported sediment. Another method to improve fish passage is to reduce the height of an existing structure. This method has been attempted on large Sabo dams in Central Honshu and Hokkaido, although some engineering problems were encountered. Progress on stream rehabilitation is incomplete. Of the 1,076 Sabo dams in this area, fish passageways have been attached to 180 dams while 32 dams have been changed into slit dams (see below). To date, these practices have been applied to 20 percent of 760 streams (Hokkaido government unpublished data).

Figure 12.7 exemplifies efforts to modify a thirty-year-old check dam in a mountain stream in southwestern Hokkaido by creating a slit in the dam wall. The central portion of the dam's panels was removed in August 2002. Within two weeks of removal, sediment removal occurred, depositing the eroded material just downstream of the modified dam. These deposits were gradually removed by ensuing torrential rainfall. Additional experience is required to establish safe procedures for dam modifications and to minimize impacts upon the environment.

Another method to rehabilitate rivers, particularly those with uniform stream morphology, is to construct low fall structures using a range of wood materials as a replacement for concrete. Three types of wood structures—log dams, wedge dams, and deflector logs—were placed to create deep pools, prevent lateral erosion, and reduce gravel transport along the upstream reach of the uniform Tobetsu River in Ishikari Catchment, central Hokkaido in 1996 (Yanai et al. 2004). An abandoned secondary channel was also excavated to reactivate flow. The influence of woody structures on stream morphology and masu salmon habitat was monitored for four years. After log placement, flow depth, mean current velocity, and substrate composition were significantly changed. Deep plunge pools formed downstream of the wedge dams and rapid currents were developed alongside deflector logs. The newly created pools will become important habitats for juvenile masu salmon. Similar rehabilitation work was carried out in 1996 in a channelized reach of the Shakotan River. Habitat complexity was improved by the addition of structures to replace the simple, uniform, stream structure with pools, riffles, and runs. An increased number of spawner and juvenile masu salmon were found in the rehabilitated reach relative to the unimproved control reach. Rehabilitation with wooden dams and deflector logs can improve habitats for masu salmon, and is a useful habitat enhancement technique in natural streams (Nagata et al. 2002). When considering the ecological and geomorphological benefits, these low fall structures should be incorporated as one of many methods to rehabilitate stream environments in Japan.

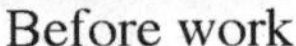

FIGURE 12.7. Improvement of an existing check dam by creating a slit, mountain stream of southwestern Hokkaido. Before (left) and after improvement (right) are compared to assess impact on the environment.

Organic Matter Processing and Human Disturbance

Organic matter processing is an integral part of ecosystem functionality. It is influenced by the nature and rate of organic matter input into streams and decomposition and transport of these materials. The construction of dams has exerted a significant impact on these relationships in Japanese rivers.

There is little information on the input of organic matter from riparian forests into headwater streams in Japan (Abe et al. 2006). From 2000 to 2002, Kochi et al. (2004) analyzed organic matter input from natural riparian forest into the headwaters of the Gokibiru stream, a second-order stream on the western coast of Hokkaido (141 degrees, 24 minutes East, 43 degrees, 23 minutes North). This area has an average annual snow accumulation of 107 centimeters. The step-pool channel is about 5.7 kilometers long, drains an area of 19.9 square kilometers, and has a mean width of 3 to 5 meters. Generally, discharge is $< 1\ m^3s^{-1}$, but flood flows occur in response to snowmelt in spring and rainstorms in summer. Ten litter traps were placed at random along this channel. Contents of the traps were collected weekly from 2000 to 2002 (other than the period from December through April, when heavy snow precluded sampling). Litter was categorized as leaves, twigs, terrestrial insects, terrestrial insect feces, and flowers. Leaves were classified visually into two categories: green leaves that fall during spring and summer months by natural loss, fragmentation by terrestrial insects, and/or storms, and senescent leaves that fall mainly in autumn.

The total annual inputs of organic matter during 2000 to 2002 at the study site ranged from 444 to $488 g/m^2$. Litter fall was greatest in October—in 2001 this accounted for 58 percent of the annual input. Although the amount of litter generated is highest in autumn, green leaves, terrestrial insect feces, and flowers appear to contribute a constant provision of food resources and/or habitat for stream dwelling macroinvertebrates. This is because the nitrogen content of green leaves, terrestrial insects and their feces, and flowers are higher than in senescent leaves, making them an important supplement for shredders that consume coarse particle organic matter during summer.

Kawaguchi and Nakano (2001) estimated that salmonids consume 51 percent of the total annual input of terrestrial invertebrates in forest reaches, showing the importance of terrestrial organic matter input on stream fauna. Nagasaka et al. (1996) also suggested that riparian forests were important in the utilization of terrestrial insects by salmonids in alder and willow riparian forests in Hokkaido. The gut content of salmonids varied between the two forest types, implying that the composition of riparian tree species has an impact on stream organisms. Decrease of tree species diversity, which results from the introduction of alien species such as the black locust, influences not only individual stream-dwelling species but also the whole stream ecosystem.

Organic matter that falls into streams from riparian forests decomposes by physical and biological breakdown. In headwater streams in Hokkaido, Yanai and Terasawa (1995) conducted decomposition experiments on the senescent leaves of nine broad-leaved tree species during autumn and winter using coarse mesh-sized bags that allowed shredder colonization. They showed that alder leaves decomposed fastest (decaying coefficient K equals 0.0129), followed by maple, birch, and willow (K equals 0.002 to 0.004) leaves, with oaks and beech leaves having the slowest rates of decomposition (K is less than 0.002). Kagaya (1990) also undertook decomposition experiments in autumn in headwater streams near Tokyo, central Japan, and reported a similar decay coefficient (K) for maple leaves of 0.002. In contrast, Kochi (2002) reported that decomposition of green leaves in summer was faster than in autumn. In this case the coefficient K for one kind of maple was greater at 0.094, and many caddis fly larvae had colonized the litter bags during summer.

Hence, during summer, decomposition of green leaves is faster than senescent leaves (Kochi and Yanai 2006a). Based on many overseas studies, temperature and leaf characteristics such as nitrogen content and toughness are inferred to affect decomposition rates in headwater streams in Japan. Motomori et al. (2001) reported that for some shredder species, carbon-nitrogen ratio and toughness affected leaf colonization. These feeder experiments suggested that although green leaves contain more nitrogen than senescent leaves, they were not as effective as a food source because of the content of secondary compounds (Kochi and Kagaya 2005).

Not all organic matter decomposes in upstream reaches. Some components are transported downstream to the river mouth. Kochi and Yanai (unpublished data) conducted feeding experiments using the *Anisogammarus pugettensis* that inhabit coastal brackish waters and found that they utilize terrestrial leaves both as a refuge from predator flounder and as a food source. This is one example of how transported coarse particulate organic matter of terrestrial origin plays an important role in coastal environments.

Kochi and Yanai (2006b) compared the decomposition of terrestrial senescent oak leaves and oak leaves that were soaked (conditioned) in freshwater streams in coastal sea areas. They found that the freshwater-conditioned leaves decomposed faster than the unconditioned oak leaves. This result indicates how the different conditioning of leaves from forest to coastal sea areas can influence the decomposition process in brackish seawater.

If dams prevent the transport of sediment from upstream reaches to the coast, how do they affect transportation and decomposition of organic matter? In dams that retain sediment, there is slow water exchange at the bottom of the reservoir. Organic matter transported from upstream accumulates on the reservoir bottom and decomposes under low oxygen levels. Decomposition experiments in a dam reservoir on a tributary of the Gokibiru River revealed that maple and oak leaves on the bottom of the dam decompose more slowly than in natural streams (figure 12.8). Furthermore, bacterial respiration of colonized leaves in the dam was about half that of leaves in natural streambeds. In addition, the nitrogen content of the leaves was lower in the lake, probably in response to the low bacterial activity (Moriwaka et al. unpublished data). This delay in decomposition inhibits healthy organic matter cycling in freshwater and promotes organic matter accumulation in dams. Therefore, whereas organic matter is transported downstream and decomposes by physical and

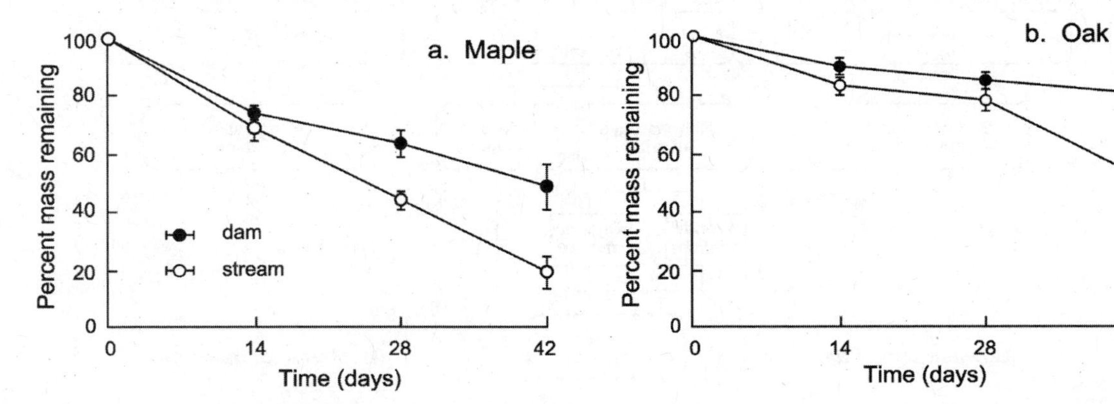

FIGURE 12.8. Decomposition of maple and oak senescent leaves in 10-millimeter mesh bags in (a) dam reservoir (-●-) and (b) natural stream (-○-). Average and standard error of five samples are shown. The experiment was started in May 2004.

biological processes in natural rivers, artificial structures such as dams inhibit this process.

The release of nutrients from salmon carcasses may have an indirect effect on leaf decomposition in upstream reaches. These nutrients may move onto leaves where they are consumed by macroinvertebrate shredders (Yanai and Kochi 2005). Hence, the restricted movement of salmon by artificial obstructions such as dams and weirs can affect decomposition of organic matter. This reinforces the need to maintain the flow of materials both from upstream forests to the coast and from the coast to upstream reaches to sustain river health.

Management of Dammed Streams

Although much is known about the structural, functional, and preventive advantages of Sabo dams, little is known about the impact of Sabo dam improvement on sedimentary, ecological, and hydrological systems for entire basins. Two factors account for this limited understanding. First, no natural reference rivers exist in Japan, making it difficult to reconstruct how sedimentary systems operated in the absence of Sabo dams. Second, it is impossible to completely disassociate and understand fluvial processes from hillslope processes, because of the continuous reworking of deposits in this high-energy geomorphic setting.

Recently, concerns for river health have been incorporated into management programs that previously were concerned solely for sediment disaster prevention. Ecosystem responses to the disruption of sediment transfer are conceptualized in figure 12.9 (central image). Natural cyclic changes in sediment volume along a channel (i.e., cycles of aggradation and degradation) result in moderate disturbances to ecosystem processes, followed by relaxation. Under these conditions, step-pool sequences are formed and reworked, and riparian vegetation is established and then reworked. Woody debris and organic matter can be exported from the riparian zone to the aquatic ecosystem. However, once Sabo dams are installed, reduced downstream sediment supply promotes degradation of the channel bed. This creates a bedrock channel with no riparian zone (left-hand side of figure 12.9). Alternatively, reduced rates of

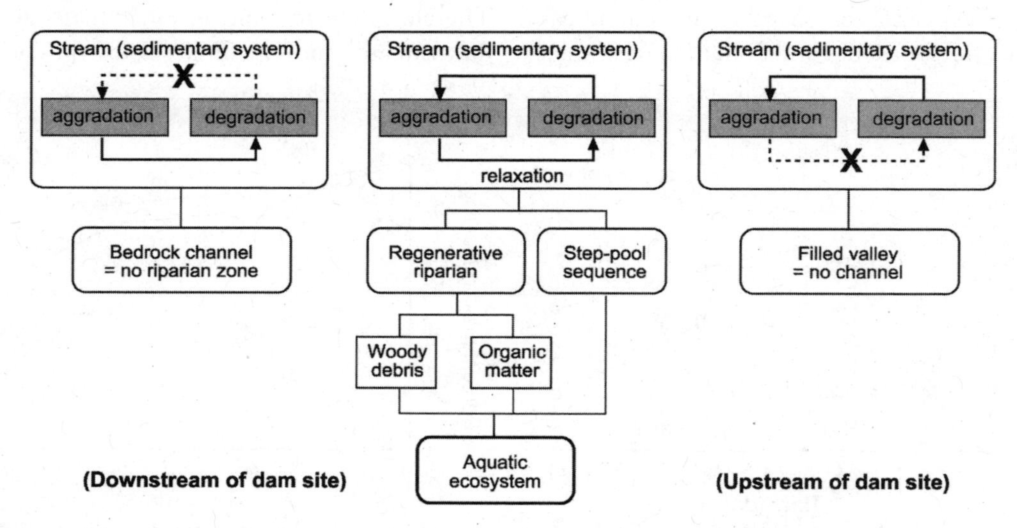

FIGURE 12.9. Change in riverbed regimes on steep sedimentary systems caused by dam construction and the ecological components yielded by the relaxation between riverbed aggradation and degradation.

sediment transport upstream of the dam site promote rapid aggradation, infilling the valley floor (right-hand side of figure 12.9).

To maintain the health of aquatic ecosystems, rehabilitation efforts must restore the balance between aggradation and degradation through improvements to the design and operation of Sabo dams. Releasing the sediment stored behind a Sabo dam promotes the recovery of biological continuity along a channel. To achieve this, several new designs have been developed and trialed in Japan, including fish ladders, cell dams, slit dams, and a bumpy obstruction system (figure 12.10).

Fish ladders are the most common form of improvement to dams, but they may cost only a little less than the original dam reconstruction (Figure 12.10a). They aim to preserve both the function of sediment storage while reinstating biological continuity without the need for a slit dam. Many kinds of ladders have been built. These commonly extend from the downstream base of a dam to above the top of the stored sediment. Although the function of the dam for sediment storage is maintained, it is often difficult to maintain aquatic habitat along the flow path of the ladder because of blockage by sediment.

Cell dams are composed of a number of round compartments that are filled with sediment from around the construction site (Figure 12.10b). Narrow spaces between each cell allow large rocks and logs to be stored behind the cell dam system, while fine sediment and water flow through. These features are used to break down debris flows in headwater reaches.

FIGURE 12.10. Recent modifications to Sabo dam operations in Japan: (a) fish ladder (b) cell dam, (c) slit dam, (d) alternate bumpy obstruction system.

Slit dams are the most popular style of dam improvement (Figure 12.10c). This relatively simple engineering procedure introduces a slit into a dam's front wall. The slit ranges in width from 2 to 10 meters, depending upon the size of sediment to be released. After engineering work, a large amount of sediment stored upstream is released from the dam pocket, restoring the passageway for fish and insects. In order to rehabilitate biological continuity, the slit dam ceases to act as a sediment storage dam. In many ways, this strategy represents a Band-Aid solution.

New designs are now being trialed to retard the destructive effects of debris flows. These designs combine engineering and geomorphic principles, increasing bed friction while maintaining sediment transfer and retaining biological continuity. For example, bumpy obstruction systems represent an alternative structure to Sabo dams, as they slow the rates of sediment movement without intercepting sediment (figure 12.10d).

Conclusion

The maintenance of riverbed regimes in mountainous streams requires a coordinated approach to the management of both hillslope and fluvial processes under variable hydrological and geological conditions. Debris flows and other forms of unpredictable hillslope processes not only disrupt human societies, but they also disturb sedimentary systems. Hence, variability and diversity are natural components of sediment supply and transport that affect sedimentary cascades. Headwater streams are characterized by heterogeneous assemblages in material size, geomorphic features, and hydrological processes. This diverse and variable set of conditions maintains riparian vegetation, aquatic fish habitats, and decomposition of organic matter.

In order to protect downstream cities and infrastructure, sediment disasters brought about by unpredictable hillslope processes such as debris flow and landslides must be negated. Sabo dams have been constructed to interrupt debris flows, while Chisan dams increase bed friction in steepland settings. These features have successfully prevented sediment disasters. However, they have induced temporary base levels and uniform hydraulic conditions in headwater streams, reducing the heterogeneity of geomorphic structure and disturbing hydrological and ecological continuity. These features cannot be removed from mountainous streams without fundamental and unrealistic reconstruction of land management systems. However, we can modify the operation of these systems, transforming their function from interception of sediment flow to reduction of sediment flow and reduction of debris flow energy. To meet the needs of sediment disaster mitigation and biological continuity, new solutions have been developed to improve Sabo-dammed streams, including fish ladders, slit dams, cell dams, and alternate bumpy obstruction systems. These site-specific operations must be framed in terms of the operation of biophysical processes at the catchment scale. Ultimately, analysis of sedimentary cascades through a source-to-sink system is a key component of sustainable land management harmonized within the bio-geo-hydro-sphere.

Acknowledgments

Two substantive reviews greatly enhanced the contribution in this chapter. Editorial comments by Emmy Macdonald and Gary Brierley improved the presentation of this chapter. Ronald DeRose is thanked for improving the early draft of this chapter. This work was partly supported by Grant-in-Aid for Scientific Research, Japan, contract no. 17208012, and Sustainability Governance Project of Hokkaido University.

References

Abe, T., M. Nunokawa, and M. Fujieda. 2006. Organic matter derived from forests and mountain stream ecosystems. *Water Science* 50:1–23.

Abrahams, A. D., G. Li, and J. F. Atkinson. 1995. Step-pool streams: Adjustment to maximum flow resistance. *Water Resources Research* 31:2593–602.

Cederholm, C. J., M. D. Kunze., T. Murota, and A. Sibatani. 1999. Pacific salmon carcasses: Essential contribution of nutrients and energy for aquatic and terrestrial ecosystems. *Fisheries* 24:6–15.

Fryirs, K. A., G. J. Brierley, N. J. Preston, and M. Kasai. 2007. Buffers, barriers and blankets: The (dis)connectivity of catchment-scale sediment cascades. *Catena* 70:49–67.

Fukushima, M. 2001. Salmonid habitat-geomorphology relationships in low-gradient streams. *Ecology* 82:1238–46.

Fukushima, M., S. Kameyama, M. Kaneko, K. Nakao, and E. A. Steel. 2007. Modelling the effects of dams on freshwater fish distributions in Hokkaido, Japan. *Freshwater Biology* 52:1511–24.

Grant, G. E., F. J. Swanson, and M. G. Wolman. 1990. Pattern and origin of stepped-bed morphology in high-gradient streams, Western Cascades, Oregon. *Geological Society of America Bulletin* 102:340–52.

Harvey, A. M. 2002. Effective timescales of coupling within fluvial systems. *Geomorphology* 44:175–201.

Helfield, J. M., and R. J. Naiman. 2001. Effects of salmon-derived nitrogen on riparian forest growth and implication for stream productivity. *Ecology* 82:2403–9.

Hirose, K., and T. Marutani. 1993. Distribution of spawning space of Oncorhynchus masou masou according to spatial changes of river bed. *Bulletin of Kyushu University Forest* 68:73–84.

Kagaya, T. 1990. Processing and macroinvertebrate colonization of leaf detritus in a mountain forest stream in Japan. *Bulletin of the Tokyo University Forests* 82:177–89.

Kasai, M., T. Marutani, and G. J. Brierley. 2004a. Channel bed adjustments following disturbance events in steep headwater settings: Findings from Oyabu Creek, Kyushu, Japan. *Geomorphology* 62:199–215.

Kasai, M., T. Marutani, and G. J. Brierley. 2004b. Patterns of sediment slug migration following typhoon-induced disturbance, Oyabu Creek, Kyushu, Japan. *Earth Surface Processes and Landforms* 29:59–76.

Kawaguchi, Y., and S. Nakano. 2001. Contribution of terrestrial invertebrates to the annual resource budget for salmonids in forest and grassland reaches of a headwater stream. *Freshwater Biology* 46:303–16.

Kochi, K. 2002. Decomposition of fresh leaves and colonization by shredders in a small mountain stream. *Japanese Journal of Ecology* 52:331–42.

Kochi, K., and T. Kagaya. 2005. Green leaves enhance the growth and development of a stream macroinvertebrate shredder when senescent leaves are available. *Freshwater Biology* 50:656–67.

Kochi, K., and S. Yanai. 2006a. Shredder colonization and decomposition of green and senescent leaves during summer in a headwater stream in northern Japan. *Ecological Research online first* No. 11284-006-0149-y.

Kochi, K., and S. Yanai. 2006b. Decomposition of leaves in coastal brackish water and their use by the macroinvertebrate *Anisogammarus pugettensis* (Gammaridea). *Marine and Freshwater Research* 57:545–51.

Kochi, K., S. Yanai, and A. Nagasaka. 2004. Energy input from a riparian forest into a forested headwater stream in Hokkaido, Japan. *Achiv für Hydrobiologie* 160:231–46.

Land, Infrastructure and Transport. 2006. *Review of sediment-related disasters*. SABO Technical Center, Dosha-saigai-no-Jittai. 3–5.

Maekawa, K., Y. Koseki, K. Iguchi, and S. Kitano. 2001. Skewed reproductive success among male white-spotted charr land-locked by an erosion control dam: Implications for effective population size. *Ecological Research* 16:727–35.

Marutani, T., G. J. Brierley, N. A. Trustrum, and M. Page. 2001. *Source-to-sink sedimentary cascades in Pacific Rim geo-systems*. Matsumoto, Japan: Sabo Work Office, Ministry of Land, Transport and Infrastructure.

Marutani, T., M. Kasai, and G. J. Brierley. 2003. Relaxation and sensitivity of steep sedimentary cascades in Japan. In *Proceedings of International Workshop for Source-to-Sink Sedimentary Dynamics in Catchment Scale*, ed. T. Araya, M. Kuroki, and T. Marutani, 63–70. Sapporo, Japan: Hokkaido University.

Motomori, K., H. Mitsuhashi, and S. Nakano. 2001. Influence of leaf litter quality on the colonization and consumption of stream invertebrate shredders. *Ecological Research* 16:173–82.

Nagasaka, Y., S. Yanai, and H. Sato. 1996. Relation between fallen insects from riparian forest and stomach content of Masu Salmon (*Oncorhynchus masou*). *Bulletin of the Hokkaido Forestry Research Institute* 33:70–77.

Nagata, M., H. Omori, and S. Yanai. 2002. Restoration of spawning and rearing habitats for masu salmon, *Oncorhynchus masou* in a channelized stream. *Fisheries Science* 68:1707–10.

Nakamura, F., and S. Kikuchi. 1996. Some methodological developments in the analysis of sediment transport processes using age distribution of floodplain deposits. *Geomorphology* 16:139–45.

Nakamura, F., T. Yajima, and S. Kikuchi. 1997. Structure and composition of riparian forests with special reference to geomorphic site conditions along the Tokachi River, northern Japan. *Plant Ecology* 133:209–19.

Nakamura K., K. Tockner, and K. Amano. 2006. River and wetland restoration: Lessons from Japan. *BioScience* 56:419–29

Niiyama, K. 1990. The role of seed dispersal and seedling traits in colonization and coexistence of Salix spp. in a seasonally flooded habitat. *Ecological Research* 5:317–32.

Oguchi, T., K. Saito, H. Kadomura, and M. Grossman. 2001. Fluvial geomorphology and paleohydrology in Japan. *Geomorphology* 39:3–19.

Shin, N., and F. Nakamura. 2005. Effects of fluvial geomorphology on riparian tree species in Rekifune River, northern Japan. *Plant Ecology* 178:15–28.

Takagi, M., and Nakamura F. 2003. The downstream effects of water regulation by the dam on the riparian tree species in the Satsunai River. *Journal of Japanese Forest Society* 85:214–21.

Takahashi, T. 1978. A mechanism of occurrence of mud-debris flows and their characteristics in motion. *Annals Disaster Prevention Research Institute, Kyoto University* 20 (B-2): 405–35.

Tsukamoto, Y. 1973. Study on the growth of stream channel (1): Relationship between stream channel growth and landslides occurring during heavy storm. *Shin-sabo* 87:4–13.

Yanai, S., and K. Kochi. 2005. Effects of salmon carcasses on experimental stream ecosystems in Hokkaido, Japan. *Ecological Research* 20:471–80.

Yanai, S., Y. Nagasaka, H. Sato, and D. Ando. 2004. Restoration of stream channel and masu salmon habitat by woody structure in a suburban stream, Central Hokkaido, Northern Japan. *Ecology and Civil Engineering* 7:13–24.

Yanai, S., and K. Terasawa. 1995. Varied effects of the forest on aquatic resources in a coastal mountain stream, in southern Hokkaido, northern Japan. (2) Autumn leaf processing of nine deciduous trees along the stream. *Journal of Japanese Forestry Society* 77:563–72.

Yoshikawa, T. 1974. Denudation and tectonic movement in contemporary Japan. *Bulletin of the Department of Geography, University of Tokyo* 6:1–14.

Yoshimura, C., T. Omura, H. Furumai, and K. Tockner. 2005. Present state of rivers and streams in Japan. *River Research and Applications* 21:93–112.

Chapter 13

Application of Integrative Science in the Management of South African Rivers

KATE M. ROWNTREE AND LEANNE DU PREEZ

> "During the last two decades . . . rivers and other aquatic ecosystems were enhanced in stature from having no rights to their own water to being one of only two sectors with a right to water; the other sector is for basic human needs."
>
> King and Brown 2006, 1

A number of fortuitous events have shaped the philosophy on which South Africa's river management policy, legislation, and practice are based. In 1992, environmentalists from the nations of the world met in Rio de Janeiro and formulated Agenda 21, the environmental blueprint that has since guided many national policies (United Nations 1993). Two years later, in 1994, South Africa gained a new government through a remarkable political transformation that enabled the development of fresh policies and fresh legislation on an essentially clean slate. In 1998, a new National Water Act (NWA) (Republic of South Africa 1998a) and a new National Environmental Management Act (NEMA) (Republic of South Africa 1998b) were passed.

Meanwhile, beginning in the 1980s, there had been growing concern from ecologists about the state of the nation's rivers (King and Brown 2006). This was exemplified by concern for the rivers that flow through the country's largest and most prestigious national park, the Kruger National Park. In the early 1990s, the Kruger Park River Research program was initiated. This program was designed to bring together researchers and managers in a cooperative framework (Breen et al. 1995) and produced a number of research outputs that can be considered as integrative river science (e.g., Van Coller et al. 1997; Heritage et al. 1997). Other groups were also active in promoting common frameworks and a common language for river science (e.g., Wadeson 1994; Rowntree and Wadeson 1997, 1999). River scientists had begun to develop methods for determining environmental flows before legislation was put in place (King and Tharme 1994; King and Brown 2006). One outcome of these new partnerships among river scientists and between scientists and river managers was the inclusion of a river ecologist in the legal team that drew up the NWA (Palmer 1999). This ensured that more than lip service was paid to the principles of Agenda 21, resulting in a strong chapter on water resource protection in the National Water Act of 1998.

Water resource protection is concerned with protecting the nation's rivers for the future. The future, however, is uncertain. In this chapter, we start with an overview of South African water legislation and its relationship to Agenda 21. The next section takes the Ecological Reserve as an example of how river scientists have developed integrated tools to meet the needs of the new legislation. We then examine some of the drivers of South African river ecosystems and make a brief assessment of how these may change in the future. The final section looks at the problem of managing for the future.

South African Water Legislation, Agenda 21, and South African River Management

South Africa has a complex political history, the legacy of which includes a backlog of environmental management issues that legislation has only begun to address in the last ten years. The country exhibits characteristics of both a developed and a developing nation with one of the world's largest disparities between rich and poor (United Nations Development Programme 2006). This poses interesting challenges for governmental sectors, not least those involved in the management of the country's natural resource base, including the nation's rivers. Human needs (in terms of water rights and development) constantly need to be balanced against protection of the environment.

Until recently, the sustainable utilization of natural resources was not a priority on the national agenda and the distribution of these resources was skewed toward a minority of the population. Prior to the Water Act 54 of 1956, under Roman–Dutch law, rivers were seen as resources belonging to the nation, implying that the public had the right to utilize them as a natural resource. African customary law also allowed unlimited public use of water. This view gradually began to change as competition grew for this relatively scarce resource (Godden 2005). The Water Act of 1956 was not explicitly discriminatory, but through linking rights to water with racially exclusive land ownership, access to water became a discriminatory issue (Palmer 1999), especially in rural areas. Furthermore, this act did not include any provision for environmental protection (Godden 2005).

Following the signing of Agenda 21 at the World Summit in 1992 (United Nations 1993) and the election of the new democratic government in 1994, legislation and management strategies have been aimed at eliminating discriminatory access rights to the country's natural resources and at instituting the concept of sustainable use. Through the NWA, the government accepted the responsibility of managing water in the public interest, promoting both equity of access to water resources and environmental protection (Palmer 1999; Godden 2005). The policy reform process provided an ideal opportunity for the inclusion of environmental issues in the new legislation in line with international obligations (Palmer 1999; Godden 2005). This new environmental awareness was reflected in the statement made by the Department of Water Affairs and Forestry (DWAF) in 1995 (Palmer 1999, 132): "Recently the needs of the environment, particularly the riverine environment, have been increasingly taken into account, and the impact of development is beginning to be taken seriously. The law needs to view the environment as the resource base from which all development leads, the foundation on which all else depends."

Several basic principles from Agenda 21 have had an important bearing on the development of South African water management policy. Three of these principles are of explicit importance to natural scientists engaged in crossdisciplinary river management, namely:

1. A focus on maintaining and protecting biodiversity
2. Recognition of the tension between human and environmental water requirements
3. Recognition of the interconnectedness of water resources and the need for holistic approaches to water management.

In terms of the protection of biodiversity, section 15.3 of Agenda 21 states: "biological resources constitute a capital asset with great potential for yielding sustainable benefits. Urgent and decisive action is needed to conserve and maintain genes, species and ecosystems, with a view to the sustainable management and use of biological resources." In response, one of the purposes of the NWA (Republic of South Africa 1998a, section 2[g]) is: "to ensure that the nation's water resources are protected, used, developed, conserved, managed, and controlled in ways which take into account, amongst other factors, protecting aquatic and associated ecosystems and their biological diversity."

Thus, within South Africa's legal framework for river management, resource protection is strongly focused on ecosystems and biological diversity. Nonbiologists engaged in river management have been encouraged to take an ecocentric approach. Geomorphologists, for example, must consider geomorphological systems primarily as drivers of physical habitat. There is no place for considering geodiversity as being important in its own right, as is the case, for example, in Australia (Rutherfurd et al. 2000).

In terms of recognizing the tension that exists between human and environmental water requirements, section 18.8 of Agenda 21 states that: "In developing and using water resources, priority has to be given to the satisfaction of basic needs and the safeguarding of ecosystems. Beyond these requirements, however, water users should be charged appropriately."

The concept of the two-part Reserve that is included in the NWA is directly in line with this principle. The Basic Human Needs Reserve addresses the backlog with respect to basic domestic water supply; the Ecological Reserve addresses the environmental sustainability of water-related activities. The maintenance of environmental health or integrity is seen from the perspective of ensuring the basic human right to a healthy environment (Godden 2005), thus prioritizing human rights with regard to natural resource protection. This is in line with the first principle of the Rio Declaration of 1992, which puts human beings at the center of concerns for sustainable development (United Nations 1993).

In terms of the recognition of the interconnectedness of river ecosystems and the need for a multidisciplinary approach to river management challenges, section 18.35 of Agenda 21 states: "Freshwater is a unitary resource. Long-term development of global freshwater requires holistic management of resources and a recognition of the interconnectedness of the elements related to freshwater and freshwater quality."

The NWA calls for the development of a national water resource strategy that provides a workable framework for the protection, use, development, conservation, management, and control of water resources for the country as a whole (NWA, chapter 2, part 1). Section 6 of the act states that: "the national water strategy must promote the management of catchments within a water management area in a holistic and integrated manner."

Furthermore, it is a fundamental environmental management principle of the South African National Environmental Management Act (NEMA) (Republic of South Africa 1998b, chapter 1, section 2[4][b]) that: "environmental management must be integrated, acknowledging that all elements of the environment are linked and interrelated, and it must take into account the effects of decisions on all aspects of the environment and all people in the environment by pursuing the selection of the best practical environmental option."

The river research community of South Africa has put much effort into developing integrated methods or procedures to support management's quest for the best practical environmental option. One example is the set of procedures put in place to determine the ecological reserve as described in the next section.

The Reserve as an Example of South African Management Frameworks

The main legal instrument for protecting South Africa's water resources is the Reserve (Republic of

South Africa 1998a). The Reserve represents the only legal right to water and comprises two components: the basic human needs reserve (BHNR) and the ecological reserve. The BHNR ensures that all South Africans have the right to a basic amount of potable water (normally taken to be 25 liters per day) for their personal use: drinking, cooking, and hygiene. The ecological reserve is defined by the NWA as "the water required to protect the aquatic ecosystems of the water resource" (NWA chapter 3, part 3). The Reserve (basic human needs and ecological reserves) refers to both the quantity and quality of the water in the resource, and will vary depending on the class of the resource.

Developing standard methods to determine and implement the ecological reserve has engaged many environmental scientists from a range of disciplines and probably represents South Africa's best example of a committed attempt to use an integrative scientific approach to achieve integrated water resource management (see Palmer 1999). The process of determining the ecological reserve has been developed and refined continually in response to new management challenges (King and Louw 1998; Kleynhans et al. 2005).

Three broad points should be noted when describing the Reserve process. First, the Reserve can be determined at three levels: comprehensive, intermediate, and rapid. Second, the degree of environmental protection versus permitted allocation to water users depends on the management class of the river (or resource unit) (figure 13.1). Third, the ecological reserve is determined in terms of environmental water requirements or the quantity and quality of water deemed to be necessary to keep the aquatic ecosystem in a condition consistent with the management class.

The three levels of ecological reserve determination depend on the size of a proposed development and the importance of the impacted river system. As the name suggests, a comprehensive reserve requires the most in-depth study and is expensive and time consuming, but the level of confidence in the results should be as high as possible, given constraints on available data and knowledge of river ecosystems. A comprehensive reserve study would be carried out when a large impoundment has been proposed or where there is pressure for a major reallocation of water in environmentally sensitive catchments. Examples include the Letaba–Olifants system and the Sabie River, both of which flow through the Kruger National Park. An intermediate reserve study is a downscaled version of the comprehensive study and is therefore cheaper and takes less time, but the level of confidence is lower. An intermediate reserve was carried out for the Kat River in the Eastern Cape; this is a river for which there is pressure for some reallocation of water, but it is not an especially water stressed catchment, nor of a high environmental status. A rapid reserve study relies on desktop analysis with limited supporting fieldwork; these studies are normally carried out as part of a regional resource assessment or at the pre-feasibility stage of a project.

The management class is determined by weighing the socioeconomic benefits of using water in a catchment against the need for environmental protection (figure 13.1). As the class changes from natural to unacceptably degraded, the level of protection and the ability of the river to support ecosystem goods and services decrease while water use and related impacts increase. Determining the need for environmental protection is itself contingent on two main considerations. First is the ecostatus of the river, described in terms of an ecological category (EC). This is a measure of the degree to which the study unit (site or river reach) has been modified by human impact relative to a natural reference condition (see table 13.1). Second is the ecological importance and sensitivity of the study area (EIS). A high EIS rating would be given to a reach with rare or endangered species present and/or to a reach in a nature conservation area.

The concept of an ecostatus is relatively new to the process of determining the ecological reserve

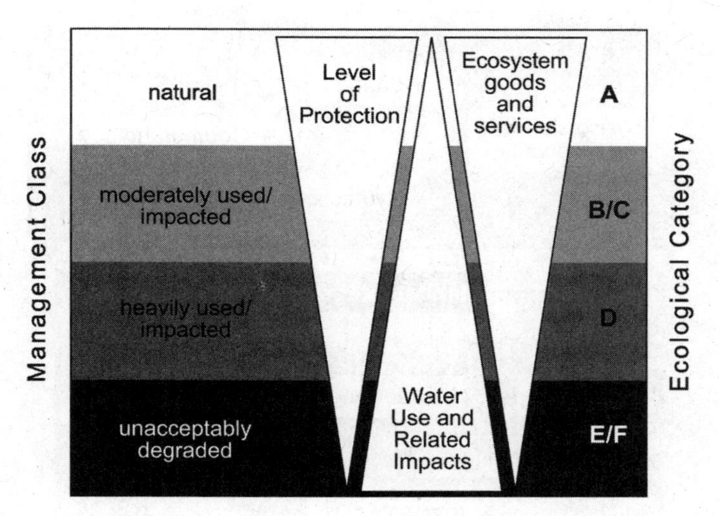

FIGURE 13.1. The relationship between management class, ecological category, level of protection, water use and related impacts and ecosystem goods and services. Ecological categories are described in table 13.1.

TABLE 13.1

Ecological categories used to describe the Ecostatus (Kleynhans et al. 2005)

Ecological category	Description
A	Unmodified, natural.
B	Largely natural with few modifications. A small change in natural habitats and biota may have taken place but the ecosystem functions are essentially unchanged.
C	Moderately modified. A loss and change of natural habitat and biota have occurred but the basic ecosystem functions are still predominantly unchanged.
D	Largely modified. A large loss of natural habitat, biota, and basic ecosystem functions have occurred.
E	Seriously modified. The loss of natural habitat, biota, and basic ecosystem functions are extensive.
F	Critical/extremely modified. Modifications have reached a critical level and the system has been modified completely with an almost complete loss of natural habitat and biota. In the worst instances the basic ecosystem functions have been destroyed and the changes are irreversible.

(Kleynhans et al. 2005) and follows thinking of ecologists such as Iverson et al. (2000) and Novotny et al. (2005). The ecostatus model is an attempt to integrate the different biophysical components of the river system to derive a single measure of its ecological state. Kleynhans et al.(2005) use Iversen et al.'s (2000) definition of ecostatus: "the totality of the features and characteristics of the river and its riparian areas that bear upon its ability to support an appropriate natural flora and fauna and its capacity to provide a variety of goods and services" (Kleynhans et al. 2005, 2–3). Ecostatus can refer to the present condition of the river (the present ecological state or PES) or to some future desired state (e.g., the recommended ecological category or REC).

The ecostatus represents the integration of a number of different components (figure 13.2). It is determined according to the ecosystem drivers (water quality, geomorphology, hydrology) that

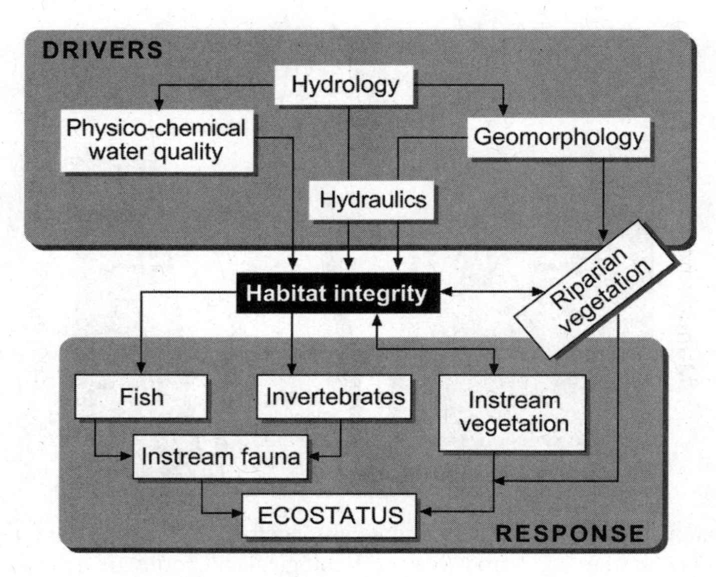

FIGURE 13.2. A driver-response model for an integrated ecostatus.

provide the habitat template of an ecosystem, and the biological responses (fish, aquatic invertebrates, and vegetation). Vegetation can be seen as both a response and a driver (as vegetation provides habitat for other organisms), and through feedback processes affects geomorphological processes such as bank erosion or bar stabilization.

In deriving the present ecostatus, the PES for each component (drivers and responses) is assessed independently. The driver status is not integrated into the ecostatus, but is used to indicate possible causes of adverse biological indicators. The health of the system is assessed from the overall state of the biological responses, which are integrated into one ecostatus value.

The PES for each component is assessed from a number of individual attributes that are each scored according to their deviation from the expected natural reference condition using a six-point system based on expert knowledge. These are referred to as metrics. The individual metrics are then weighted according to their perceived importance in a given river system. Again, this is a matter of expert judgement. Metrics for each component are combined using a multi-criteria decision approach (MCDA) into a series of assessment indices: the hydrological driver assessment index (HAI), the geomorphological driver assessment index (GAI), the physico-chemical driver assessment index (PAI) for water quality, the fish response assessment index (FRAI), the macroinvertebrate response assessment index (MIRAI), and the vegetation response assessment index (VEGRAI). The indices are designed to allow the development of consistent rating systems that are aggregated in a theoretically justifiable way to derive the ecostatus, an important part of the process for determining the ecological reserve (figure 13.3).

Having determined the PES, the recommended ecological category (REC) is decided on the basis of trends in each component. The specialist must weigh how realistic it is to either improve upon or maintain the present conditions. Resource quality objectives (RQOs) are set to either maintain or achieve this category. RQOs are measures specified to achieve a desired level of

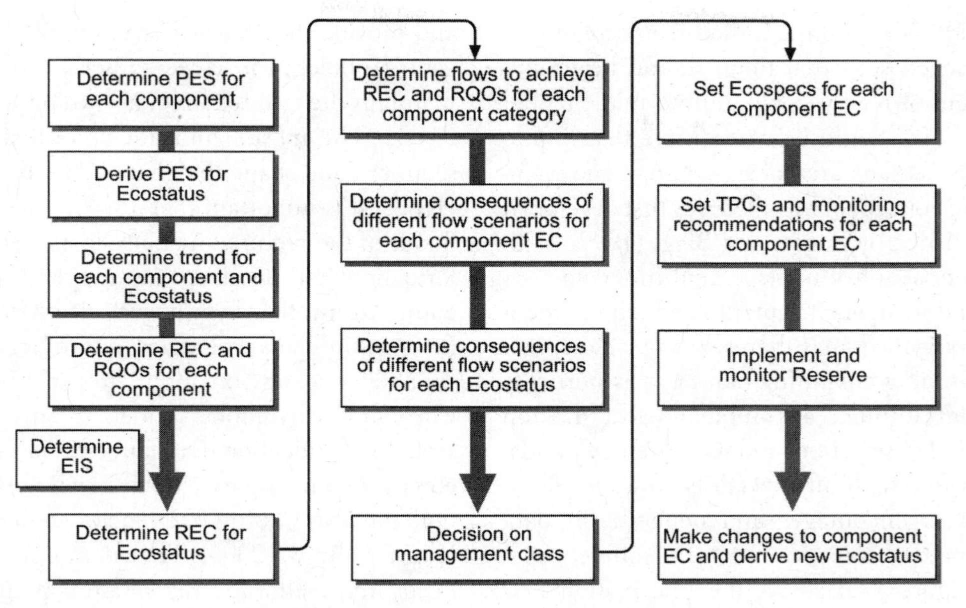

FIGURE 13.3. The ecological reserve process. The first column shows how the recommended ecological category (REC) for the ecostatus is derived. The second column summarizes the process of determining flow requirements and the third column relates to implementation and monitoring.

protection and are based on clear numerical or descriptive goals for the water quantity and quality, biota, and habitat of the resource (Rogers and Biggs 1999). These recommendations are then used together with the EIS rating to determine the REC for the ecostatus.

The flow that is required to keep the river in this recommended category is determined following the steps shown in the second column of figure 13.3. The impacts of alternative flow scenarios on each component of the ecostatus model are evaluated, and the consequent impact on the ecostatus foresighted. The outputs from this process are then fed into the management class decision process. The minister makes the final decision on the management class. In doing so, he or she must weigh the environment's needs against the social and economic benefits of water use. It is at this point that stakeholders are involved in the decision-making process.

High levels of uncertainty cloud decisions on flow requirements. King and Brown (2006) provide a useful overview of the challenges facing the assessment of environmental flows. No ecological specialist can claim to fully understand the observed ecological system. Predicting the future is even more difficult. Moreover, the data on which decisions are based are always inadequate, short term, and highly localized. Monitoring is therefore a critical part of the ecological reserve process. Monitoring is needed first to ensure that the ecological reserve is being implemented according to the recommendations and second to see if the recommended flow requirements actually meet the ecological needs of the system. Monitoring for the second requirement is based on setting "ecospecs" for each component, together with thresholds of probable concern (TPCs). An ecospec is a specification related to a measurable attribute and a TPC is a statement of the degree of change that is not acceptable. For example, the geomorphologist may set an ecospec that there should be no encroachment of reed vegetation onto mid-channel bars. The TPC could be when more than 5 percent of the surface area of bars suffers encroachment. The concept of the TPC was first developed in South Africa by Rogers and Biggs (1999) as an approach to adaptive management in a data-poor context and in an environment in which the management objectives may themselves be changing.

Determining environmental flows is normally carried out through a combination of desktop studies, field observation, and measurement, and a workshop in which all specialists come together. The process encourages and depends on data sharing and discussions between the different specialists. Figure 13.4 shows the information networks that are important to the geomorphologists. A similar diagram could be constructed for each specialist.

Rowntree and Wadeson (1998) described a geomorphological framework for assessing water requirements for maintaining physical habitat related to channel morphology. While the biological specialists are most concerned with the low or intermediate flows that occur for most of the time and provide the "normal" habitat, the task of the geomorphologist is to recommend flows that will maintain the river channel according to specified RQOs. The most significant flows that transport sediment and shape the channel are flood events. Hence, it is important that a sufficient number of floods of the required magnitude are retained. Information on flood magnitude and frequency comes from the hydrologist. Translating these flows into depths, velocities, and other hydraulic variables depends on input from a hydraulician who applies hydraulic models to surveyed cross sections. Verification of required flood frequencies comes from the riparian botanist who links vegetation zones to wetting frequencies (Boucher 2002; King and Brown 2006). The geomorphologist also needs information on vegetation condition to understand trends in channel stability. The fish and invertebrate ecologists provide the geomorphologist with information on their required habitat. This informs the choice of RQOs that geomorphologists will aim to achieve. In turn, the geomorphologist provides the ecologists with an assessment of present and future habitat conditions that can be related to attributes of channel morphology and bed sediment.

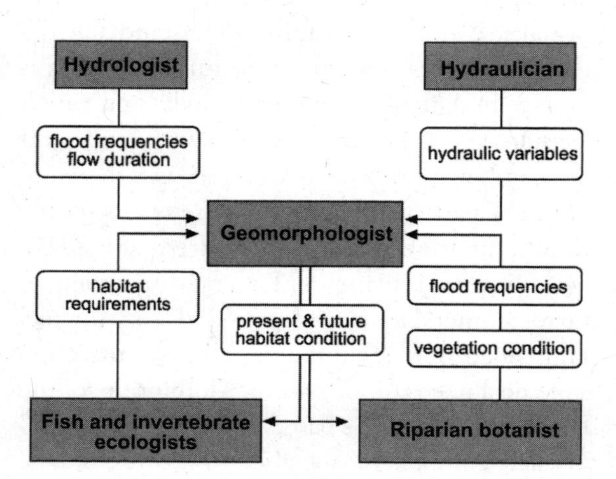

FIGURE 13.4. Information networks important to a geomorphologist in an ecological reserve study.

Future Fluvial Geomorphologies

The purpose of the ecological reserve is to provide protection of the water resource, specifically its ecosystems, for future generations. It is therefore important that in setting the ecological reserve due consideration is given to how ecosystem drivers may change in the future. This section provides an overview of changes to the main landscape scale drivers of South Africa's water related ecosystems. The focus will be on their implications for future fluvial geomorphologies. The main drivers considered here are climate change, urban development, rural landscape changes, and water-related policy implementation (figure 13.5).

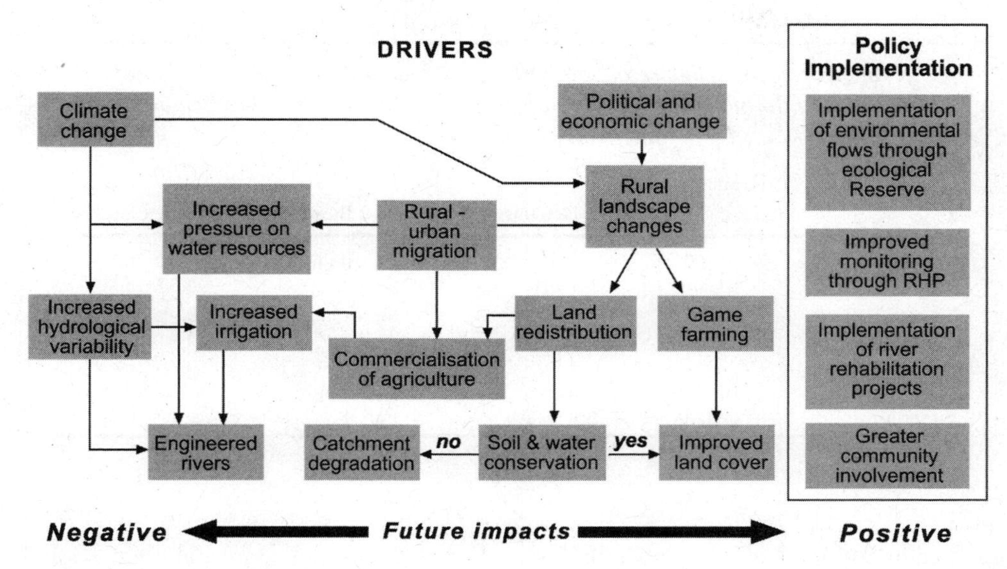

FIGURE 13.5. Drivers of future South African rivers. Negative consequences can be reduced by the successful implementation of policies to protect the nation's water resources.

Climate Change

Future rivers of South Africa will reflect the climate of the future. Climate determines hydrological and sediment processes directly through the rainfall-runoff response and indirectly through the vegetation response. The present climate of South Africa is characteristically dry and both temporally and spatially variable. Table 13.2 summarizes the main climatic zones and associated biomes as presented by Palmer and Ainslie (2005). They classify 70 percent of the country as semi-arid or drier and less than 10 percent as humid or super-humid. The aridity is reflected in the small area that supports perennial streamflow, with intermittent or even ephemeral rivers draining large areas of the country. Annual runoff coefficients across the country are low, averaging 10 percent and seldom exceeding 25 percent except in the smallest mountain catchments (Alexander 1985). Most rivers are characterized by highly variable regimes with long periods of low flows punctuated by severe floods. Ephemeral rivers dominate the western interior.

Although it is difficult to discern climate change within South Africa's already variable climate, at least one group of climate modelers predicts that over the next half century the summer rainfall regions of eastern South Africa will have substantially wetter summers, while parts of the Western Cape may experience wetter winters (Hewitson 1999; Meadows and Hoffman 2003). These changes may be accompanied by seasonal shifts of rainfall and increased variability. Other models suggest a decrease in rainfall throughout South Africa, though the effect will be felt earlier in the west (Schulze and Perks 2000; Turpie et al. 2002). Hoerling et al. (2006) predict an increased frequency of droughts over southern Africa. All authors agree that stream flow will become even more variable. The frequency of extreme flood events may well increase, and increased sediment yields are likely to result from the reduced vegetation cover that will accompany the more frequent droughts.

The National Botanic Institute predicts large changes to many of the country's biomes as a result

TABLE 13.2

Climatic zones and biomes of South Africa (excluding Lesotho and Swaziland). The listing of the biomes reflects their approximate relationship to the climatic zones (after Palmer and Ainslie 2005).

Climatic zone	Mean annual rainfall (in mm)	% Country	Biome	% Country
Desert	< 200	22.8	Succulent Karoo	6.79
Arid	201–400	24.6	Nama-karoo	24.5
Semiarid	401–600	24.6	Savannah	34.4
Subhumid	601–800	18.5	Thicket	3.4
			Fynbos	6.5
Humid	801–1000	6.7	Grassland	24.3
Super-humid	< 1000	2.8	Forest	0.12

of carbon dioxide–induced climate change, including significant reductions in the area of grasslands and an almost complete loss of forest and succulent thicket (Turpie et al. 2002). These predicted changes to land cover will surely affect catchment hydrology and streamflow, but how so is less certain. Schulze and Perks (2000) warn of problems of prediction due to uncertainties in models, incomplete understanding of the physical climate system, and an inability to predict biotic and human response to climate change.

How will channel geomorphology respond to these changes? Experience from the Kruger National Park has shown that rivers flowing through the park are highly dynamic and that geomorphological processes are driven both by extreme events, such as the flood of 2000 that tore out large trees from the riparian zone and caused sand-bed reaches to become bedrock controlled, and by more frequent, moderate flood events that promote the rebuilding of sedimentary features. Van Coller et al. (1997) have demonstrated that the interplay between sediment dynamics and riparian vegetation plays an important role in determining the emergent path of channel change following these extreme events. These moderate flow events are considered to be an important component of the ecological reserve (King and Brown 2006), being especially important for both riparian vegetation and channel geomorphology (Boucher 2002; Dollar and Rowntree 2003). If future hydrological regimes become more variable, with more frequent high-magnitude events and fewer moderate floods, the natural ability of channels to recover after being reset by large floods will decrease, and channel morphology will be driven more by extreme events. How do scientists identify emergent ecosystem trajectories that express adaptation to a new environment? How should managers modify ecological flow requirements in response to these changing conditions? These are questions that still need to be addressed through integrative river research.

Urban Development

The population classified as urban in 2001 was 57.5 percent, increasing at a rate of 2.4 percent since 1996 (Statistics South Africa 2003). Increased demands for water resulting from urban development can only be met by providing sufficient storage and transfer systems. Most of South Africa's rivers have already been dammed. Interbasin schemes (IBTs) transfer water between catchments from areas of surplus to areas of high demand, often using river channels as the conduit for transferring water. Many South African rivers are highly regulated. Increased scour below dams and redistribution of sediment downstream have been noted in some instances. For example, the channel of the Mhlutuze River in KwaZulu Natal has been changed from a sand bed channel to a cobble bed channel for 20 kilometers below the dam wall (Freeman and Rowntree 2005). Massive channel enlargement has been observed in some receiving streams of IBTs, such as the Skoonmakers River in the Eastern Cape (Rowntree and Du Plessis 2003).

The ecological reserve aims to provide ecological flows that will reduce the impacts of future schemes, but given the changes to sediment dynamics and flood frequencies that accompany these large-scale engineering schemes, it is unrealistic to aim to maintain the pre-dam channel geomorphology. However, research into the impacts of dams and interbasin transfer schemes can be used to anticipate future channel geomorphologies and the likely impact on ecosystem dynamics.

Rural Landscape Changes

Although 57 percent of South Africa's population lives in urban areas (Statistics South Africa 2003), less than 1 percent of the landscape is urban (data from Midgley et al. 1994). While urban areas are significant pollution sources, the impact of urban areas on fluvial geomorphology is highly localized. It is therefore changes to the rural landscape that impact most on catchment-scale processes. The political geography of South Africa changed dramatically in 1994 with the end of Apartheid and the installation of a democratically elected government. Three important processes that are changing the rural landscape are the Land Reform Program, legal changes that impact on farm labor, and various poverty-reduction initiatives.

The Land Reform Program aims to bring about a redistribution of land ownership to the advantage of the previously disenfranchised black and colored population. Concerns have been raised that unless the new owners are given sufficient support in terms of access to infrastructure and extension services, land degradation and increased catchment sediment yields may result. The environmental outcome of land reform programs will depend on the development of parallel programs to support sustainable agriculture based on good soil and water conservation practice.

Commercial farms under white ownership have come under economic pressure due to changes in employment laws and stock theft. As a consequence, many farms in marginal dryland areas have been transformed from livestock farms to game farms. The prospect is that game farming could increase land cover and reduce erosion potential, thus reducing catchment sediment yields and channel sedimentation.

Water-Related Policy Implementation

Numerous water-related policies and programs aim to protect water resources and, specifically, water-related ecosystems. These include the implementation of environmental flows (discussed above) through the National Water Act of 1998, improved monitoring through the River Health Program, the implementation of various river and catchment rehabilitation schemes and the greater inclusion of local communities in the governance of land and water resources. This last process is probably the most imperative, as without the cooperation and guidance of local people, legislation to protect the environment will be ineffective.

Various poverty-relief schemes have started to tackle environmental problems. For example, the Department of Water Affairs and Forestry (DWAF) is responsible for Working for Water (http://www.dwaf.gov.za/wfw/), an alien plant eradication program aimed at increasing streamflow by reducing water consumption by fast-growing alien trees such as *Acacia mearnsii*. Working for Wetlands (http://www.dwaf.gov.za/wfw/Wetlands) is another DWAF initiative that aims to rehabilitate eroded wetlands and restore their natural hydrological function. The Department of Agriculture has instigated a LandCare program (http://www.nda.agric.za/docs/Landcarepage/landcare.htm) that aims to tackle land degradation, especially in the former homeland areas (Rowntree 2006). Together, these programs aim to promote an attitude among local groups of good environmental stewardship, thereby improving the ecological integrity of the catchment areas and riparian zones. All three programs should restore some natural biodiversity, help to stabilize flows, and decrease catchment sediment yields.

To be sustainable these rehabilitation schemes need to be based on a good understanding of the underlying geoecological processes. Working for Water is based on the premise that removal of the alien trees will increase both streamflow and biodiversity. Only limited experimental research is available to substantiate these claims (Rowntree and Beyers 1999; Dye and Jarmaine 2004), but direct observations of improved biodiversity and increased flow of headwater streams lend credence to the success of the program. The Working for Water program has brought together ecologists, forest hydrologists, and engineers to gain a better understanding of the hydrological impacts of invasive alien plants (Gorgens and van Wilgen 2004).

Tooth and McCarthy (2007) point to the need to base wetland restoration programs on a good understanding of long-term geomorphological and sedimentological processes. To date, Working for Wetlands has largely taken a structuralist approach, raising the water table in eroded wetland streams by means of gabion weirs, as well as reducing water loss by the removal of alien trees.

Integrative Science and the Future of South African River Management

South Africa's policies and legislation have been put in place to protect the water resource for future generations. They are implemented through programs and methods that have been developed by crossdisciplinary teams of environmental scientists. Both the legislation and the associated methods have been held up as models for good environmental management. Are these approaches, however, really allowing us to apply integrative science that can accommodate the complexity of our river systems? Are they able to take into account the uncertain future? These questions are considered in more detail below.

River management requires an integrated approach that brings together a number of the traditional scientific disciplines. Integration can play itself out, however, in a number of ways, some of which could potentially—albeit inadvertently—compromise certain disciplines. As noted by Pawson and Dovers (2003), it is too easy to assume that crossdisciplinarity will emerge when representatives of different disciplines get together, yet as members of different disciplinary cultures they use particular discursive practices (see chapters 1 and 3).

In South Africa, biologists have spearheaded the development of river management frameworks. As a consequence, these frameworks have been developed within a biological research paradigm. Other disciplines have been called upon to develop methods and protocols in line with existing biological assessment frameworks. It can therefore be asked, "Has the central place of river biology compromised the contributions from other sciences? Has the leading role of biologists resulted in the application of incorrect assumptions and inappropriate conceptual frameworks, ultimately leading to injudicious management practices?"

Fluvial geomorphology is one of the subdisciplines for which this assertion may be true. Not only does fluvial geomorphology assess abiotic, rather than biotic, system components, but there is also a disparity in the spatial and temporal scales that need to be considered. While biologists prefer to conduct surveys on a site, or at most, a reach-specific basis, fluvial geomorphologists must consider the catchment as a whole and must consider longer time periods of change. Rowntree and Wadeson (1997, 1999) developed a spatially explicit hierarchical model for South Africa that was designed to provide a framework linking biota and geomorphology at different scales. Dollar et al. (2007) demonstrated how such a framework could be applied to address the challenges of scale in crossdisciplinary river research.

The sampling design and assessment frameworks used by crossdisciplinary teams tend to favor the paradigms of biologists. This is exemplified by the National River Health Program (RHP), a monitoring initiative designed to gauge the health of all major South African river systems, from which State of River reports are intended for wide distribution to the private and public sectors in an effort to increase public knowledge and appreciation of river systems (see http://www.csir.co.za/rhp/). Although the RHP claims to assess river health through monitoring the status of components of the river system, including macroinvertebrate communities, fish assemblages, riparian plant communities, water quality, and river geomorphology, the program is driven by a method known as the South African scoring system (SASS) (see Dickens and Graham 2002). SASS is a biomonitoring method that uses the abundance and diversity of its macroinvertebrate species as a measure of water quality. One of the biotopes (i.e., habitats) included in this assessment is "stones-in-current," where the stones can be kicked in order to release the biota into the water column where they can be netted. For this reason a riffle (with shallow, fast-flowing water over relatively loose stones) is purposefully sought out during the site-selection process, despite the fact that they are not common features in many reaches of South African rivers.

Fish biologists also play a key role in the RHP. Fish sampling methods, such as fish shocking and the use of sein nets, means that easy access is necessary to get the equipment onto the site. Sites are often close to bridges or roads and, as for SASS, have to include a riffle feature as well as a pool. Such sites are anathema to geomorphologists, who sees bridges and roads as system perturbations.

Geomorphologists are seldom part of the site-selection process in the RHP and our experience is that sites selected by biologists are not always representative of the geomorphology of the larger reach in which they are situated. Furthermore, a geomorphological assessment should take place at the scale of the reach, not an isolated site, as selected for SASS assessments. Because teams of scientists work together, it is not always possible for geomorphologists to get a good impression of a reach due to constraints of time and, in certain areas, safety.

The issue of site selection underlines the problem of taking into account the different scales at which different river scientists work (see chapter 3). An invertebrate biologist doing a SASS assessment works within and around a single riffle. A fish biologist extends the scale to include upstream of downstream pools. A geomorphologist scales up the assessment to include the entire reach. Moreover, reach-scale processes can only be interpreted within the context of the upstream river network and the catchment that it drains.

The reference condition concept can be used to demonstrate that not only is there a practical bias toward the biologists' paradigm, but also a theoretical one. Many South African river management procedures (for example, the determination of ecological water requirements for the ecological reserve) require that a reference condition be derived for each component of the system (i.e., fish, macroinvertebrates, riparian vegetation, geomorphology, and hydrology). Dallas (2000, 2), quoting Reynoldson et al. (1997), defines the ecological reference condition as "the condition that is representative of a group of least impacted or minimally disturbed sites organized by selected physical, chemical, and biological characteristics." Biological reference conditions are established by undertaking sampling in a number of the least impacted sites within a larger geographically defined sampling unit, namely the ecoregion.

Problems arose when following this approach to establish a geomorphological reference condition due to the different spatial and temporal scales involved (Du Preez and Rowntree 2006). River reaches cannot be isolated from upstream processes, so few reaches can be considered minimally disturbed. River morphology in any reach also represents the effect of long-term processes and recovery from major events, which are a part of the natural system. Geomorphologists working in the Kruger National Park observed major changes in channel morphology during the floods of 2000, with sediment being stripped from some alluvial reaches. After the flood the morphological units within the active channel (below the bankfull level) readjusted quickly to the normal range of flows, but those above the bankfull level showed little change from their post flood condition (Heritage et al. 2004).

Du Preez and Rowntree (2006) defined the reference condition as "the geomorphological system that supports the natural ecosystem, where a system is a set of components connected through flows of energy and matter to accomplish a set function." (Du Preez and Rowntree 2006, 43). The GAI (geomorphological assessment index) is based on the extent that the capacity of the geomorphological system to support the natural ecosystem has been degraded due to changes to system connectivity, the reach sediment balance, the perimeter resistance of the channel (especially riparian vegetation), and the channel morphology. Rather than seeing the reference condition as being a deterministic channel morphology that can be predicted from reach and catchment controls, the reference condition is conceptualized as a geomorphological system composed of flows, storages, and morphological responses that are dynamic and indeterminate.

Scientists increasingly question how accurately we can predict certain complex systems (see chapters 1 and 14). This is not simply because of a lack of information, the interpretation emanating from the reductionist paradigm of conventional science, but rather because certain systems have characteristics that make them inherently unpredictable. The geomorphological concept of indeterminacy of response (also termed system "singularity" or "uniqueness") states that channels have many different ways of responding to the same stimulus for change, whereas its converse, the equifinality problem, states that different change stimuli can produce similar results (Thornes 1977). System characteristics that confound prediction include such elusive yet intuitively important concepts such as chaos, system resilience, and system resistance (e.g., Gribbin 2005, Phillips 1992a, 1992b).

River scientists working in South Africa have developed a number of models that could be used to predict channel change, assuming that present-day drivers remain the same. For example, qualitative rule-based models have been developed as part of the Kruger National Park River Research Program to predict the response of systems to natural and anthropogenic factors affecting water supply (Jewitt et al. 1998). Heritage et al. (1997) developed a sediment transport and storage model for predicting annual change at the channel type scale while Birkhead et al. (2000) presented various geomorphological change models as part of a project aimed at developing a decision support system for water resource management.

The methods that are being developed to support river management in South Africa are based on the premise that scientists can predict future channel condition, or "postdict" a reference condition based on present-day observations of processes and impacts (causes). If certain systems are inherently unpredictable, however, then the reductionist principle of cause and effect is invalid and systems may not behave (or have behaved) in the way one might expect at all spatial and temporal scales. Trofimov and Phillips (1992) identified various factors that confound prediction in geomorphological systems. These include the need to determine the geomorphological memory of a system; whether the system has passed through any bifurcation points on its trajectory that in essence reset the system, making previous states irrelevant; and the complex response of geomorphological systems to perturbations. These concepts could help to explain the geomorphological changes observed in the rivers of the Kruger National Park.

As uncertainty is inherent to the behavior of certain systems, and environmental changes may shift the boundary conditions of the future, it is difficult (if not impossible) to make accurate predictions of future river conditions (see chapter 5). However, one of the general objectives of integrated environmental management set out in chapter 5 of the National Environmental Management Act (section 23[2][b]) is to "identify, *predict*, and evaluate the actual and potential impact on the environment, the risks and consequences and alternatives and options for mitigation of activities, with a view to minimizing negative impacts, maximizing benefits, and promoting compliance with the principles of environmental management."

In light of the above statements regarding uncertainty, this is a weighty expectation. The act does, however, recognize the potential limitations of predictive tools. Section 24 of the same act, which pertains to the implementation of the objectives of integrated environmental management, states that "procedures for the investigation, assessment, and communication of the potential impact of activities must, as a minimum, ensure *(amongst other things)* reporting on *gaps in knowledge, the adequacy of predictive methods and underlying assumptions*, and *uncertainties* encountered in compiling the required information" (24[7][e], parentheses and emphasis added).

Although South African river scientists are required to attach a confidence value to their assessments and predictions in an ecological reserve study, official reporting on the adequacy of predictive methods and uncertainties, as stipulated by the NEMA, does not take place. The predictive tools that have been developed thus far in South Africa have often not been tested rigorously. In the case of geomorphological models, the spatio-temporal scale of change is likely to be significantly greater than that of the management framework, making testing unrealistic.

However, all is not necessarily lost. Trofimov and Phillips (1992, 203) draw a useful distinction between prediction ("the ability to foretell future evolution and outcomes") and forecasting ("estimating or calculating possible, likely, or desired future outcomes"). By these definitions, they state that while prediction for many systems may be impossible, some degree of forecasting may be attainable. The key seems to be the honest assessment of our statements about the future and the admission that in most cases we achieve a degree of educated forecasting, but probably never true prediction. An example of forecasting can be seen in ecological reserve studies. Our understanding of channel response to changing system drivers has been built into indices such as the GAI. These indices (the GAI, MIRAI, PAI, FRAI, and VEGRAI) are used to assess the impact of different water-use scenarios on the future ecostatus of the river. The results are used to help stakeholders make an informed decision about their desired management class.

Conclusion

The application of integrative river science to protecting water resources is not straightforward or easily achieved. In South Africa, many river scientists have applied their expertise to develop methods that comply with the National Water Act of 1998 and international agendas. In some ways, the need to develop assessment models may have taken priority over the advancement of river science. Strong financial support for applied research programs has come from the Water Research Commission. The process by which an ecological reserve is recommended for any South African river system brings together a team of scientists who together represent the different components of the river ecosystem. This team shares its cumulative knowledge, applying expert judgment to make decisions that will affect the river's future management. The methods used have evolved since the mid 1990s when King and Tharme (1994) first presented their review of instream flow methodologies. These methods have been continually adapted and modified in an attempt to achieve an answer that balances scientific integrity with the need for a recommendation that can be implemented in practice. The problem of implementing the recommendations of a reserve study is highlighted by the fact that not one of the comprehensive and intermediate studies has yet gone to the full implementation stage. Scientific integrity is often compromised by the lack of real understanding of the complexity of river systems, and points to the need for further integrative research by teams of river scientists to develop process-response models of river systems.

It is not always the case that environmental management approaches reflect contemporary research and scientific thinking (Doyle et al. 1999). Pressure is placed on scientists to maintain the integrity of their various scientific disciplines while providing practical, feasible answers to questions posed by the environmental management sector. This challenge was well put by Rogers and Biggs (1999) who pointed to the tendency of models used as management tools to be based on an uncritical acceptance of weak science. This criticism could be directed at the indices forming the basis of the ecostatus model as these have yet to undergo rigorous scientific testing.

In the scramble for recognition in the crossdisciplinary arena, there is a danger that river scientists could lose touch with their disciplinary base. One problem of promoting integrative, crossdisciplinary science is that, in the effort to integrate, the individual disciplines may lose their sense of identity. South Africa is a large country, and the fluvial geomorphology community is small, geographically dispersed, and itself suffers from a lack of integration. In trying to work with other disciplines, we may have forgotten how to work together to further our scientific understanding of how rivers work.

An enhanced theoretical understanding is required to better embrace the uncertain future of the complex systems that comprise the rivers of South Africa. With these advances, system characteristics such as chaos, resilience, sensitivity, and resistance could be factored into the current assessment methods. This would allow a degree of forecasting based on a systems' propensity for change, rather than being based simply on uncertain future system inputs.

Acknowledgments

The authors wish to thank two reviewers for their constructive comments on this chapter.

References

Alexander, W. J. R. 1985. Hydrology of low-latitude southern hemisphere land masses. In *Perspectives in southern hemisphere limnology*, ed. B. R. Davies and R. D. Walmsley, 75–83. The Hague, Netherlands: Dr. W. Junk Publishers.

Birkhead, A. L., G. L. Heritage, C. S. James, K. H. Rogers, and A. W. van Niekerk. 2000. Geomorphological change models for the Sabie River in the Kruger National Park. *Water Resources Commission Report* No. 782/1/00, Pretoria, South Africa.

Boucher, C. 2002. Flows as determinants of riparian vegetation zonation patterns in selected southern African rivers. *Proceedings of the Environmental Flows for River Systems incorporating the Fourth International Ecohydraulics Symposium: An international working conference on assessment and implementation Cape Town, South Africa, 3–8 March 2002*. http://www.uidaho.edu/ecohydraulics/faculty/klausjorde/c-town/Proceedings/Boucher/ (accessed March 13, 2007).

Breen C. M., H. Biggs, M. C. Dent, A. Gorgens, J. O'Keeffe, and K. H. Rogers. 1995. Designing a research programme to promote river basin management. *Water Science and Technology* 32:103–9.

Dallas, H. F. 2000. *The derivation of ecological reference conditions for riverine macroinvertebrates*. NAEBP Report Series No.12. Pretoria, South Africa: Institute for Water Quality Studies, Department of Water Affairs and Forestry.

Dickens, C. W. S., and P. M. Graham. 2002. The South African Scoring System (SASS) Version 5 rapid biological method for rivers. *African Journal of Aquatic Sciences* 27:1–10.

Dollar, E. S. J., and K. M. Rowntree. 2003. Geomorphological research for the conservation and management of Southern African rivers, Volume 2: Managing flow variability: The geomorphological response. *Water Research Commission Report* 849/1/03. Pretoria, South Africa.

Dollar, E. S. J., C. S. James, K. H. Rogers, and M. C. Thoms. 2007. A framework for interdisciplinary understanding of rivers and ecosystems. *Geomorphology* 89:147–62.

Doyle, M. W., D. E. Miller, and J. M. Harbor. 1999. Should river restoration be based on classification schemes or process models? Insights from the history of geomorphology. *ASCE International Conference on Water Resources Engineering, Seattle, Washington, 1999.*

Du Preez, L., and K. M. Rowntree. 2006. Assessment of the geomorphological reference condition: An application for resource directed measures and the River Health Programme. *Water Resources Commission Report* No. 1306/1/06. Pretoria, South Africa.

Dye, P., and C. Jarmaine. 2004. Water use by black wattle (*Acacia mearnsii*): Implications for the link between removal of invading tress and catchment stream response: Working for water. *South African Journal of Science* 100:40–44.

Freeman, N. M., and K. M. Rowntree. 2005. Our changing rivers: An introduction to the science and practice of fluvial geomorphology. *Water Research Commission Report* No. TT 238/05. Pretoria, South Africa.

Godden, L. 2005. Water law reform in Australia and South Africa: Sustainability, efficiency and social justice. *Journal of Environmental Law* 17:181–205.

Gorgens, A. H. M., and B. W. van Wilgen. 2004. Invasive alien plants and water resources in South Africa: Current understanding, predictive ability and research challenges. *South African Journal of Science* 100:27–33.

Gribbin, J. 2005. *Deep simplicity: Bringing order to chaos and complexity*. New York, NY: Random House.

Heritage, G. L., A. R. G. Large, B. P. Moon, and G. Jewitt, 2004. Channel hydraulics and geomorphic effects of an extreme flood event on the Sabie River, South Africa. *Catena* 58:151–81.

Heritage, G. L., A. W. van Niekerk, B. P. Moon, L. J. Broadhurst, K. H. Rogers, and C. S. James. 1997. The geomorphological response to changing flow regimes of the Sabie and Letaba river systems. *Water Resources Commission Report* No. 376/1/97. Pretoria, South Africa.

Hewitson, B. 1999. Deriving Regional Precipitation Scenarios from General Circulation Models. *Water Resources Commission Report* No. 751/1/99. Pretoria, South Africa.

Hoerling, M., J. Hurrell, J. Eischeid, and A. Phillips. 2006. Detection and attribution of twentieth-century northern and southern African rainfall change. *Journal of Climate* 19:3989–4008.

Iversen, T. M., B. L., Madsen, and J. Bøgestrand. 2000. River conservation in the European community, including Scandinavia. In *Global perspectives on river conservation: Science policy and practice*, ed. P. J. Boon, B. R. Davies, and G. E. Petts, 79–104. Chichester, UK: Wiley.

Jewitt, G. P. W., G. Heritage, D. Weeks, J. Mackenzie, A. H. M Görgens, J. O'Keeffe, and K. Rogers. 1998. Modelling abiotic-biotic links in the Sabie River. Report to the KNPRRP and Water Research Commission. *Water Resources Commission Report* No. K777/98. Pretoria, South Africa.

King, J. M., and C. Brown. 2006. Environmental flows: Striking the balance between development and resource protection. *Ecology and Society* 11(2):Article 26. http://www.ecologyandsociety.org/vol11/iss2/art26/ (accessed December 16, 2007).

King, J. M., and D. Louw. 1998. Instream flow assessments for regulated rivers in South Africa using the building block methodology. *Aquatic Ecosystem Health and Management* 1:109–24.

King, J. M., and R. E. Tharme. 1994. Assessment of the instream flow incremental methodology and initial development of alternative instream flow methodologies for South Africa. *Water Research Commission Report* No 295/1/94. Pretoria, South Africa.

Kleynhans, C. J., M. D. Louw, C. Thirion, N. J. Rossouw, and K. Rowntree. 2005. River ecoclassification: Manual for ecostatus determination (version 1). *Department of Water Affairs and Forestry/Water Resources Commission* Report No. KV 168/05. Pretoria, South Africa.

Meadows, M. E., and T. M. Hoffman. 2003. Land degradation and climate change in South Africa. *The Geographical Journal* 169:168–78.

Midgley, D. C., W. V. Pitman, and B. J. Middleton. 1994. Surface water resources of South Africa 1990. *Water Resources Commission Report* No. 228. Pretoria, South Africa.

Novotny, V., E. Bartosova, N. O'Reilly, and T. Ehlinger. 2005. Unlocking the relationship of biotic integrity waters to anthropogenic stresses. *Water Research* 39:184–98.

Palmer, A. R., and A. M. Ainslie. 2005. Grasslands of South Africa. In *Grasslands of the world*, ed. J. M. Suttie, S. G. Reynolds, and C. Battello, 77–120. Rome, Italy: Food and Agricultural Organization of the United Nations.

Palmer, C. G. 1999. Application of ecological research to the development of a new South African water law. *Journal of the North American Benthological Society* 18:132–42.

Pawson, E., and S. Dovers. 2003. Environmental history and the challenges of interdisciplinarity: An antipodean perspective. *Environment and History* 9:53–75.

Phillips, J. D. 1992a. The end of equilibrium? *Geomorphology* 5:195–201.

Phillips, J. D. 1992b. Nonlinear dynamical systems in geomorphology: Revolution or evolution? *Geomorphology* 5:219–29.

Republic of South Africa 1998a. *National Water Act*, Act No. 36 of 1998. The Department of Water Affairs and Forestry. Government Gazette 19182. Pretoria, South Africa: Government Printer.

Republic of South Africa 1998b. *National Environmental Management Act*, Act No. 107 of 1998. The Department of Environmental Affairs and Tourism, Government Gazette 19519. Pretoria, South Africa: Government Printer.

Reynoldson, T. B., R. H. Norris, V. H. Resh, K. E. Day, and D. M. Rosenberg. 1997. The reference condition: A comparison of multimetric and multivariate approaches to assess water-quality impairment using benthic macroinvertebrates. *Journal of the North American Benthological Society* 16:833–52.

Rogers, K., and H. Biggs. 1999. Integrating indicators, endpoints and value systems in strategic management of the rivers of the Kruger National Park. *Freshwater Biology* 41:439–51.

Rowntree, K. M. 2006. Integrating catchment management through LandCare in the Kat Valley, Eastern Cape Province, South Africa. *Physical Geography* 27:435–46.

Rowntree, K. M., and G. Beyers. 1999. An experimental study of the effect of Acacia Mearnsii (black wattle trees) on streamflow in the Sand River, Zwartkops River Catchment, Eastern Cape. Report to the Department of Water Affairs and the Water Research Commission. *Water Resources Commission Report* No KV123/99. Pretoria, South Africa.

Rowntree, K. M., and M. Du Plessis. 2003. Geomorphological research for the conservation and management of Southern African rivers, Vol 1 Geomorphological Impacts of River Regulation. *Water Research Commission Report* No. 849/2/03. Pretoria, South Africa.

Rowntree, K. M., and R. A. Wadeson. 1997. A hierarchical geomorphological model for the assessment of instream flow requirements. *Geoöko plus* 4:85–100.

Rowntree, K. M., and R.A. Wadeson. 1998. A geomorphological framework for the assessment of instream flow requirements. *Aquatic Ecosystem Health and Management* 1:125–41.

Rowntree, K. M., and R. A. Wadeson. 1999. A hierarchical geomorphological model for the classification of selected South African river systems. *Water Research Commission Report* No. 497/1/99. Pretoria, South Africa.

Rutherfurd, I. D., K. Jerie, and N. Marsh. 2000. *A rehabilitation manual for Australian streams, volumes 1 and 2*. Canberra, Australia: Cooperative Research Centre for Catchment Hydrology, and the Land and Water Resources Research and Development Corporation.

Schulze, R. E., and L. A. Perks. 2000. *Assessment of the impact of climate change on hydrology and water resources in South Africa*. Report to South African Country Studies for Climate Change Programme ACRUcons Report 33.

Statistics South Africa 2003. *Census 2001: Investigation into appropriate definitions of urban and rural areas for South Africa: Discussion document*. Report No. 03-02-02 (2001). Pretoria, South Africa: Statistics South Africa.

Thornes, J. B. 1977. Hydraulic geometry and channel change. In *River channel changes*, ed. K. J. Gregory, 91–100. Chichester, UK: Wiley.

Tooth, S., and T. S. McCarthy. 2007. Wetlands in drylands: Geomorphological and sedimentological characteristics, with emphasis on examples from southern Africa. *Progress in Physical Geography* 31:3–41.

Trofimov, A. M., and J. D. Phillips. 1992. Theoretical and methodological premises of geomorphological forecasting. *Geomorphology* 5:195–201.

Turpie, J. H., H. Winkler, R. Spalding-Fecher, and G. Midgley. 2002. *Economic impacts of climate change in South Africa: A preliminary analysis of unmitigated damage costs*. Cape Town, South Africa: Southern Waters Ecological Research & Consulting & Energy Y & Development Research Center, University of Cape Town.

United Nations 1993. *Report of the United Nations conference on environment and development. Rio de Janeiro, 3–14 June 1992*, Vol. I, Agenda 21. New York, NY: United Nations.

United Nations Development Programme (UNDP). 2006. *Beyond scarcity: Power, poverty and the global water crisis*. New York, NY: United Nations Development Programme.

Van Coller, A. L., K. H. Rogers, and G. L. Heritage. 1997. Linking riparian vegetation types and fluvial geomorphology along the Sabie River within the Kruger National park, South Africa. *African Journal of Ecology* 35:194–212.

Wadeson, R. A. 1994. A geomorphological approach to the identification and classification of instream flow environments. *South African Journal of Aquatic Science* 20:1–24.

PART IV

Managing the Process of River Repair

> The most neglected and least understood aspect of conservation is interaction with citizens to determine and incorporate their value systems into management.
>
> Rogers and Bestbier 1997, 23

Ecological rehabilitation is about people and process. Finding a way to link environmental and social goals offers the most likely pathway to rehabilitation success. Unless applications are "owned" by the communities involved, prospects for long-term, sustainable river health are likely to be compromised. What we seek to achieve through river management is dependent on the values we place on rivers and the approaches to management that we adopt. Discipline-bound knowledge can only provide partial answers to problems faced. Cross-disciplinary science is required to guide the process of river repair.

As shown in the international case study chapters, diverse perspectives and agendas drive the process of river repair. Place-based knowledge coupled with political, cultural, and societal perspectives vary significantly across the globe. As such, a mix of top-down and bottom-up approaches and governance structures has been adopted in different parts of the world. Despite the specific nature of these arrangements, the process of river repair is a collective responsibility that requires appropriate visions, effective implementation, and societal "will" to improve river health. With this comes a commitment to maintaining and documenting the process within an adaptive management framework.

One reason why past efforts to improve river health were relatively ineffective was the deterministic outlook with which rivers were viewed and management applications were performed. Chapter 14 documents uncertain futures and the need to restore uncertainty in the management process through acceptance of river diversity, complexity, and variability. To achieve this, system-specific understanding of river dynamics and change, and the incorporation of societal values and perspectives, must be framed within an adaptive management approach. Guidance on ways in which this can be achieved is explored in chapter 15, with direct references to contributions from each chapter in this book.

Reference

Rogers, K. H., and R. Bestbier. 1997. *Development of a protocol for the definition of the desired state of riverine systems in South Africa*. Pretoria, South Africa: Department of Environmental Affairs and Tourism.

Chapter 14

Restoring Uncertainty: Translating Science into Management Practice

MICK HILLMAN AND GARY J. BRIERLEY

> Learning how to deal with uncertainty and adapt to changing conditions is becoming essential in a world where humanity plays a major role in shaping biospheric processes from genetic levels to global scales.
>
> Olsson et al. 2004, 75

Uncertainty in river management occurs across the biophysical, social, and institutional domains. This chapter argues that uncertainty is unavoidable and should not be regarded as a barrier to action or a reason for ill-considered efforts to get things done, but rather as part and parcel of managing rivers; if it is not embraced, uncertainty should at least be accepted and acknowledged. We start by examining the dimensions and sources of uncertainty and their broad implications for management.

The inherent biophysical diversity, variability, and complexity of river systems make their behavior unpredictable. Key drivers and relationships that shape ecosystem structure and function vary from system to system and change over time. Hence, similar disturbances may produce different responses in different places and at different times. The resilience of each system is influenced by its unique history and configuration. Ultimately, the maintenance of healthy ecosystems, and indeed our own health, is contingent upon the desire and capacity to work with the uncertainties of biophysical interactions in nature. Application of generic and prescriptive management tools or practices in a quest for certainty fails to address these concerns.

An awareness of these biophysical uncertainties must be viewed in relation to contrasting human attitudes toward, and relationships and interactions with, the environment. The diversity and changing nature of these relationships introduce a separate tier of uncertainty in efforts to manage river health. A third tier of uncertainty lies in the management structures and institutional arrangements that frame the search for a balance between economic development and environmental repair. In this chapter, we explore these various biophysical, social, and institutional dimensions of uncertainty in the management of river systems. Ultimately, we regard uncertainty as something that should be embraced rather than feared.

At its broadest level, two types of uncertainty can be differentiated: *ontological* uncertainty due to the inherent and intrinsic variability of complex ecosystems: what we might call the unknowable; and *epistemological* uncertainty due to some limitation on the knowledge of those investigating or managing those systems: what might be called the unknown (Van Asselt and Rotmans 2002; Wheaton et al. 2008). This latter type of uncertainty can be further classified into structural and methodological concerns, as shown in table 14.1.

In box 14.1, two case studies presented at the Australian Stream Management Conference in 2005 are used to illustrate these various types and dimensions of uncertainty.

How we see and act upon these various types of uncertainty is dependent upon our view of the world and of river systems in particular. An engineering worldview emphasizes demands for control and stability, underpinned by a positivist perspective that defines science as the search for, and prediction of, empirical regularities to derive universal truths (Van Asselt and Rotmans 2002). The positivist paradigm has come under sustained attack following what Funtowitz and Ravetz (1993) refer to as a skeptical crisis, in which critics argued that scientific knowledge was in reality nothing more than a set of conventions deemed to be true by a particular group (Feyerabend 1975; Van Asselt and Rotmans 2002). Rather than viewing uncertainty as a problem, post-normal perspectives see uncertainty as part of the inherent beauty of an environment within which the diversity, dynamics, and vitality of rivers resonate with human nature and spiritual values (Stoffle and Arnold 2003; Jackson 2006). Such thinking recognizes that ontological uncertainty in aquatic ecosystems may be a necessary precondition for biodiversity—for example, flow variability promotes a range of habitats and niches (Montgomery and Bolton 2003). In this mindset, complexity and uncertainty are considered as part of ongoing evolutionary processes. Whereas science was previously understood as steadily advancing certainty in knowledge, enhancing the capacity to predict and control the natural world, it is now seen as loaded with inherent

TABLE 14.1

Sources of ontological and epistemological uncertainty (adapted from Van Asselt and Rotmans 2002)

Sources of ontological uncertainty	Sources of epistemological uncertainty	
	Structural	Methodological
1. *Inherent natural randomness*—the chaotic, nonlinear, and unpredictable nature of natural processes	1. *Irreducible ignorance*—"we cannot know," the processes are just too complex	1. *Immeasurable in practice*—"we know what we don't know"—lack of data or information with which to characterize the full range of natural variability due to problems of temporal and spatial resolution, upscaling, and averaging
2. *Value diversity*—differing attitudes, uses, and goals	2. *Indeterminacy*—"we will never know," we understand the basic principles but are unable to make sound predictions	2. Lack of observations and measurements—"could have, should have, would have, but didn't"
3. *Behavioral diversity*—actions that may differ from stated values	3. *Reducible ignorance*—"we do not know what we do not know, but we may in the future"	3. *Inexactness*—related to lack of precision, lack of accuracy, measurement, and calculation; "we know roughly"
4. *Societal randomness*—unpredictable values, actions, and so on, expressed in social and institutional processes	4. *Conflicting evidence*—"we don't know what we know"—knowledge is not fact but interpretation, and interpretations frequently contradict and challenge each other	
5. *Technological surprise*—unanticipated developments or outcomes		

Box 14.1. Illustration of the types of uncertainty presented in table 14.1 based on case studies of the response to woody debris in sandy bed streams (after Borg et al. 2005; Brooks and Cohen 2005).

Initial Hypothesis

Engineering theory predicts that logs encourage the scouring of long, semipermanent pools suitable for fish habitats.

Findings

- In practice, scouring and pool formation is highly variable across time and location.
- Engineering methods are not very useful when considering stochastic realities of biologically relevant geomorphic processes.

Identified sources of biophysical uncertainty.

- Effects of other geomorphic/hydraulic features (e.g., lateral sandbars) promoted by woody debris.
- Pools scour or infill depending on flow conditions.
- Engineering predictions based on peak flows and flume modeling.

Examples of Types of Uncertainty

- *Ontological*: Inherent nonlinearity and unpredictability of geomorphic and hydraulic processes.
- *Ontological (value and behavioural diversity)*: Conflicting perspectives on the value of woody debris—bank erosion, mobilization of logs in floods versus promotion of biodiversity and aesthetics values—not derivable from stakeholder position.
- *Structural*: Interpretations of multiple possible causal factors in rate of scour/infilling.
- *Structural*: Links between pool formation and suitability for particular breeding cycles.
- *Methodological*: Problems of measuring the full range of natural variability at appropriate scales.
- *Methodological*: Sampling is usually done once or twice per year at coarse scour pool depths.

uncertainties, so that ignorance can no longer be expected to be conquered (Funtowitz and Ravetz 1991).

In management terms, programs increasingly attempt to acknowledge and work with uncertainty, striving to enhance natural recovery mechanisms and the physical and ecological integrity of living, variable, and evolving systems (Everard and Powell 2002). This recognition is reflected in recent river management initiatives in Europe, such as the Loire Vivante program in France and the Space to Move program in the Netherlands (Piégay et al. 2005). However, this new approach to uncertainty brings with it new tensions in the relationship between science, management, and the broader community. For those who retain a controlling top-down outlook on managing rivers, problems occur when black-box science, institutional capacity, or political will fail to provide firm and assured explanations of environmental behavior and respond accordingly. This narrow perspective fails to appreciate the real challenges of integrative river science and management.

As Sarewitz (2004) argues, the uncertainty that fuels environmental controversies should be understood not as a failure of scientific understanding but as the lack of coherence among competing scientific understandings, amplified by the various political, cultural, and institutional contexts within which science is carried out. Many managers, policy makers and decision makers are fearful of admitting to uncertainty, believing that the public may view this as a fatal sign of weakness (Newson and Clark, forthcoming). As uncertainty is considered to represent a risk, and river managers are risk-averse, managers view uncertainty as something negative or undesirable and prefer to entrench themselves in "rituals of verification" aimed at minimizing liability rather than giving due regard to strategies that work with the inherent variability and complexity of natural systems (Newson and Clark 2008).

Ironically, at a personal level, we usually confront such uncertainties in our daily lives with a degree of equanimity, making decisions about complex problems with incomplete information (Pollack 2003). The remainder of this chapter argues for the adoption of such an approach in river management and for restoring uncertainty as part of a more harmonious relationship with natural systems. As a starting point, we consider the scientific, societal, and managerial dimensions of uncertainty in river management practice. We appraise the characteristics and application of various management tools developed to facilitate environmental decision making in the face of biophysical and social uncertainty. We argue for prudence in the application of such tools, ensuring that they address concerns for equity and precaution as key principles of sustainability.

Sources of Uncertainty in the Management of River Systems

Biophysical, social, and managerial sources of uncertainty can be identified, each with its own inherent limitations that may compromise the performance of management. These various forms of uncertainty are considered in turn.

Sources of Biophysical Uncertainty

The evolution of the natural world has been characterized by experimentation, blind alleys, extinctions, and shifts in direction rather than by steady and consistent progress. Particular spatial and temporal combinations of events have generated distinctive and often unique responses. Often it has been major, extreme events that have prompted shifts in evolutionary trajectory—the occurrence of which are unpredictable, let alone the consequences. Within any given branch of science there are inherent uncertainties about the way natural systems operate. These uncertainties are confounded many times over when ecosystems are viewed in a holistic manner. As any river system is made up of a complex web of biophysical interactions, integrative, whole-of-ecosystem management approaches both introduce and increase possibilities for uncertainty. For example, while incidental increases in discharge may instigate trivial responses in some biophysical relationships (e.g., bedload mobility), the consequences may be of fundamental concern for other attributes (e.g., habitat viability, nutrient flux, food web processes).

Key drivers that fashion physical and ecological integrity, and the functional relationships that sustain ecosystem behavior, are themselves subject to change over time (see chapters 4, 5, and 6). More important, the limiting factors that constrain ecosystem processes may vary markedly from system to system. Unravelling cumulative impacts of differing forms of disturbance events over time requires system-specific, cross-scalar insights. The nonlinear nature of biophysical interactions ensures that surprising responses are inevitable. Pronounced variability in critical threshold conditions that may alter ecosystem behavior pose significant uncertainties for a host of management applications. In the woody debris studies (box 14.1), the relationship between hydraulic conditions and geomorphic features resulted in large variations in pool scour between sites, as opposed to the predictions of engineering models. This in turn made the link between flows, channel structure, and habitat complex, site-specific, and uncertain. In light of these considerations, system-specific insights must emphasize the inherent range of natural variability that may be experienced, framing appraisals of future adjustments in terms of more or less likely trajectories of change. Looking to the past to flag prospective futures highlights that new directions are eminently possible (chapter 5).

These circumstances mean that the best that can be achieved in scientific investigations of environmental interactions are inferences with variable degrees of reliability—*not* certainty. The nondeterministic nature of biophysical interactions ensures that endeavors to work with nature must recognize, and recognize explicitly, the inherent indeterminacy of ecosystems, and hence the uncertainty of predictions about their future states. Stochastic approaches to the appraisal of river futures provide a more realistic basis upon which to launch programs of river repair, using foresighting exercises to appraise prospective future states for any given system, such as the geomorphic and biotic response to the introduction of woody debris (box 14.1) or the provision of environmental flows (Stewardson and Rutherfurd 2005). Many layers of social and cultural uncertainty must be added to this information base to frame management decision making.

Sources of Social and Cultural Uncertainty

Economic development, population growth, and technological advances have driven social and cultural transformations that pose significant challenges to sustainability agendas, management of biodiversity, and regional sensitivity to global environmental change. Indicators of economic development, such as gross domestic product, along with community attitudes and government policies, shape what are seen as justifiable, economically viable conservation and rehabilitation programs. In practice such choices are often framed as rational or technical questions of efficiency, masking the high number of political and ethical choices that underpins environmental decision making.

Changing demographics within any given region influence the land-use pressure placed upon environmental systems, promoting increased societal awareness of, and management concern for, environmental degradation (chapter 8). This has been augmented by the apparent proliferation and localized impacts of natural hazards and extreme events. Greater appreciation of risk has been accompanied by growing recognition of a broader spectrum of recreational, cultural, and spiritual values as core business in river management, contrasting starkly with a long-standing singular focus in developed and postcolonial societies on the consumptive use-value of water (Pigram 2006; Jackson 2005; Gibbs 2006). In the case of woody debris (box 14.1), this increased range of values means that community concerns now encompass the habitat and aesthetic benefits of introduced logs, along with more traditional concerns over bank stability and flood hazards (Brooks and Cohen 2005). Nor are these values necessarily derivable from knowing the occupation of the landholder or their particular use of the river (chapter 8). Attention is being drawn to the importance of social connection and place-identity as factors to be considered in the assessment of river health and in developing broad-based catchment visions and plans, particularly where rapid demographic and land-use change is occurring (see chapter 8).

However, for many in river management and the broader society, the promise of technofix as a driver of economic progress or an environmental savior continues to loom large (Higgs 2003). For others, this belief in technologically based, cost-effective, and environmentally friendly solutions is frowned upon as undermining genuine efforts to alter attitudes and approaches. For the techno-skeptics, overzealous, technically proficient restoration framed within the "device paradigm" detracts from the real work of protecting ecosystems (Higgs 2003). From this viewpoint, incorporation of multiple perspectives, use-values, and knowledges is a prerequisite for moving beyond marquee corporate-style restoration projects and toward focal restoration, a goal shaped by historical and cultural associations and dialogue between people and their environment (Clifford 2001).

These conflicting views of what constitutes river health and repair combine with recognition of scientific limitations and the importance of local knowledge to generate a "messy" or "fuzzy" social context for river management (Clark 2002; Lachapelle et al. 2003). Differing interest groups and competing perspectives can at times create a climate of mutual suspicion in which "wicked" decisions (Lachapelle and McCool 2005) appear necessary: between people and their places, between social justice and environmental protection, and between biophysical and social priorities. For example, provision of environmental flows has to balance such priorities in the interests of long-term sustainability (Hassall and Associates et al. 2003). These tensions are compounded by short-term policy shifts as governments—or even individual ministers or bureaucrats—come and go, making the matching of what is physically possible with what is socially and politically acceptable a test for all concerned. Engagement with this challenge in decision-making processes, which range from command-and-control perspectives to authentic participatory comanagement, itself constitutes a major source of process uncertainty.

Sources of Management (Institutional) Uncertainty

The level of uncertainty in scientific data, social perceptions, the political climate, and the risks involved for stakeholders, determine both how research is undertaken and how information is used in management. It is very important to distinguish the inherent and intrinsic uncertainty in both biophysical and social systems (table 14.1) from the ambiguity and confusion resulting from misinformation or mismanagement. While concern has been expressed about the proliferation of management agencies, each with its own agendas and power plays, inefficiencies of overarching bodies are equally evident in a breakdown of direction and communication. Nor should inherent uncertainty be equated with the confusion, unrealistic expectations, and lack of information that often accompanies periods of institutional transition, affecting not just employees but also the local/regional community, and prospectively impacting upon the environment itself (Morrison et al. 2004; Hillman and Howitt, forthcoming). Associated loss of trust, goodwill, and continuity, while more intangible than a loss of financial or human resources, is still a significant barrier to equitable and effective natural resource management (Lachapelle et al. 2003). The costs and benefits of institutional change and the resultant regimes are distributed unevenly and potentially unjustly with little recognition of the resultant dislocation and disruption. A random generation of winners and losers from institutional change is no foundation for success in the new regime. A key challenge in the face of this uncertainty in institutional arrangements lies in avoiding a managerial impasse of similar scope to the "tragedy of the commons," whereby responsibility for the process of environmental repair falls through the cracks between governments, corporations, and individual actions.

In the face of social and biophysical diversity and variability, a placeless universalism—whether articulated through generic principles, models, or training programs—only serves to increase the sense of mistrust and irrelevance. River management institutions have generally been set up to deal with some hypothetical biophysical and social norm, compromising their capacity to manage for diversity and variability. Recognition of surprise and natural variability in institutional planning and decision-making is the basis of both ecosystem-centered and adaptive management approaches to river rehabilitation (see chapter 15). In most cases, variability and diversity are products both of nature and of human activity. It is sometimes hard to separate the two, but one implicit or explicit goal has often been to reduce this variability in the interests of notional development for human needs (Powell 2000). Managing against variability not only induces ecological damage, but also promotes social and institutional conflict. Ultimately, institutions that try to manage against variability end up being themselves managed by such variability, as flood, drought, erosion, and siltation dictate the form, pace, and place of rehabilitation work within reactive management programs (Mance et al. 2002; Hillman 2006).

Notwithstanding these concerns, effective, and equitable environmental decision making is contingent upon both sound and useful scientific information and a coherent and workable set of decision rules to set priorities and allocate resources (Rogers 2003). This requires the integration of biophysical information and assessments with sociocultural characteristics. This has become something of a holy grail in natural resource management (Hassall and Associates et al. 2003; Natural Resources Commission 2005). Within institutional structures, a number of factors combine to create uncertainties in knowledge sharing and the effective use of scientific information. Substantial differences occur between stakeholders as to the relative importance of scientific rigor, of ownership through transparent processes of information collection, or of producing outcome-oriented information with immediate application irrespective of the process itself. It is clearly a difficult task to present complex information to stakeholders in a digestible form without oversimplifying the issues. Adopting a lowest-common-denominator approach runs the risk of alienating some stakeholders or emphasising divisions between those with local knowledge and those with a broader theoretical background. In this case, how much do stakeholders need to know? Perhaps a bottom line is Mitchell's view (1998) that effective adaptive management requires at least an awareness of which part of the knowledge base to apply to the question at hand.

Inherent and institutional uncertainties of river management also present the ongoing risk that some problems, and indeed places, will be placed in the "too hard" basket. A more general criticism suggests that government-initiated participatory processes focus on issues that are most likely to be solved, rather than those of most concern to the participants (Sarkissian et al. 1997). From this perspective, the process may be dominated by a "solution in search of a problem," which appears attractive to some participants and, more important, is within their reach (De Bruijn and Ten Heuvelhof 1999, 182). In this instance the pressure is on agencies to achieve some gains on the ground rather than become bogged down in intractable quests for long-term solutions. This contrasts with the pressure on community stakeholders to get something done about the latter.

In the next section, the implications of these sources of uncertainty are considered through an examination of the characteristics and application of biophysical assessment tools in river management. While promising much as a basis for decision making, these tools often conceal ethical and political choices and foster a false sense of institutional assurance and security. Alternatively, the acknowledgment of a degree of uncertainty may generate a lack of confidence and allow interest groups to "spin the facts" by selective application (Moss 2007, 5). However, this also makes them a valuable lens through which to consider how biophysical, social, and managerial uncertainties are bound up in river management.

The Assessment of Condition in River Management: Characteristics and Uncertainty

Uncertainty in science lies not just in the attainment of empirical knowledge, but also in the integration of such knowledge for meaningful application to the real world. Approaches to data gathering and knowledge generation strive to reduce the bounds of uncertainty in decision making. In this chapter, we use the phrase *biophysical assessment tools* (BATs) as the generic term to refer to a range of techniques employed to capture information on the condition of a river system, to communicate this understanding to stakeholders, and to use these insights to guide management (chapter 7). The role of such tools has frequently been ignored or downplayed in the discourse of river management, despite their centrality to environmental decision making and their key role in catchment-scale approaches to stream rehabilitation (Dunn 2004). Moreover, historical and community research indicates that the absence of, or disagreement over the usefulness of, such tools can lead to the perception or reality of favoritism, ad hoc priority-setting, and fragmented planning (Hillman 2006; Regan et al. 2006). This problem may be particularly severe in river management, where capturing variability and complexity is a major challenge (Hilden 2000; Hillman 2005).

The diverse characteristics of biophysical assessment tools contribute to difficulties of comparison, reliability, and validity that reflect inherent, structural, and methodological sources of biophysical uncertainty and contribute to the management uncertainty discussed in this chapter. These characteristics are framed in the context of scientific, community, and management norms and expectations. Moreover, outputs of assessment techniques are applied (or not) in the face of contested social understandings and interests and through geographically and historically framed institutional structures. This places BATs at the nexus of uncertainty in river management. Hence, analysis of the characteristics and use of such tools provides a means to examine the relationship between biophysical, social, and management dimensions. In this section, we analyze the nature and use of BATs in relation to three key characteristics—scale, indicators, and methodology. Finally, we consider their application in environmental decision making.

Scale

The variety of scales at which ecological assessments are performed raises questions of representation, inclusion, and exclusion, not least in attempting (usually) to "scale-up" reach- or site-based assessment for use in catchment or regional decision making and priority setting. Some techniques target specific ecological communities such as macroinvertebrates or riparian vegetation across a range of spatial scales, while others are scale-specific. Managers need to consider these trade-offs. While using a particular scale provides a level of scientific consistency, there may be broader community concerns about the arbitrary choice of a particular unit of study. This is particularly the case where biophysical characteristics used to determine homogeneity do not coincide with social perceptions or socioeconomic units. This may in turn create uncertainty over the appropriate scope of management intervention, as suggested by Bond and Lake (2003, 198): "The spatial extent of restoration is rarely set from the perspective of target species or communities, instead being driven more often by human perceptions of important scales, or by issues of economic and social convenience."

However, if the choice of scales can be made on a holistic basis, aiming for biophysical rigor and representativeness and matching social understanding expectations, opportunities for community engagement and local input are enhanced, especially in relation to unique, special, or remnant features. This choice in turn requires institutional flexibility and transdisciplinary cooperation.

Indicators

Applications of differing indicator schemes for the assessment of river health typically create an inconsistent and patchy mosaic of information (Fairweather 1999). Factors such as precision, richness, usefulness, and connectivity influence the selection of indicators. In some cases, indicators are target-specific and/or driven pragmatically by available data (Abell et al. 2002). Other techniques attempt to use broader, more integrative indicators such as diversity, naturalness, connectivity, and representativeness. There are important and difficult trade-offs for managers in selecting those characteristics that are readily available, those that are understandable for the broader community, and those that best capture the broader functioning of the riverine ecosystem. Using secondary integrative indicators provides a broader approach but makes the analysis vulnerable to criticism of what may be counted as reliable, valid, or rigorous data or indicators. This may lead to a failure to address primary causes of degradation. For instance, major debates have occurred over the assessment of representativeness and naturalness in SERCON (Boon et al. 2002), while a proposed indicator of connectivity was seen as too "rubbery" for inclusion in an appraisal of sub-catchment health in the Hunter Valley, Australia (Sanders 2003).

Methodology

Assessment methodologies are primarily oriented to the reliability and validity of the data produced, and usually less to social goals such as community involvement and learning, although there are examples of both sets of goals being given priority (Newcastle City Council 2004; Townsend et al. 2004). An emphasis on aerial photos, geographic information systems, and formal surveys provides a degree of certainty, although neither the actual techniques nor the basic data used are widely understood or accessible. In contrast, approaches using subjective but scientifically uncertain assessment alongside primary data are potentially more accessible to local input. In practice, however, these techniques often become the province of "expert panels" due to resource and time constraints (Hukkinen 2003; Fraser and Lepofsky 2004). This limits the potential for critical insights to be made. Rather, prepackaged knowledge is *subsequently* presented for consideration by the broader community and represented as a "bolt-on" in decision support systems (McCann 2001). Such approaches foster the "exclusionary impact of science" (Lane et al. 2004, 112). Ultimately, they lead to hostile managerial and community responses to uncertain predictions rather than an acceptance of such uncertainty and of surprises down the track.

Output and Application

The output and application of the techniques analyzed are also substantively different in practice, varying from colored maps of ecological condition or targeting of particular areas for on-ground works, weighted scores, or indices, or providing a matrix comparing risk/stress against condition. Such outputs are often regarded as a starting point for a more holistic assessment of biophysical and socioeconomic criteria for management decisions. Without such publicly available criteria, the way is open to a wide range of uncertainties and inequities, including ad hoc and nontransparent decisions, favoritism, and allocation of resources to the loudest voices. On the other hand, the uncritical application of such criteria risks imposing decontextualized and abstract principles at the expense of place-based concerns. In addition, the rigid application of decision rules in river management creates sacrificial lambs in parts of the catchment seen as beyond repair or in areas where the incompatibility of rehabilitation of environmental values with socioeconomic costs is too great (Sanders 2003). Such sacrifice may be absolute, where a sub-catchment or reach is written off, or may be relative, where a lower standard of river health is required, as provided in the principle of "derogation" within the European Water Framework Directive (Carter and Howe 2006).

Given their potential to create confusion and conflict in river management, these areas of uncertainty must be addressed at both technical and institutional levels. There are several technical questions that must be addressed in the use of assessment tools.

- How effectively do these tools appraise key drivers, relationships, and limiting factors?
- How representative are tools and their application?
- How integrative are these tools, particularly in promoting a crossdisciplinary perspective?
- Do these tools assist in the development of system-specific conceptual models?

Unfortunately, these technical questions in the use of assessment tools are often conflated with the social and institutional dimensions of river management. Several key points must be addressed here, including:

- Appraisal of whether the selected approach considers "the things that matter";
- The potential socioeconomic costs of uncertainty, such as a "wasted" environmental flow (Stewardson and Rutherfurd 2005); and
- Whether the tool is accessible and relevant to the broader community.

Underpinning these points is the central procedural question of who gets to decide what matters, meaning that transparency and participation are critical criteria. While integrative river science may tell us what is physically possible or likely—for instance, that a river in a moribund state of decline is naturally on the way out—this is no guarantee of the social acceptability of allowing or facilitating that trajectory. The wide range of community views on these issues point to the importance of negotiating the use of indicators and providing a rationale that has meaning across sections of the population (Nancarrow et al. 2002; Ferreyra and Beard 2007). Ultimately, a key question is whether the inevitable trade-offs in developing indicators reflect a physically and socially framed approach to assessment and decision making, or are merely indicative of disciplinary choice and managerial convenience. The latter are more likely to promote uncertainty and disagreement through failure to address the diversity and variability of particular localities.

Uncertainty and Sustainability

For river managers dealing with the range of uncertainties in research, assessment, and decision making, the sustainability principles of precaution and equity are of fundamental importance, as highlighted below.

Precaution

Uncertainty over the scientific knowledge or the outputs of assessment tools may be seen as a rationale for inaction or alternatively for ill-considered, nonstrategic, and unmonitored management intervention in the political interests of getting something done, preferably within the current electoral cycle. This knee-jerk political and institutional response can be tempered by recognition that acting in the face of uncertainty still requires the best use of available information, anticipation of likely outcomes, and strategic monitoring of the results. A precautionary approach would favor the trialing and testing of new approaches to both mitigate further harm and to facilitate social learning as part of an adaptive management approach (Pahl-Wostl 2002; Tippett et al. 2005). The precautionary principle also promotes the articulation of an ecosystem-based approach that is often missing from technocentric approaches to sustainable development (Bosselmann 1999) and to river management itself. According to Eckersley (2004), an authentic precautionary approach provides a means of representing nonhuman species and future generations who are incapable of participating in contemporary dialogue or decisions that pose them potentially serious risks.

While precaution is often viewed as acting in the face of scientific uncertainty to prevent irreversible environmental damage (Rajeswar 2001), the principle can also be invoked to prevent economic development where there is a serious but unproven risk to ecological values or where assessment of current values is ambiguous—and particularly where the absence of certain or sound science is invoked in favor of such development. This in effect involves a shifting of the burden of proof toward the proponent of a development and away from the need for others to conclusively demonstrate that such action is harmful. Such reverse onus of proof is now applied in the granting of new allocation licences for water extraction in the United Kingdom and as part of the European Habitats Directive. This is particularly significant where ecological communities are facing cumulative impacts from a series of developments (Krieg and Faber 2004).

Equity

The naturally uneven distribution, variability, and biophysical complexity of aquatic ecosystems underlie the importance of developing a just approach to river management. In the absence of certainty over outcomes, a new policy or management plan is more likely to be assessed in terms of judgments about fairness in the decision-making process itself (Nancarrow et al. 2002; Jorgansen et al. 2006). However, it is critical to avoid universal or stereotyped assumptions about what criteria are used in such judgments. In practice these vary between individuals, catchments, and cultures (chapter 8). This in turn requires an acknowledgment of the need to understand community perceptions of a "fair go" within the varying biophysical opportunities and constraints across regions and catchments. Moreover, differing notions of justice and fairness exist both *between* and *within* particular interest groups such as irrigators, recreational users, and conservationists. Cross-cultural differences are also significant in many situations in assessing how equitably an organization allocates resources (Conner 2003). New programs such as the European Water Framework Directive and the Australian National Water Initiative (see chapters 11 and 9, respectively) are predicated upon recognition of these multiple concerns and the need for transparency and input from a wide range of community voices at varying scales (Lane et al. 2004; Connelly and Richardson 2005).

For this reason, attention to the representativeness, integration, and applicability of assessment tools becomes central. If, as we have argued, such tools do embed particular biases and scales, or favor particular ecosystem components, the likelihood of perceived inequity is increased. The potential for such inequity compounds the likelihood that uncertainty will generate real conflict and acts against the promotion of a precautionary adaptive approach. In river management, uncertainty about the distribution of impacts or contested views of river health does not provide a mandate to walk away from issues of justice. Indeed, it is under those very conditions of uncertainty that the perception of just process becomes critical to community acceptance of management actions. Recognizing the hidden ethical components of scientific tools and understanding provides a sound basis to promote considerations of fairness in management decisions.

Living with Uncertainty in the Era of River Repair

According to a survey of river restoration practitioners by Wheaton et al. (2006), the dominant strategies for dealing with perceived uncertainty are constraint (do nothing) or denial (know nothing). In contrast, the perspective outlined in this chapter suggests that scientists and managers have a responsibility to restore uncertainty to river management. Such thinking emphasizes the need for management practices to move beyond the application of deterministic efforts to control or limit the inherent diversity, variability, and complexity of the natural world. Insights into biophysical processes will always contain some doubt—the key shift requires greater capacity to cope with a sense of the unknowable as well as the unknown (see box 14.1). Beyond this biophysical uncertainty, due regard must also be given to social uncertainties, perhaps most simply represented by our capacity to cope with surprise and change. Looking beyond these inevitable biophysical and social uncertainties, avoidable ambiguities and confusion of management processes should themselves be minimized. Due regard must be given to concerns for precaution, justice, and equity, ensuring that strategies for social inclusion entail genuine commitment to hear, respond to, and learn from multiple voices in open and transparent approaches to adaptive management. Figure 14.1 provides the outline of a broad approach to managing for uncertainty based on these principles and some lessons from this and previous chapters. This systems-based approach is intended as a set of generic principles and stages, since providing a prescriptive model or exemplar runs contrary to the key messages of this chapter.

Such recognition promotes robust, proactive frameworks that minimize damage brought about by unforseen circumstances. Fragmented, discipline-bound thinking cannot adequately address the concerns for ecosystem health that are raised by complexity and variability. A different kind of science, addressing a different set of questions, is required, containing crossdisciplinary, cross-scalar knowledge that views rivers within their landscape and societal context. This science considers systems as wholes, thereby providing a richer explanation of process relationships and their inherent uncertainties. These applications move beyond the placelessness and universalism that pervade a generalized approach to modeling in the tradition of positivist science. In contrast, integrative science and connection to place are sound platforms upon which to develop a sustainable precautionary approach to environmental management. This approach, built upon an honest and authentic understanding of reality, provides a sound platform for dialogue in environmental decision making.

Conclusion

The complexity and variability of social-ecological systems, and their governance, ensure that management cannot be based on accurate prediction of adjustments toward a particular future stage or end point. However, acceptance of uncertainty and the potential for surprising outcomes does not

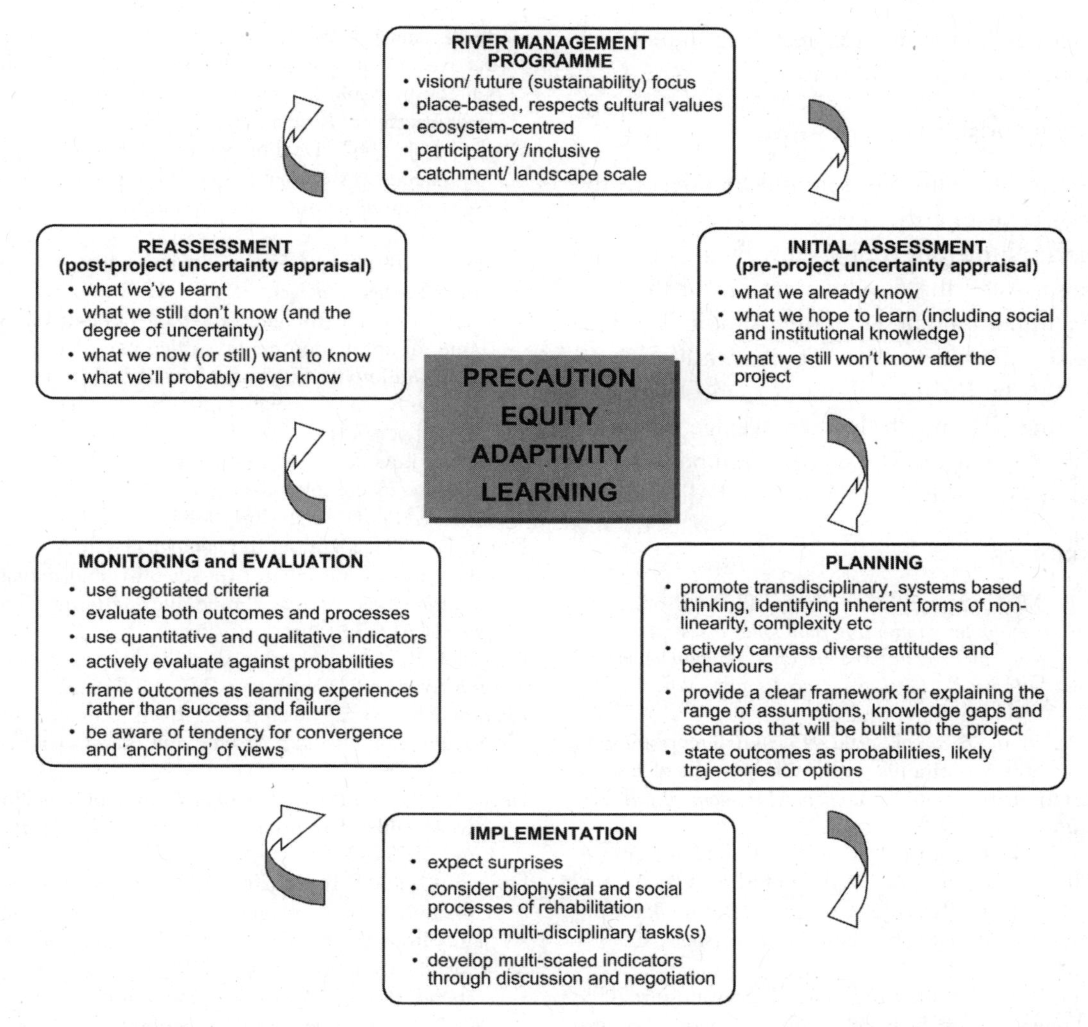

FIGURE 14.1. Overview of stages in an uncertainty-based river management system. The diagram is framed as an environmental management system that operates broadly in a clockwise manner, recognizing (and promoting) iterative reappraisal and reassessment at any stage, between any components.

imply a failure of existing knowledge systems. On the contrary, to be certain about the future is to be wrong. We cannot conquer uncertainty; rather, we have to learn to live with it, recognizing its importance as a determinant of the structure and function of aquatic ecosystems. Expecting surprise, accepting change, and managing risk are core components of approaches to river science and management that embrace uncertainty. Although we cannot anticipate all future events, we can manage and rehabilitate landscapes in ways that allow for uncertainty. Planning for resilience should allow systems a greater ability to deal with and recover from surprise and future change (Hildebrand et al. 2005). Explicit recognition of diverse forms of uncertainty, and their prospective impact upon system performance and environmental management practice, is a key component in the establishment of resilient institutions. Building on this premise, adaptive capacity is the key to coping with vulnerability and risk, restoring uncertainty within river management practice.

Acknowledgments

We thank Malcolm Newson and Stephen Darby for their constructive review comments on this chapter. Numerous discussions with our colleagues assisted the development of our thinking on this important and contentious topic. We wish to make special mention of conversations and comments by Kirstie A. Fryirs, Susan Owen, and Ines Winz. We gratefully acknowledge guidance from river managers who commented on this contribution, especially Eric Hatfield and Allan Raine.

References

Abell, R., M. Thieme, E. Dinerstein, and D. Olson. 2002. *A sourcebook for conducting biological assessments and developing biodiversity visions for ecoregion conservation: Volume II: Freshwater ecoregions.* Washington, DC: World Wildlife Fund.

Bond, N. R., and P. S. Lake. 2003. Local habitat restoration in streams: Constraints on the effectiveness of restoration for stream biota. *Ecological Management and Restoration* 4:193–98.

Boon, P. J., N. T. Holmes, P. S. Maitland, and I. R. Fozzard. 2002. Developing a new version of SERCON (System for Evaluating Rivers for Conservation). *Aquatic Conservation: Marine and Freshwater Ecosystems* 12:439–55.

Borg, D. I., I. D. Rutherfurd, and M. Stewardson. 2005. Merging the fish, the wood and the engineer: The complex physical response to additions of woody debris. In *Proceedings of the 4th Australian Stream Management Conference*, ed. I. D. Rutherfurd, I. Wisniewski, M. J. Askey-Doran, and R. Glazik, 95–102. Hobart, Australia: Department of Primary Industries, Water and Environment.

Bosselmann, K. 1999. Justice and the environment: Building blocks for a theory on ecological justice. In *Environmental justice and market mechanisms,* ed. K. Bosselmann and B. J. Richardson, 30–57. The Hague, Netherlands: Kluwer Law International.

Brooks, A., and T. Cohen. 2005. Wood reintroduction in a multi-objective river rehabilitation project: The Upper Hunter River Rehabilitation Initiative. In *Proceedings of the 4th Australian Stream Management Conference,* ed. I. D. Rutherfurd, I. Wisniewski, M. J. Askey-Doran, and R. Glazik, 134–40. Hobart, Australia: Department of Primary Industries, Water and Environment.

Carter, J., and J. Howe. 2006. The Water Framework Directive and the Strategic Environmental Assessment Directive: Exploring the linkages. *Environmental Impact Assessment Review* 26:287–300.

Clark, M. J. 2002. Dealing with uncertainty: Adaptive approaches to sustainable river management. *Aquatic Conservation: Marine and Freshwater Ecosystems* 12: 347–63.

Clifford, S. 2001. Meandering: Nature, culture and rivers. *Water Science and Technology* 43:17–26.

Connelly, S., and T. Richardson. 2005. Value-driven SEA: Time for an environmental justice perspective? *Environmental Impact Assessment Review* 25:391–409.

Conner, D. S. 2003. Socially appraising justice: A cross-cultural perspective. *Social Justice Research* 16:29–39.

De Bruijn, J. A., and E. F. Ten Heuvelhof. 1999. Scientific expertise in complex decision-making processes. *Science and Public Policy* 26:179–84.

Dunn, H. 2004. Defining the ecological values of rivers: The views of Australian river scientists and managers. *Aquatic Conservation: Marine and Freshwater Ecosystems* 14:413–33.

Eckersley, R. 2004. *The green state: Rethinking democracy and sovereignty.* Cambridge, MA: MIT Press.

Everard, M., and A. Powell. 2002. Rivers as living systems. *Aquatic Conservation: Marine and Freshwater Ecosystems* 12:329–37.

Fairweather, P. G. 1999. State of environment indicators of "river health": Exploring the metaphor. *Freshwater Biology* 41:211–20.

Ferreyra, C., and P. Beard. 2007. Participatory evaluation of collaborative and integrated water management: insights from the field. *Journal of Environmental Planning and Management* 50:271–96.

Feyerabend, P. 1975. *Against method: Outline of an anarchistic theory of knowledge.* London, UK: New Left Books.

Fraser, J., and J. Lepofsky. 2004. The uses of knowledge in neighbourhood revitalization. *Community Development Journal* 39:4–12.

Funtowitz, S. O., and J. R. Ravetz. 1991. A new scientific methodology for global environmental issues. In *Ecological economics: The science and management of sustainability,* ed. R. Costanza, 136–52. New York, NY: Columbia University Press.

Funtowitz, S. O., and J. R. Ravetz. 1993. The emergence of post-normal science. In *Politics and morality: Scientific uncertainty and decision-making,* ed. R. von Schomberg, 85–123. Dordrecht, Netherlands: Kluwer.

Gibbs, L. M. 2006. Valuing water: Variability and the Lake Eyre Basin, central Australia. *Australian Geographer* 37:73–85.

Hassall and Associates, H. Ross, and Mary Maher and Associates. 2003. *Social impact assessment of possible in-*

creased environmental flow allocations to the River Murray System. Canberra, Australia: Murray-Darling Basin Commission Living Murray Initiative.

Higgs, E. 2003. *Nature by design: People, natural process and ecological restoration*. Cambridge, MA: MIT Press.

Hilden, M. 2000. The role of integrating concepts in watershed rehabilitation. *Ecosystem Health* 6:41–50.

Hilderbrand, R. H., A. C. Watts, and A. M. Randle. 2005. The myths of restoration ecology. *Ecology and Society* 10(1): Article 19. http://www.ecologyandsociety.org/vol10/iss1/art19/ (accessed December 14, 2007).

Hillman, M. 2005. Justice in river management: Community perceptions from the Hunter Valley, New South Wales, Australia. *Geographical Research* 43:152–61.

Hillman, M. 2006. Situated justice in environmental decision-making: Lessons from river management in Southeastern Australia. *Geoforum* 37:695–707.

Hillman, M., and R. Howitt. Forthcoming. Institutional change in natural resource management in New South Wales, Australia: Sustaining capacity and justice. *Local Environment*.

Hukkinen, J. 2003. From groundless universalism to grounded generalism: Improving ecological economic indicators of human/ environmental interaction. *Ecological Economics* 44:11–27.

Jackson, S. 2005. Indigenous values and water resource management: A case study from the Northern Territory. *Australasian Journal of Environmental Management* 12:136–46.

Jackson, S. 2006. Compartmentalising culture: The articulation and consideration of indigenous values in water resource management. *Australian Geographer* 37:19–31.

Jorgansen, B. S., G. J. Syme, and B. E. Nancarrow. 2006. The role of uncertainty in the relationship between fairness evaluations and willingness to pay. *Ecological Economics* 56:104–24.

Krieg, E. J., and D. R. Faber. 2004. Not so black and white: Environmental justice and cumulative impact assessments. *Environmental Impact Assessment Review* 20:667–94.

Lachapelle, P. R., and S. F. McCool. 2005. Exploring the concept of "ownership" in natural resource planning. *Society and Natural Resources* 18:279–85.

Lachapelle, P. R., S. F. McCool, and M. A. Patterson. 2003. Barriers to effective natural resource planning in a "messy" world. *Society and Natural Resources* 16:473–90.

Lane, M. B., G. T. McDonald, and T. H. Morrison. 2004. Decentralisation and environmental management in Australia: A comment on the prescription of the Wentworth Group. *Australian Geographical Studies* 42:103–15.

McCann, E. J. 2001. Collaborative vision or urban planning as therapy? The politics of public-private policy making. *Professional Geographer* 53:207–18.

Mance, G., P. J. Raven, and M. E. Bramley. 2002. Integrated river basin management in England and Wales: A policy perspective. *Aquatic Conservation: Marine and Freshwater Ecosystems* 12:339–46.

Mitchell, D. 1998. The interdependence of adaptive management and environmental monitoring. In *Water: Wet or dry: Proceedings of the Water and Wetlands Management Conference. St. Leonards, New South Wales*, 149–54. Sydney, Australia: New South Wales Nature Conservation Council.

Montgomery, D. R., and S. M. Bolton. 2003. Hydrogeomorphic variability and river restoration. In *Strategies for restoring river ecosystems: Sources of variability and uncertainty in natural and managed systems*, ed. R. C. Wissmar and P. A. Bisson, 39–80. Bethesda, MD: American Fisheries Society.

Morrison, T. H., G. T. McDonald, and M. B. Lane. 2004. Integrating natural resource management for better environmental outcomes. *Australian Geographer* 35:243–58.

Moss, R. H. 2007. Improving information for managing an uncertain future climate. *Global Environmental Change* 17:4–7.

Nancarrow, B. E., C. S. Johnston, and G. J. Syme. 2002. *Community perceptions of roles, responsibilities and funding for natural resources management in the Moore catchment*. Perth, Australia: CSIRO Land and Water.

Natural Resources Commission. 2005. *Standard for quality natural resource management*. Sydney, Australia: Natural Resources Commission.

Newcastle City Council. 2004. *Creek assessment instructions*. Unpublished manual.

Newson, M. D., and M. J. Clark. 2008. Uncertainty and the sustainable management of restored rivers. In *Uncertainties in river restoration*, ed. S. Darby and D. Sear. Chichester, UK: Wiley.

Olsson, P., C. Folke, and F. Berkes. 2004. Adaptive comanagement for building resilience in social-ecological systems. *Environmental Management* 34:75–90.

Pahl-Wostl, C. 2002. Towards sustainability in the water sector: The importance of human actors and processes of social learning. *Aquatic Sciences* 64:394–411.

Piégay, H., S. E. Darby, E. Mosselman, and N. Surian. 2005. A review of techniques available for delimiting the erodible river corridor: A sustainable approach to managing bank erosion. *River Research and Applications* 21:773–89.

Pigram, J. J. 2006. *Australia's water resources: From use to management*. Collingwood, Australia: CSIRO Publishing.

Pollack, H. N. 2003. *Uncertain science . . . Uncertain world*. Cambridge, UK: Cambridge University Press.

Powell, J. M. 2000. Snakes and cannons: Water management and the geographical imagination in Australia. In *Environmental history and policy: Still settling Australia*, ed. S. Dovers, 47–73. South Melbourne, Australia: Oxford University Press.

Rajeswar, J. 2001. Conservation ethics versus development: How to obviate the dichotomy? *Sustainable Development* 9:16–23.

Regan, H. M., M. Colyvan, and L. Markovchick-Nicholls. 2006. A formal model for consensus and negotiation in environmental management. *Journal of Environmental Management* 80:167–76.

Rogers, K. 2003. Adopting a heterogeneity paradigm: Implications for management of protected savannas. In *The Kruger experience: Ecology and management of savanna heterogeneity*, ed. J. T. Du Toit, K. Rogers, and H. C. Biggs, 41–58. Washington, DC: Island Press.

Sanders, N. 2003. *Rapid assessment of environmental, social and economic values in the Hunter River catchment*. Sydney, Australia: New South Wales Healthy Rivers Commission and 4Site Co. Pty. Ltd.

Sarewitz, D. 2004. How science makes environmental controversies worse. *Environmental Science and Policy* 7:385–403.

Sarkissian, W., A. Cook, and K. Walsh. 1997. *Community participation in practice*. Western Australia: Institute for Science and Technology Policy, Murdoch University.

Stewardson, M., and I. Rutherfurd. 2005. Quantifying uncertainty in estimating environmental flow requirements. In *Proceedings of the 4th Australian Stream Management Conference*, ed. I. D. Rutherfurd, I. Wisniewski, M. J. Askey-Doran, and R. Glazik, 553–61. Hobart, Australia: Department of Primary Industries, Water and Environment.

Stoffle, R. W., and R. Arnold. 2003. Confronting the angry rock: American Indians' situated risks from radioactivity. *Ethnos* 68:1–20.

Tippett, J., B. Searle, C. Pahl-Wostl, and Y. Rees 2005. Social learning in public participation in river basin management: Early findings from HarmoniCOP European case studies. *Environmental Science and Policy* 8:287–99.

Townsend, C., G. Tipa, L. Teirney, and D. Niyogi. 2004. Development of a tool to facilitate participation of Maori in the management of stream and river health. *EcoHealth* 1:184–95.

Van Asselt, M. B., and J. Rotmans 2002. Uncertainty in integrated assessment modelling: From positivism to pluralism. *Climatic Change* 54:75–105.

Wheaton, J. M., S. E. Darby, and D. A. Sear. 2008. The scope of uncertainties in river restoration. In *River restoration: Managing the uncertainty in restoring physical habitat*, ed. S. E. Darby and D. Sear, 21–39. Chichester, UK: Wiley.

Wheaton, J. M., S. E. Darby, D. A. Sear, and J. A. Milne. 2006. Does scientific conjecture accurately describe restoration practice? Insight from an international river restoration survey. *Area* 38:128–42.

Chapter 15

River Futures

Gary J. Brierley, Kirstie A. Fryirs, and Mick Hillman

> Conservation needs to be built on a foundation of individual awareness of and concern for nature. . . . What we need now is imagination, combined with a burning sense of the value of nature, its importance to human life, and its place in a sustainable economy. Much can be done. It is on our vision, and our energy, that the future of nature depends.
>
> Adams 2003, 9

The chapters in this book have highlighted a major shift in approach to the management of river systems, prompting significant hope for improvement in river health across the globe. Ecosystem values are now firmly placed alongside strategies to meet human needs, in what has been called an era of river repair. Increased recognition of the inherent diversity, variability, and complexity of river systems has prompted moves away from a "one size fits all" approach to river management. Emerging approaches to river management increasingly embrace place-based programs that respect the values and attributes of each river system. Visionary programs strive to enhance the resilience and self-sustaining capacity of social-ecological systems. In this chapter, we discuss measures that can be applied to maximize the effectiveness of management efforts that facilitate the process of river repair. We start with an overview of key findings from the case study chapters of this book that describe management actions to improve river health in differing parts of the world. We then summarize the development and application of integrative scientific insights that have accompanied this transition in management practice. Uptake of scientific guidance is then framed in relation to institutional and socioeconomic constraints. Finally, we consider the relationship between our capacity to enact processes of river repair and the societal and political will to actively utilize this capacity.

The Emerging Process of River Repair

In recent decades there has been a fundamental shift in the way rivers are perceived and managed across the globe. As noted in chapter 1, management approaches increasingly consider ecosystem values, moving beyond the deterministic era of command and control that used engineering approaches to regulate and stabilize channels. Today, concerns for rehabilitation lie at the heart of river management, as illustrated by the various case studies documented in this book. Inevitably, this transition toward incorporation of ecosystem values has occurred at different times in different places. In general terms, however, appreciation of river diversity, dynamic, and complexity has enhanced our understanding of controls upon river condition and health (see chapter 7).

As we enter the era of river repair, different countries and continents have placed different emphases on the importance of conservation and rehabilitation activities relative to other river management concerns (e.g., water supply, navigation, sediment disasters). This reflects different pressures that operate in highly varied landscape and social settings. Each case study is embedded within a unique societal context with a specific history of human-disturbance and economic development. This in turn is overlayed on different biophysical settings and histories of river adjustment. The importance of place and environmental history cannot be understated when drawing comparisons between management approaches adopted in various parts of the world. While some examples documented in the case studies are further along the trajectory of river repair, all have adopted an approach that is heavily reliant on the use of best available crossdisciplinary science. Notwithstanding this broad shift, an ethos of control and regulation still pervades many practices and institutions. Past approaches and attitudes are not easily shifted, and generational change may be required before genuine commitment to the process of river repair is initiated. Rather than dwell on these examples, this book has presented case studies that demonstrate a future-focus and effective use of science in river management.

Contemporary Australian practice shows an increase in awareness of the damage done to river systems since European colonial settlement in the late eighteenth century (chapter 9). With low population density across a vast island continent, river management regimes that are heavily dependent upon volunteer involvement have developed within a bottom-up participatory approach and a conservation-first ethos. River science has guided management practice for several decades. Increasingly, such applications have a crossdisciplinary and catchment-scale focus.

Rivers in the United States have experienced a set of postcolonial pressures similar to Australia's (chapter 10). In this instance, however, a regulatory, top-down approach framed the development of water resources. For example, the U.S. Army Corps of Engineers and the Bureau of Reclamation established a comprehensive program of river regulation, marked by dam construction for all major river systems in the western States. However, the shift toward the era of river repair has prompted the initiation of numerous high-profile, large-scale river rehabilitation initiatives, accompanied by an emerging ethos of stewardship.

European river management has to deal with a long history of human disturbance (chapter 11). In this old-world society there is a long tradition of regulation through channelization and stabilization. In recent years, however, ecosystem values at the heart of the European Water Framework Directive (WFD) have provided a strong policy framework that guides the process of river repair, in many cases spanning national boundaries. Increasing emphasis upon "space to move" or "erodible corridor" concepts apply principles of hydrosystem and/or geoecology science in the process of river rehabilitation. This case study demonstrates how a strong policy framework with a coherent focus can promote commitment to an improvement in river condition across a range of societies.

For many hundreds of years, Japanese river management practice has been driven by disaster management and minimization of risk in a highly sensitive, densely populated, tectonically active landscape setting (chapter 12). Extensive reforestation programs and Sabo dam construction have been applied to control sediment transfer in river systems in a highly regulated, top-down, engineering-dominated approach to river management. In recent years, however, the incorporation of geoecological values has modified this traditional engineering approach and control ethos.

Recent river management practice in South Africa has been framed around a strong emphasis on water law, reflecting land-use pressures, population growth, rapid political change, and the resultant concerns for equitable access to water resources (chapter 13). A strong emphasis on crossdisciplinary science in environmental flow allocation and adaptive management practices through the water reform process has sought to minimize environmental damage.

These case studies demonstrate that transitions in river management practice *have* occurred and an era of river repair *has* begun across the globe. Crossdisciplinary science, with an emphasis upon ecosystem health, has played a key role in this transition.

Pressures and demands for conservation and/or rehabilitation activities vary in differing parts of the world. The case studies highlight the societal, cultural, historical, and environmental differences in old-world societies (Europe, Japan) relative to new-world (i.e., postcolonial) societies (United States, South Africa, Australia). These histories have shaped the nature and extent of system resilience, responses to disturbance, and prospects for recovery. Underpinning these differences are variable relationships to the environment—our capacity to live with risks, leave river systems alone, and allow them to recover naturally. Increasingly, ecosystem science and values are integrated in giving rivers a voice for themselves, as highlighted in the following section.

Using Coherent Scientific Information to Guide the Process of River Repair

Ultimately, ecosystem values underpin the process of river repair. Our efforts to capture the diversity, variability, and complexity of river systems as a template for management activities can be expressed in terms of the effectiveness with which we are able to "think like an ecosystem" (Leopold 1949). Multiple crossdisciplinary components and complex interactions make up this constantly changing template, as outlined in part II of this book. Chapter 4 asserted the primacy of the catchment scale as a basis for management actions, framing the spatial organization of river systems in terms of gradient-induced biophysical relationships. Framing rivers in their landscape context situates local-scale phenomena in context of their interactions with adjacent compartments. While characteristic biophysical patterns may be evident from catchment to catchment within a given ecoregion, shaped largely by geological, climatic, and land use considerations, it is the catchment-specific nature of these interactions that should underpin management practice. Principles outlined in chapter 4 also highlight the importance of landscape heterogeneity as a platform from which to appraise relationships between habitat availability and habitat viability. Attention is drawn to the importance of naturalness as a basis to consider the complexity of any given system, noting that some systems are naturally simple (i.e., relatively homogenous). In general terms, however, protection of biodiversity can be promoted by increasing the range of available habitat, striving to increase the resilience of the system in light of prevailing and future pressures and stressors.

Spatial considerations in chapter 4 are complemented by analysis of temporal considerations in chapter 5. Focus is placed on the need to frame river rehabilitation practices within the evolutionary trajectory of a given river system. The effectiveness of management activities is likely to be enhanced if they work with change and are able to communicate different prospective scenarios. To do this, we must understand the behavioral regime and sensitivity of a river system, appreciating how the system adjusts to disturbance events in the short term, and how these adjustments fit within the longer-term evolutionary sequence.

Prospects for success in river rehabilitation activities are markedly enhanced when they address the causes rather than the symptoms of environmental degradation. To achieve this, we must understand key controls on ecosystem functionality for any given system (chapter 6) and undertake meaningful and integrative assessments of river condition (chapter 7). From such insights, threshold conditions that bring about change in system state can be identified. Based on these insights, conceptual models of ecosystem performance can be developed, enabling foresighting exercises to "vision" system adjustments to disturbances such as climate change or management intervention (chapters 2 and 5). The question that must be addressed in such exercises is: How will we know, before it's too late, that we are approaching the limits to ecosystem viability? Such exercises provide a proactive platform for adaptive management practice.

Managing the Process of River Repair

It is one thing for scientists to develop insights into ecosystem structure and function, and quite another to see them used effectively in management practice. Scientists have a responsibility to adequately explain complex and uncertain ideas to stakeholders, decision makers, managers, and the public. Practical uptake of scientific insights in the process of river repair is likely to be maximized when information relates to a particular place and practitioners own that knowledge. In a sense, our relationship to place provides a basis to link both biophysical and social connectivity, as outlined in chapter 8.

We now have excellent understanding of the types of initiatives that must be taken to kick-start and sustain rehabilitation efforts. However, effective rehabilitation practice extends beyond technical competence and efficiency to embrace a range of social, cultural, political, moral, and aesthetic qualities. Meaningful engagement among managers, stakeholders, researchers, and the community is required to facilitate this process. Collectively, we must ask ourselves what is manageable, and then prioritize where to place effort and limited resources. An adaptive management framework provides the most appropriate basis to facilitate these discussions and applications, as discussed below.

What Is Manageable?

The success of river repair reflects the combined outcome of two primary considerations: our capacity to bring about improvements in river health on the one hand, and societal/political will to undertake and sustain actions on the other. Many historical factors have brought about irreversible changes to river condition by altering ecosystem interactions and instigating species extinctions. We live with the legacy of past actions that have modified ground cover and prevailing biophysical fluxes—that is, water, sediment, and nutrient movement. Similarly, the contemporary character and behavior of river systems is fashioned by the legacy of direct human impacts, such as dam and weir construction, channelization of rivers, and even rehabilitation projects themselves.

Another factor that is beyond our immediate control is population growth and distribution. Population density, relative to the life-supporting capacity of any given region, exerts a primary influence upon environmental health. Although these trends change over time and are largely associated with socioeconomic development, we live with the legacy of past trends—relatively stable populations in the developed world but significant growth in many developing countries. Urbanization and shifts in agricultural practices and land use, combined with the uneven impacts of climate change and sea-level rise, ensure that the degree and nature of human disturbance to the natural environment is forever changing. While global environmental changes will make some areas increasingly marginal, others will become more viable and desirable. Access to water will play a major role in development prospects. Alternatively, in some instances forced migrations may be instigated as part of adaptation.

While it is essential to understand how those global, and to a large extent uncontrollable, factors influence the contemporary health of river systems, it is even more important to consider those issues that we are able to affect through our actions. These considerations fashion how we go about the process of river repair, if we decide to do so. The effectiveness of this process is heavily contingent upon societal engagement and attitudes toward the environment. Major ecological rehabilitation will not be achieved unless society approves the goals and objectives (Cairns 2000). Unfortunately, at this stage, our efforts to move beyond reactive, piecemeal responses are hindered by social, economic, and institutional impediments. The key step in moving beyond this impasse lies in reformation of societal attitude and political will.

Inevitably, actions to enhance river repair reflect economic capacity and the social priority placed upon these activities. We also need to recognize what we cannot afford *not* to do. In many places, proactive measures are urgently required to maintain biodiversity through conservation measures and improve river health through rehabilitation initiatives before inherent values are lost or compromised. Working with stakeholders to minimize the negative consequences of conflicting land uses is essential. As the case studies in this book illustrate, scientific insights can effectively guide our ability to apply strategic initiatives to target vulnerabilities in the system, highlighting those issues that we are most able to address with a given resource over a given timeframe. While practitioners often expect a quick fix, this guidance often indicates that this is not possible. Just as important, benefits will not persist unless their distribution is equitable and society has sufficient regard for the rehabilitated ecosystem to protect its integrity. In practice these two criteria are strongly interdependent—neither technically feasible goals nor scientifically valid goals are achievable in the absence of societal acceptance.

Visioning and Prioritization of Conservation and Rehabilitation Initiatives

As noted in chapter 2, the articulation of a clearly defined vision is a fundamental underpinning of environmental management through which a shared sense of purpose is accompanied by a collective sense of responsibility. The history of river management is littered with examples of reactive and piecemeal actions characterized by limited success and wasteful expenditure. As river rehabilitation is now a multibillion dollar industry, there are increasing demands from society and policy makers to show that management initiatives are working (Wissmar and Bisson 2003). Unfortunately, countless on-the-ground actions have poorly articulated goals, no commitment to maintenance and/or auditing, and little sense of how these actions fit within a broader-scale, longer-term vision (Bernhardt et al. 2005). Competing or conflicting goals dilute or negate the effectiveness of our actions, wasting time and money, eroding goodwill, and limiting ongoing engagement with subsequent actions.

Such lack of coherent planning limits our capacity to achieve the bigger-picture outcomes that underlie long-term sustainability and biodiversity management. A logical, socially acceptable, and affordable plan of action is required, embedding short-to-medium-term measurable target conditions within longer-term goals. Among the many issues to be addressed are agreement on what the problem is, knowledge of where and when to intervene, and assessment of the measures that need to be applied to improve the situation. Therefore, key considerations to be addressed within the visioning process include the environmental values that we wish to protect, the values that we should aim to enhance, and the trends that we should endeavor to promote or reverse.

However harsh it may appear at first sight, justice in river management is not about achieving equality of outcome or even about equal distribution of resources. Such thinking flies in the face of biophysical diversity. Rather, targeted activities promote conservation and rehabilitation successes in selected areas, while other areas may be subjected to environmental deterioration. Tensions and dilemmas may arise between conflicting ideals and priorities in rehabilitation works (Hillman and Brierley 2005). First, preventing is better, and cheaper, than curing. Urgent care for charismatic species must be balanced with longer-term preventative measures that protect or enhance ecosystems. The question needs to be asked: Is it more appropriate to spend huge amounts on saving the last remaining individuals of a species on the brink of extinction, or to invest in protecting habitat that is used by many other species? Single-interest rehabilitation projects are likely to be counterproductive to the management of the system as a whole. A second dilemma is in determining whether it is better to invest in purchasing and managing small patches of good quality habitat or in rehabilitating larger tracts of currently degraded habitat. While it makes good sense to focus efforts on high-profile problem sites in social and community terms through creating local jobs and fostering social capital, spending a fortune on sites where the prospects for sustained environmental success are limited makes little sense in economic terms. A third dilemma focuses on the flexibility and adaptability of management systems to address changes in what are highly dynamic and uncertain systems. Programs should be sufficiently adaptive and flexible (i.e., nonprescriptive) such that they can respond to new insights and/or opportunities as they arise. However, such perspectives should *not* be misinterpreted as excuses to deviate from longer-terms goals should anything untoward happen along the way—the way we manage crisis situations is a key determinant of prospects for sustained success.

With these dilemmas in mind, prioritization frameworks must adopt a spatially integrated approach to river repair, with an appropriate balance of conservation and rehabilitation initiatives. Often the acute cases are more immediate and appealing to the local community and hence attract both more demands for action and more funding and support for that action. Prompt attainment of such tangible outcomes through short-term rehabilitation measures will engender sustained support for the process of river repair. However, these small-scale initiatives should be framed explicitly in the context of longer-term, ecosystem-scale visions (see chapters 2 and 8). Prioritizing rehabilitation projects with a conservation-first approach is the most environmentally efficient means to allocate resources (Brierley and Fryirs 2005). A conservation-first approach promotes measures that protect relatively intact systems within any given ecoregion. In some instances, this may simply require these areas to be left alone rather than subjecting them to development pressures. Building out from these remnants often presents the most logical biophysical approach to enhance ecosystem recovery.

A first priority in management plans should be to provide a clear statement of conservation goals (Hooper et al. 2005; Abell et al. 2007). Although habitat protection for rare, threatened, and endangered species remains a key focus for conservation programs, it is now widely recognized that there are too many species to save them one at a time. Conservation measures alone are unable to protect ecosystem functions, services, and habitat for many species (Hilderbrand et al. 2005). Hence, emphasis has now shifted toward an ecosystem approach to environmental management. In these applications, measures that address concerns for habitat loss and fragmentation are balanced by broader-scale measures that support ecosystem functionality, striving to enhance the resilience of the system and its prospects for recovery. Such applications give due regard to both the availability and the viability of habitat. In designing rehabilitation initiatives, target conditions for any given reach should be framed in terms of an appropriate reference, recognizing that flexibility is required as inherently dynamic systems may adopt multiple future scenarios and trajectories. Indeed, there are inherent dangers in striving to lock any ecosystem into a particular state.

Hobbs (2003) applies the medical triage concept to ecosystems and landscapes, considering that ecosystems, like humans, have the capacity to recover from stress and disturbance, and there are differing degrees of recoverability depending on the extent of the "injury" and the resilience of the "patient." Conservation managers face the same decisions about whether to invest heavily in acute emergency treatment versus longer-term and preventative care. Like doctors lacking sure diagnoses, managers must also consider uncertainty and lack of information; but awareness that we cannot be certain of the future should not inhibit our preparedness to act *now*, recognizing that precautionary actions may be the most appropriate path to follow. Strategic, preventative initiatives must be sufficiently flexible to be adapted when the system behaves (or responds) in unexpected ways. This requires the adoption of an adaptive management approach.

Adaptive Management

Adaptive management of environmental systems is all about effective and sustainable governance of conservation and rehabilitation activities, recognizing the need for flexibility to cope meaningfully with ongoing social and environmental changes (Walters 1997; Rogers 2003). In theory, carefully nurtured partnerships between science, management, and the community promote pragmatic, goal-orientated management programs that learn and grow in each cycle (Rogers 1998). Such strategies are proactive, retaining sufficient flexibility to enable social-ecological systems to bend rather than break when subjected to external pressures. Rather than attempting to hold the system in its existing (or a particular) state, working relationships address surprises, conflicts, and risks as they arise. Adaptive management programs recognize that we don't always know what the consequences of our actions are going to be and we aren't always going to get things right. However, if appropriate experimentation and documentation procedures are in place, we should be able to learn from experience, communicating findings to others in a process of learning by doing. Such strategies recognize explicitly that managers can only make contingency plans for extreme events, not prevent or even necessarily anticipate them, and that surprising outcomes are inevitable. Given its timescale, complexity, and transdisciplinary nature, coping with uncertainty is a goal of this process, rather than attempting to remove it or using it as an excuse for inaction (Clark 2002; see chapter 14). Adoption of an adaptive management approach should not be viewed as an alternative to predictive thinking. Rather, activities should be framed within a specific plan for learning, moving beyond trial-and-error procedures.

In this section, we consider various issues that determine the effectiveness of adaptive management programs and the capacity to cope and work with change. We start by considering the importance of governance structures with which to implement environmental management practices. This is followed by a discussion of approaches to monitoring and information management. Finally, we consider the importance of large-scale rehabilitation experiments and demonstration sites as tools with which to guide adaptive management processes.

Governance and Adaptive Management

Contrasting opportunities and limitations are presented by top-down and bottom-up approaches to environmental governance, reflecting dogmatic, authoritarian regimes on one hand, and the challenge of participatory politics on the other hand. Ultimately, a meaningful middle ground must be sought, linking visionary and substantive policy and legal frameworks enacted by effective leaders with genuine on-the-ground engagement and commitment to the process of river repair by the community. Key issues of equity, justice, ownership, and responsibility underpin the prospects for sustained success in the design and implementation of environmental management practices. Effective management of ecosystems therefore requires flexible governance that is responsive to social and environmental change (Olsson et al. 2004). Effective institutions provide genuine leadership and guidance in managing for sustainability, fostering supportive, equitable, and productive relationships among managers, stakeholders, researchers, and the community. Breaking down social and institutional constraints that arise from agency structure/mentality, personality clashes, or divergent agendas is required.

The process of river repair requires patience and ongoing commitment. Expectations of overnight transformation are generally unrealistic. In many instances, substantive success will only be achieved over long timeframes—timeframes that are well beyond those of political cycles. Effective and sustained leadership is required to guide this process. Such leadership goes beyond the "strong man," error-free steadfast political stereotype to encompass a genuine integrative capacity, humility, and flexibility in the face of experience, surprise, and mistakes.

Ownership of what we seek to achieve in river management, and the way we go about it, is vital to sustained success. Participatory frameworks promote stewardship of river systems, enhancing the view of environmental management as a collective responsibility. Viewed in this light, the process of ecological rehabilitation has inherent democratic capacity (Higgs 2003). The combination of value to nature and value to community gives rehabilitation activities the capacity to enhance participatory politics. Appropriate mechanisms must be set in place to adequately resource such processes, fostering community leaders while minimizing the burn-out of champions, with appropriate arrangements for succession planning.

A corollary of this is that sustained success in rehabilitation efforts will only be achieved if it is accompanied by the development of more secure career structures in this field. Greater use could be made of apprenticeships or traineeships to deliver on-the-ground training with industry or agency practitioners. Efforts to enhance professionalism among river practitioners may require the accreditation of individuals and institutions. While volunteer commitment will remain an integral part of river rehabilitation, this should not be at the expense of adequately paid and secure jobs in the field.

In terms of governance structures, it is also important to consider the power balance that determines institutional resilience. As noted by Nietzsche (1909–1913, section 8, number 466, in Levy 1964): "The overthrow of opinions is not immediately followed by the overthrow of institutions; on the contrary the new opinions dwell for a long time in the desolate and haunted house of their predecessors, and conserve it even for want of a habitation." Challenges lie in promoting visionary, just, and appropriate resources institutions with which to promote and sustain the process of river repair.

Monitoring, Information Management, and Knowledge Transfer

When viewed as an experiment as part of adaptive management, failure can be just as valuable to the process as success, provided we learn from it (Kondolf 1995). To date, our efforts to monitor and audit system responses to management treatments have been paltry, compromising our capacity to learn what works where, and why (Bernhardt et al. 2005). Appropriate processes must be set in place to frame what we are trying to achieve (visions and goals), implement measures effectively, and appraise their effectiveness. Monitoring requirements must be realistic and achievable within organizational constraints. Appropriate, measurable parameters must be selected. Pre-project data and control sites are required. Feedback should be used to reframe goals, priorities, and the needs of monitoring processes themselves. Documentation and communication must be maintained throughout the process, setting aside adequate time for analysis and assessment.

However, caution and care are required in appraising the effectiveness of our actions. In many instances, perceived benefits may simply reflect what we measure and where we measure things (chapter 14). For example, improvements in river condition at particular treatment sites may come about, in part, at the expense of deteriorating river conditions elsewhere (e.g., Brooks et al. 2006).

Appropriately informed and responsive governance structures are thus required to act upon findings from monitoring exercises while sustaining and building upon successes already achieved. Capacity for successful implementation of adaptive management processes is most forcefully tested during times of stress or crisis, such as instances of major flooding. Ironically, these are the very instances that reactive and politically driven management strategies seek to address. In many ways, threats to long-term ecosystem health are most pronounced when management responses seek to protect human values and assets from natural variability. Ideally, management structures are sufficiently robust to hold the course, enacting initiatives and contingencies that have been designed for these very instances (Rogers 2006).

While we can intellectualize about all management being an experiment, the reality is that there are often adverse consequences if the experiment goes wrong and the agency or individual involved is blamed—not least a loss of confidence by the community itself. This requires that we do not create false expectations and recognize explicitly that what we do now may not be considered "right" in the future. Our mistakes will only continue to be a problem if we keep repeating them (Hobbs 2003). In particular, appropriate questions must be asked about the representativeness and transferability of understanding derived elsewhere.

Use of Large-Scale Rehabilitation Experiments and Demonstration Sites as Tools for Adaptive Management

Rehabilitation initiatives can be viewed as experiments in adaptive management. Working collectively with managers in the design and application of large, team-based, crossdisciplinary experiments provides an opportunity to test the efficacy of conceptual models and improve methods, technology, and approaches to enquiry. Such projects can be used to showcase what can be achieved through concerted and sustained efforts. Prospectively, these experiments promote rapid learning by testing a range of techniques, expanding the rehabilitation toolbox while learning about the system (Shields et al. 2003). By extension, demonstration sites provide a vehicle to highlight the design, implementation, and maintenance of on-the-ground treatments, reforming practice by promoting policy change and developing networks with the potential to change attitudes within both the government and community.

Site selection is a key issue in the design and implementation of large-scale, crossdisciplinary rehabilitation experiments and demonstration sites. The resource needs for such initiatives are often very high, requiring carefully crafted and targeted activities. Given the near-universal need for river repair initiatives, strategic decisions must be made about the representativeness of selected sites and the selection of strategically located and topical applications. Should we prioritize high-profile sites where the requirements to enhance ecosystem performance are both demanding and risky, or less costly sites that already show signs of recovery such that the prospects for rehabilitation success are enhanced? Other factors that influence this decision include prospects for local ownership of actions taken, maintenance of outcomes, and mechanisms to facilitate transferability of lessons learned.

Conclusion

We live in an age of immediacy, crisis, and alarmism. Every day, the media draws attention to threats (and occasionally opportunities) brought about by global changes such as security concerns, economic prospects, climate change, natural hazards, demographic trends, and land- and water-use pressures. With each passing day, the human imprint upon planet Earth increases. As the global population continues to increase, and our technological capacity appears forever greater, we exert an increasingly profound influence upon the world around us. Unsustainable lifestyles and levels of consumption threaten the viability of ecosystems and human survival itself. Given the ever-increasing interdependence between ecosystems and human communities, their sustainability must be managed simultaneously as a transdisciplinary endeavor (Keough and Blahna 2006). Environmental security is a prerequisite for human survival.

Adaptive management frameworks provide the most appropriate basis with which to facilitate the process of river repair. What we need now is the commitment and social will to adequately resource and sustain such initiatives, remembering that healthy river futures require ongoing maintenance. Approaches to environmental management, and the priorities that we place on these measures, reflect societal choice and opportunity. Choice reflects the values that we care about—what we seek to achieve in our relationship to the natural world and to each other. The choices we make reflect our needs, values, and perceptions, as we frame spiritual, recreational, aesthetic, and cultural connections in relation to consumptive needs. Opportunity refers to what it is possible to achieve through conservation and rehabilitation efforts at any given locality. Ultimately, the process of river repair entails the negotiation of societal priorities. Unfortunately, confronting images of degraded rivers and aquatic ecosystems across most of the planet reminds us of the choices we have made to date as a society.

Efforts to "think like an ecosystem" promote a perspective in which the environment is not "out there"—rather, we're within it. Environmental degradation and loss of biodiversity affect all of us, though some are affected more than others. The longer our unsustainable practices continue, the greater the extent and degree of environmental degradation and biodiversity loss. Healthy and sustainable river systems will not be achieved by passively waiting for someone else to fix a problem. Ultimately, the process of river repair is about the actions we take. The future of rivers is in our hands.

References

Abell, R., J. D. Allan, and B. Lehner. 2007. Unlocking the potential of protected areas for freshwaters. *Biological Conservation* 134:48–63.

Adams, W. M. 2003. *Future nature: A vision for conservation*. London, UK: Earthscan.

Bernhardt, E. S., M. A. Palmer, J. D. Allan, G. Alexander, K. Barnas, S. Brooks, J. Carr, S. Clayton, C. Dahm, J. Follstad-Shah, D. Galat, S. Gloss, P. Goodwin, D. Hart, B. Hassett, R. Jenkinson, S. Katz, G. M. Kondolf, P. S. Lake, R. Lave, J. L. Meyer, T. K. O'Donnell, L. Pagano, B. Powell, and E. Sudduth. 2005. Synthesizing U.S. river restoration efforts. *Science* 308:636–37.

Brierley, G. J., and K. A. Fryirs. 2005. *Geomorphology and river management: Applications of the River Styles framework*. Oxford, UK: Blackwell.

Brooks, A. P., T. Howell, T. B. Abbe, and A. H. Arthington. 2006. Confronting hysteresis: Wood based river rehabilitation in highly altered riverine landscapes of southeastern Australia. *Geomorphology* 79:395–422.

Cairns, J. J. 2000. Setting ecological restoration goals for technical feasibility and scientific validity. *Ecological Engineering* 15:171–80.

Clark, M. J. 2002. Dealing with uncertainty: Adaptive approaches to sustainable river management. *Aquatic Conservation: Marine and Freshwater Ecosystems* 12:347–63.

Higgs, E. 2003. *Nature by design: People, natural process and ecological restoration*. Cambridge, MA: MIT Press.

Hildebrand, R. H., A. C. Watts, and A. M. Randle. 2005. The myths of restoration ecology. *Ecology and Soci-*

ety 10(1): Article 19. http//www.ecologyandsociety.org/vol10/iss1/art19/ (accessed December 16, 2007).

Hillman, M., and G. J. Brierley. 2005. A critical review of catchment-scale stream rehabilitation programmes. *Progress in Physical Geography* 29:50–70.

Hobbs, R. J. 2003. Ecological management and restoration: Assessment, setting goals and measuring success. *Ecological Management and Restoration* 4 (supplement): S2–S3.

Hooper, D. U., F. S. Chapin III, J. J. Ewel, A. Hector, P. Inchausti, S. Lavorel, J. H. Lawton, D. M. Lodge, M. Loreau, S. Naeem, B. Schmid, H. Setälä, A. J. Symstad, J. Vandermeer, and D. A. Wardle. 2005. Effects of biodiversity on ecosystem functioning: A consensus of current knowledge. *Ecological Monographs* 75:3–35.

Keough, H. L., and D. J. Blahna. 2006. Achieving integrative, collaborative ecosystem management. *Conservation Biology* 20:1373–82.

Kondolf, G. M. 1995. Five elements for effective evaluation of stream restoration. *Restoration Ecology* 3:133–36.

Leopold, A. 1949. *A Sand County Almanac, and sketches here and there*. Oxford, UK: Oxford University Press.

Levy, O. 1964. *The complete works of Friedrich Nietzsche*. New York, NY: Russell and Russell.

Olsson, P., C. Folke, and F. Berkes. 2004. Adaptive comanagement for building resilience in social-ecological systems. *Environmental Management* 34:75–90.

Rogers, K. 1998. Managing science/management partnerships: A challenge of adaptive management. *Conservation Ecology* 2(2): R1. http://www.ecologyandsociety.org/vol2/iss2/resp1/ (accessed December 15, 2007).

Rogers, K. H. 2003. Adopting a heterogeneity paradigm: Implications for management of protected savannas. In *The Kruger experience: Ecology and management of savanna heterogeneity*, ed. J. T. du Toit, K. H. Rogers, and H. C. Biggs, 41–58. Washington, DC: Island Press.

Rogers, K. H. 2006. The real river management challenge: Integrating scientists, stakeholders and service agencies. *River Research and Applications* 22:269–80.

Shields Jr., F. D., R. R. Copeland, P. C. Klingeman, M. W. Doyle, and A. Simon. 2003. Design for stream restoration. *Journal of Hydraulic Engineering* 129:575–84.

Walters, C. J. 1997. Challenges in adaptive management of riparian and coastal ecosystems. *Conservation Ecology* 1:xi-xii.

Wissmar, R. C., and Bisson, P. A. 2003. Strategies for restoring river ecosystems: Sources of variability and uncertainty. In *Strategies for restoring river ecosystems: Sources of variability and uncertainty in natural and managed systems*, ed. R. C. Wissmar and P. A. Bisson, 1–10. Bethesda, MD: America Fisheries Society.

About the Contributors

Angela Arthington
Australian Rivers Institute, Griffith University, Nathan, Queensland, 4111, Australia.
a.arthington@griffith.edu.au

Angela is professor of freshwater ecology in the faculty of environmental science at Griffith University in Brisbane, Queensland, Australia, where she works within the Australian Rivers Institute. She leads research programs and projects in the eWater Cooperative Research Centre (CRC), Tourism CRC, and the Reef and Rainforest Research Centre. Her research is primarily concerned with fish ecology in rivers and their floodplains, flow-ecological relationships, and determination of environmental flow methods for tropical and arid-zone rivers. Angela is coauthor of *Freshwater Fishes of North-Eastern Australia* (CSIRO Publishing, 2004), and is a member of the Freshwater Cross-cutting Network of DIVERSITAS, a global biodiversity science program.

Andrew Boulton
School of Environmental and Rural Science, University of New England, Armidale, NSW, 2350, Australia.
aboulton@une.edu.au

Andrew's interests lie in river ecology and management, especially groundwater–surface water interactions, and the effects of catchment land use on aquatic fauna and instream processes. He has worked on rivers in Australia, France, and Arizona, taught at several universities, and is increasingly aware of the importance of social sciences and knowledge structures in ecosystem management.

Laurent Bourdin
Agence de l'Eau Rhone Méditerranée et Corse, 2-4 allée de Lodz, 69 363, Lyon cedex, 07, France.
laurent.bourdin@eaurmc.fr

Laurent works at a water agency that is a basin-scale French state administration focused on water management and improvements to aquatic ecosystems. His work is based on the physical component of freshwater ecosystems in the French Rhône basin. He is in charge of promoting and planning river rehabilitation activities, notably to implement the Water Framework Directive.

Gary J. Brierley
School of Geography, Geology and Environmental Science, The University of Auckland, 10 Symonds Street, Private Bag 92019, Auckland, New Zealand.
g.brierley@auckland.ac.nz

Gary is chair of physical geography in the School of Geography, Geology and Environmental Science at the University of Auckland. His primary research focus lies in the use of principles from fluvial geomorphology for river management activities. His work as a fluvial geomorphologist builds upon undergraduate training in England, postgraduate training in Canada, and postdoctoral experience in Australia. At Macquarie University, Sydney, he worked for more than ten years on river responses to human disturbance. He is codeveloper of the River Styles framework, a physical template for crossdisciplinary management applications, and he is coauthor of the book *Geomorphology and River Management* (Blackwell Publishing, 2005). He is section editor (Environment and Society) for *Geography Compass* and is on the editorial board of the journals *Geomorphology*, *Catena*, and the *Open Geology Journal*.

Bruce Chessman
NSW Department of Environment and Climate Change, PO Box 3720, Parramatta, NSW, 2124, Australia.
bruce.chessman@dnr.nsw.gov.au

Bruce is a principal research scientist in the New South Wales Department of Environment and Climate Change. His main interests are in freshwater biodiversity conservation and the assessment and management of human impacts on freshwater ecosystems.

Carola Cullum
Centre for Water in the Environment, Department of Animal, Plant and Environmental Sciences, University of Witwatersrand, Private Bag 3, Wits 2050, South Africa.
cullum@biology.biol.wits.ac.za

After twenty years as a strategic marketing research consultant, Carola returned to the University of Auckland to study biological and geographical science. She is now preparing a doctoral thesis under the supervision of Kevin Rogers at the University of the Witwatersrand. Her thesis work is developing a hierarchical framework for the analysis and management of the savannas of Kruger National Park.

Leanne du Preez
Department of Geography and Catchment Research Group, Rhodes University, PO Box 94, Grahamstown, 6140, South Africa.
L.dupreez@ru.ac.za

Leanne is a Mellon Foundation lecturer in the Geography Department at Rhodes University in Grahamstown, South Africa. She has participated in various river management projects, including a number of ecological reserve studies and the South African National River Health Programme. She is currently completing her PhD thesis, which critically assesses the role of fluvial geomorphology in the interdisciplinary milieu of river management in South Africa. Her research interests include fluvial geomorphology, catchment processes, and theoretical geomorphology.

Kirstie A. Fryirs
Department of Physical Geography, Division of Environmental and Life Sciences, Macquarie University, North Ryde, NSW, 2109, Australia.
kfryirs@els.mq.edu.au

Kirstie is a lecturer in the Department of Physical Geography at Macquarie University. She is a fluvial geomorphologist who has worked on various river systems in Australia. Her research focuses on disturbance responses of rivers, their sensitivity to change, their condition, and their recovery potential. She also works on sediment budgets and the (dis)connectivity of sediment movement through catchments. Kirstie's work has been extensively applied in river management practice. She is codeveloper of the River Styles framework and coauthor of the book *Geomorphology and River Management* (Blackwell Publishing, 2005).

James Grove
Department of Civil Engineering, Monash University, Clayton, Victoria, 3800, Australia.
James.Grove@eng.monash.edu.au

As a research fellow, James has concentrated on the catchment scaling of geomorphic processes. He has applied this work to audit and assess stream condition. When not in Australia, he also enjoys working on arctic river systems.

Gertrud Haidvogl
Institute for Hydrobiology and Aquatic Ecosystem Management, University of Natural Resources and Applied Life Sciences, Max Emanuelstraße 17, A-1180, Vienna, Austria.
gertrud.haidvogl@boku.ac.at

Gertrud Haidvogl is an environmental historian with a research focus on historical interactions between riverine landscapes and societies and their ecological impacts. Gertrud is also involved in a range of river management and river restoration projects.

Mick Hillman
Department of Environment and Geography, Manchester Metropolitan University, All Saints, Manchester, M15 6BH, United Kingdom.
m.hillman@mmu.ac.uk

Mick is senior lecturer in environmental management and policy in the Department of Environment and Geography at Manchester Metropolitan University. He has an academic and practice background in community development and environmental science. His primary research focus is on the application of transdisciplinary perspectives to environmental management, with a particular focus on issues of social and environmental justice. His doctoral research at Macquarie University in Sydney examined the interaction of biophysical and social dimensions of river management in the Hunter Valley in New South Wales.

Richard Hobbs
School of Environmental Science, Murdoch University, Perth, Western Australia, 6150, Australia.
rhobbs@murdoch.edu.au

Richard is an Australian Professorial Fellow at Murdoch University in Western Australia. His research interests span basic ecology, restoration ecology, landscape ecology, and conservation biology. He is particularly interested in the management and repair of degraded and altered ecosystems. He has worked mostly in the fragmented agricultural landscapes of Western Australia, but also has ongoing long-term research interests in the dynamics of serpentine grasslands in Northern California. He is currently editor in chief of the journal *Restoration Ecology*.

Jochem Kail
Department of Hydrobiology, Institute of Biology, Faculty of Biology and Geography, University of Duisburg-Essen, Weinheimstr. 20, 53229 Bonn, Germany.
jochem.kail@uni-due.de

Jochem is a fluvial geomorphologist with special interest in stream restoration (reference conditions, restoration measures, use of large wood) and implementation of the Water Framework Directive. He is interested in the link between hydromorphology and stream biota, and the impact of climate change on hydromorphology of streams and rivers. After being involved in several research projects at the University of Duisburg-Essen, he now works as a consultant at the umweltbüro-essen in the field of river research and management.

Daniel Keating
Department of Physical Geography, Division of Environmental and Life Sciences, Macquarie University, North Ryde, NSW, 2109, Australia.
dkeating@els.mq.edu.au

Dan is currently a research officer with the Department of Physical Geography at Macquarie University. He has an earth and environmental sciences background. Since completing a MSc in 2002, he has held a number of technical positions at the university and has been project manager for the Upper Hunter River Rehabilitation Initiative (UHRRI).

Shun-ichi Kikuchi
Graduate School of Agriculture, Hokkaido University, Kita 9 Nishi 9, Kita-ku, Sapporo Hokkaido, 060-8589, Japan.
kikku@for.agr.hokudai.ac.jp

Shun-ichi is a geoecologist and assistant professor of laboratory of Earth Surface Processes and Land Management in the Graduate School of Agriculture at Hokkaido University. His interests lie in ecological interactions between vegetation dynamics and earth surface disturbances, especially the impacts of flooding, debris flows, landslides, and volcanic activity. For more than ten years he has studied vegetation responses to natural disturbance in Hokkaido. He has used this work to underpin effective management and restoration of ecosystems. He recently started field research on revegetation process following volcanic disturbance at Mt. Ruapehu in New Zealand.

Kaori Kochi
Graduate School of Science and Engineering, Saitama University, Shimo-okubo 255, Sakura-ku, Saitama 338-8570, Japan.
k-kochi@r5.dion.ne.jp

As an adjunct lecturer, Kochi is interested in the diversity and decomposition of coarse particulate organic matter entering streams from riparian forests and organic matter circulation between upstream forests and coastal areas. She thinks the effect of organic matter such as green leaves and flowers that is deposited in spring and summer on macroinvertebrate shredders is of considerable importance in rivers in temperate zones.

G. Mathias Kondolf
Department of Landscape Architecture and Environmental Planning, 300 Wurster Hall, University of California, Berkeley, CA 94720, USA.
kondolf@berkeley.edu

Matt's research focuses on human impacts on rivers, emphasizing concerns for their resilience and their rehabilitation. He has particular interest in salmon-bearing rivers and Mediterranean-climate rivers, studying the effects of human alterations such as dams and gravel mining. He is coeditor (with Hervé Piégay) of the recently published book entitled *Tools in Fluvial Geomorphology* (Wiley, 2003).

Garreth Kyle
Department of Biological Sciences, Division of Environmental and Life Sciences, Macquarie University, North Ryde, NSW, 2109, Australia.
gkyle@bio.mq.edu.au

Garreth has recently submitted his PhD at Macquarie University in Sydney, where he currently works as a researcher. His thesis examined the functional and structural composition of plant communities in the context of riparian rehabilitation. He has a keen interest in the practical side of riparian rehabilitation and has been involved in further work examining the potential factors that affect revegetation success. Garreth's undergraduate degree was completed at the University of Otago in Dunedin, New Zealand.

Michelle Leishman
Department of Biological Sciences, Division of Environmental and Life Sciences, Macquarie University, North Ryde, NSW, 2109, Australia.
Michelle.Leishman@mq.edu.au

Michelle is a plant ecologist with research interests in plant functional traits, seed and seedling biology, invasive plants, and restoration ecology. She believes that it is critical to use our understanding of the underlying processes of ecosystems to inform restoration ecology.

Tomomi Marutani
Laboratory of Earth Surface Processes and Land Management, Graduate School of Agriculture, Hokkaido University, Sapporo, Hokkaido, 060-8589, Japan.
marut@for.agr.hokudai.ac.jp

Tomomi has an interest in mitigating sediment disasters at the catchment scale, considering both hillslope and fluvial processes. He leads a collaborative international project that examines catchment-scale sedimentary cascades in various Pacific-Rim countries, including New Zealand, Australia, China, and Korea. His interest is now extending to sustainability of catchment systems, entailing analysis of biological and hydrological continuities.

Sarah Mika
School of Environmental and Rural Science, University of New England, Armidale, NSW, 2350, Australia.
smika@une.edu.au

Sarah is a postgraduate student at the University of New England, Armidale, NSW, Australia. Her research focuses on the processes of vertical connectivity in rivers and techniques to restore these mechanisms.

Larissa A. Naylor
Department of Geography, University of Exeter, Cornwall Campus, Penryn, Cornwall, TR10 9EZ, United Kingdom.
l.a.naylor@exeter.ac.uk

Larissa is a biogeomorphologist. Her work focuses on the interactions between ecology, geology and geomorphology in fluvial and coastal settings. She has a particular interest in understanding scale linkages in geomorphic systems and how these relate to those of cognate disciplines. She has considerable experience in interdisciplinary and applied research including wetland restoration, climate change adaptation, and integrated catchment management in Britain and Canada.

David Outhet
NSW Department of Water and Energy, 10 Valentine Street, Parramatta, NSW, 2150, Australia.
David.Outhet@dnr.nsw.gov.au

David is the principal geomorphologist in the NSW Department of Water and Energy. He mainly provides staff training and advice to NSW Catchment Management Authorities on River Styles mapping and the translation of fluvial geomorphology research into practical river rehabilitation applications.

Margaret A. Palmer
Chesapeake Biological Laboratory, University of Maryland Center for Environmental Science, Solomons, MD 20688, USA.
mpalmer@umd.edu

Margaret is a restoration ecologist whose work focuses on the link between structure and function in stream ecosystems. Her work also includes riverine-coastal interactions, especially in response to land use and climate change.

Hervé Piégay
University of Lyon, UMR 5600, CNRS Environment Ville Société, Site ENS-Lsh, 15 Parvis René Descartes, BP 7000, 69342, Lyon cedex 07, France.
hpiegay@free.fr

Hervé is a research director at CNRS, where he leads an interdisciplinary research group working on biological and physical interactions within the LTER Rhône catchment. He is a specialist in fluvial geomorphology, with particular interests in the interactions between channels and their floodplains. He has significant experience in interdisciplinary and applied research on river restoration, maintenance, and planning. He is coeditor (with G. Mathias Kondolf) of the book *Tools in Fluvial Geomorphology* (Wiley, 2003) in which methodological and technical approaches to geomorphology are introduced and exemplified to a large audience of scientists (ecologists, geologists, engineers, managers) and practitioners.

Kate M. Rowntree
Department of Geography and Catchment Research Group, Rhodes University, PO Box 94, Grahamstown, 6140, South Africa.
K.Rowntree@ru.ac.za

Kate is head of the Geography Department at Rhodes University in Grahamstown, South Africa. She has been instrumental in the development of river management frameworks at the national level. She has also contributed significantly to participatory river management endeavors in South Africa. She has worked extensively with the South African Water Research Commission and has contributed to various river management-related studies undertaken by the South African Department of Water Affairs and Forestry. Her research interests include fluvial geomorphology for river management, catchment processes, soil erosion and land degradation, and integrated catchment management.

Darren Ryder
School of Environmental and Rural Science, University of New England, Armidale, NSW, 2350, Australia.
dryder2@une.edu.au

Darren is a senior lecturer in ecosystem rehabilitation and aquatic ecology at the University of New England, Armidale, NSW, Australia. His research is focused on the development and application of functional ecological indicators such as organic matter and nutrient dynamics, and ecosystem metabolism to underpin effective management and rehabilitation of freshwater ecosystems.

Mark Sanders
Boffa Miskell Ltd, PO Box 110, Christchurch, 8140, New Zealand.
mark.sanders@boffamiskell.co.nz

Mark has worked as a scientist and manager on research-based river and wetland management projects in New Zealand and Australia, since the early 1990s, although he has mucked about in streams and ponds for as long as he can remember. He has an overriding interest in bridging the gap between science and management, and in working with the wider community to achieve integrated ecosystem management. He now works as an ecological consultant in Christchurch, New Zealand.

Laurent Schmitt
Faculty of Geography, History, History of Art, Tourism, University of Lyon, UMR 5600 CNRS, 5 av. Pierre Mendès-France, 69676 BRON cedex, France.
laurent.schmitt@univ-lyon2.fr

Laurent is a fluvial geomorphologist. His research topics focus on river classification and restoration, functional links between hydro-geomorphology and hydro-ecology, integrated management of periurban hydrosystems, and fluvial palaeo-dynamics.

Alexandra Spink
Department of Physical Geography, Division of Environmental and Life Sciences, Macquarie University, North Ryde, NSW, 2109, Australia.
aspink@els.mq.edu.au

Alex is an environmental scientist who completed a MSc(Hons) at Macquarie University. Her thesis was an interdisciplinary project examining the history of river rehabilitation in the upper Hunter catchment since 1950 and the associated geomorphic and social changes that occurred. She currently works for the Department of Primary Industries, Water and the Environment in Tasmania.

Martin Thoms
School of Resource, Environmental and Heritage Sciences, Applied Ecology Research Group, School of Environmental Sciences, University of Canberra, ACT, 2601, Australia.
Martin.Thoms@canberra.edu.au

Martin's primary research interests are in the interdisciplinary study of riverine landscapes that brings together fluvial geomorphology, landscape ecology, and freshwater ecology. The setting for most of his research has been the large floodplain rivers of the Murray Darling Basin and those of the Lake Eyre Basin in central Australia. He also has active research collaborations in South Africa and the United States. Currently, Martin is the regional editor of *River Research and Application* and is on the editorial boards of *Geomorphology* and the *Open Geology Journal*.

Ellen Wohl
Department of Geosciences, Colorado State University, Ft. Collins, CO 80523-1482, USA.
ellenw@cnr.colostate.edu

Ellen received her BS in geology from Arizona State University and her PhD in geosciences from the University of Arizona in 1988. Since graduating, she has been on the faculty at Colorado State University. Her research focuses on rivers in bedrock canyons, mountain rivers, and human interactions with river systems.

Seiji Yanai
Department of Environmental Design, Hokkaido Institute of Technology, 4-1, 7-15, Maeda, Teine, Sapporo, 006-8585, Japan.
yanai@hit.ac.jp

Seiji is a professor at the Hokkaido Institute of Technology. His research focuses on interactions between terrestrial and aquatic ecosystems from forests to rivers, estuarine, and coastal areas. He believes that healthy watershed ecosystems should be restored, facilitating material flows by anadromous fish from the ocean by modifying or removing dams in Japan.

INDEX